Climate Change and Individual Moral Duties

Anna Luisa Lippold

Climate Change and Individual Moral Duties

A Plea for the Promotion of a Collective Solution

BRILL | mentis

Gedruckt mit Unterstützung des Förderungsfonds Wissenschaft der VG WORT

Cover illustration: Environmental Pollution (Photo by Pete Linforth)

Bibliographic information published by the Deutsche Nationalbibliothek

The Deutsche Nationalbibliothek lists this publication in the Deutsche Nationalbibliografie; detailed bibliographic data available online: http://dnb.d-nb.de

Internet: www.mentis.de

Cover design: Anna Braungart, Tübingen
Production: Brill Deutschland GmbH, Paderborn

ISBN 978-3-95743-185-1 (paperback)
ISBN 978-3-95743-742-6 (e-book)

Contents

Acknowledgments

Over the past three and a half years that I have been working on this book, many people have supported and encouraged me. Here, I would like to at least acknowledge those who have been especially important.

First of all, I would like to express my gratitude to my supervisor, Klaus Steigleder, who has not only strongly supported me throughout the process of writing this book but who has constantly been a source of motivation, patience, immense knowledge and discussion since we first met during my master studies in 2013. Thank you to Corinna Mieth for agreeing to be my second supervisor and providing me with an opportunity to present my hypotheses in her research colloquium. My sincere thanks also go to Helmut Pulte and members of the colloquium of the Faculty of Philosophy for commenting on my first thoughts on individual climate duties back in 2016.

Many thanks go to the doctoral research colloquium of Klaus Steigleder, in particular Friederike Asche, Daniela Berner-Zumpf, Lukas Hambach, Nora Henn, Johannes Graf Keyserlingk, Itzik Pazuelo, Alina Pfleghardt, Raoul Scheppart, Vandad Sohrabi and Julia Weinheimer, for feedback and lively discussions on several chapters of my book. Further, I am grateful to Marissa Hünnefeld for proof-reading and improving the wording where necessary.

I would like to express my gratitude to the Konrad-Adenauer-Stiftung e. V. (KAS) for their financial and ideational support via a doctoral scholarship and in particular to Gernot Uhl, Daniela Bister and Johannes Gold for their support.

Special mention goes to Aleksandra Brand for personal exchange during countless lunch and coffee breaks. Also, I would like to thank my friends Lea, Julia and Lea who have supported me from a distance.

I am deeply indebted to my parents, Susanne and Dirk, who have supported me in innumerable ways throughout the years of studying and research. Thank you for being a constant source of encouragement.

At last, there is Marius, my love and best friend, who has commented on my thoughts from a non-philosophical perspective, who has celebrated the ups and cushioned the downs of writing with me, who always knows when it is time to go for a run and who is my refuge for happiness. Thank you for almost unbelievable support – this book is dedicated to you!

List of Tables

List of Acronyms

ACE	Action for Climate Empowerment
AR4	IPCC's Fourth Assessment Report
AR5	IPCC's Fifth Assessment Report
ATAG	Air Transport Action Group
CBA	Cost-Benefit-Analysis
CCS	Carbon Capture and Storage
CCU	Carbon Capture and Utilization
CDR	Carbon Dioxide Removal
CEO	Chief Executive Officer
CH_4	methane
CIEL	Centre for International Environmental Law
CO_2	carbon dioxide
CO_2eq	carbon dioxide equivalent
CORSIA	Carbon Offsetting and Reduction Scheme for International Aviation
DESA	Department of Economic and Social Affairs
DRV	Deutscher Reiseverband [German Travel Association]
EC	European Commission
ECA	Emission Control Areas
EEA	European Economic Area
EU	European Union
EU ETS	European Union Emissions Trading Scheme
GAP on ESC	Global Action Program on Education for Sustainable Development
GDP	Gross domestic product
GHG	Greenhouse gases
GMST	Global Mean Surface Temperature
Gt	Gigatons [unit]
HDI	Human Development Index
IAEA	International Atomic Energy Agency
IATA	International Air Transport Association
ICAO	International Civil Aviation Organization
ICS	International Chamber of Shipping
IEA	International Energy Agency
IMO	International Marine Organization
INDCs	Intended Nationally Determined Contributions
IPCC	Intergovernmental Panel on Climate Change
LDC	Least Developed Countries

LEL	Lower Explosive Limit
LLDC	Landlocked Developing Countries
LOSC	UN Law of the Sea Convention
MCC	Mercator Research Institute on Global Commons and Climate Change
MEP	Member of the European Parliament
MEPC	Marine Environmental Protection Committee
micro-CHP	Micro combined heat and power
MP	Member of the Parliament
MRO	Maintenance, repair and overhaul services
MSC	Marine Stewardship Council
MW	megawatt [unit]
N_2O	dinitrogen monoxide
NASA	National Aeronautics and Space Administration
NAZCA	Non-State Actor Zone for Climate Action
NDCs	Nationally Determined Contributions
NESS	Necessary element of a sufficient set
NGO	Non-governmental organization
OECD	Organisation for Economic Co-operation and Development
OPEC	Organization of Petroleum Exporting Countries
PGC	Principle of Generic Consistency
ppm	parts per million [unit]
RCP	Representative Concentration Pathway
RF	Radiative Forcing
RTK	Revenue Ton Kilometres, measuring the share in international transport performances
SIDS	Small Island Developing States
SR15	IPCC's Special Report on 1.5 degrees
SUV	Sport Utility Vehicle
t	tonne [unit]
TED	Technology, Entertainment, Design (e.g. talk or conference)
UAE	United Arab Emirates
UEL	Upper Explosive Level
UK	United Kingdom
UN	United Nations
UNCTAD	United Nations Conference on Trade and Development
UNDP	United Nations Development Program
UNEP	United Nations Environment Programme
UNESCO	United Nations Educational, Scientific and Cultural Organization
UNFCCC	United Nations Framework Convention on Climate Change

UNWTO	World Tourism Organization (Specialized Agency of the UN)
US	United States
VAT	Value added tax
vol%	volume percent or percentage by volume [unit]
WMO	World Meteorological Organization
WWF	World Wide Fund for Nature

CHAPTER 1

Introduction and Outline

The scientific consensus has never been clearer: human activities are heating up the planet while the window to limit global climate change to an acceptable level is slowly but steadily closing. Effective and far-reaching policy measures that accelerate climate action across state, sub-state and non-state actors are either missing or gain moment far too slowly. Climate change is not simply a somewhat slight inconvenience on peoples' lives – climate change is the single biggest threat to human well-being.[1] Climate change is not a regional challenge – climate change will increasingly hit human beings globally. In fact, climate impacts are already imposing upon humans across the planet. In light of this, it is surprising that even civil society remains largely inactive.

The Paris Agreement sent an urgently needed, clear, global signal concerning the direction of future climate policy. And yet, the very long road to the Paris Agreement and also the lack of ambitious results at three conferences since Paris reveal the complexity of international governance, the persistent disagreement over the distribution of burdens as well as the ineffectiveness and inefficiency regarding decision making processes. Over the years, major effort has been made on improving climate science and policy but there is comparatively little effort that explores the role of individual agents with regard to global climate action. Rather than bypassing climate policy on any level, this book asks: What should individuals do to supplement the global fight against climate change?

Typically, normative advice to individual agents focuses on reducing personal emissions. Environmental organisations in particular invite people to ride a bike, use public transport, switch to green electricity, use energy efficient light bulbs, refrain from flying or switch to a vegetarian diet. Suggestions are often stimulated by catchy slogans of the style 'going green is easy', 'every little bit counts' or 'plant a tree, get air for free'. This is the dominant moral narrative of the public discourse.

When it comes to climate change, most citizens are torn between the conviction that fighting climate change is worth the global effort on the one hand and doubts about their personal ability to make a meaningful difference on the

1 In contrast to other big threats such as famine, nuclear war or poverty there are no emergency interventions or diplomatic last-minute solutions that could stop climate change. Rather, climate change increases the risk of famine, war and poverty.

 | DOI:10.30965/9783957437426_002

other. Possible thoughts or comments are: 'everyone else will continue to have a non-vegetarian diet anyways', 'my driving the car does not make too much of a difference after all' or 'regardless of whether I take the plane or not, the plane will take off anyways'. With regard to climate change, many individual agents are, what David Zoller calls, "complicit social reformers" (Zoller, 2015, p. 995): Principled people who are convinced that collective harms are worth the fight, who may even attempt to adjust their personal behaviour accordingly. Principled people who turn around, move on with their daily lives and thereby contribute to the same collective harm they are actually protesting against. Individual agents might spend their holidays overseas while accurately separating the waste, refraining from the use of disposable plastic bags, and using ecological, organic shampoo. Most individuals probably fly home with the unpleasant feeling that flying for a holiday might not have necessarily been beneficial to the environment but because 'every bit counts' they actually end up believing they have done their part to global climate action. However, waste separation, reusable bags and organic shampoo will not stop climate change.

Recently, communication scientists discussed how climate change communication should be shaped in order to get the message about climate science across to individual agents (see e.g. Moser, 2016; Rayner & Minns, 2015; Salama & Aboukoura, 2018; Trolliet, Barbier, & Jacquet, 2019; van der Linden, Maibach, & Leiserowitz, 2015). One of them is David Fenton, a communication and campaign specialist who has consulted former National Aeronautics and Space Administration (NASA) climate scientist James Hansen with regard to holding his famous TED talk a few years ago. In summer 2018, he was asked in an interview why presenting scientific facts about climate change was not an effective strategy to convey the urgency to take action immediately. Part of Fenton's answer was: "Facts don't work by themselves. Facts are important and they're essential and they're the only ethical thing to communicate, which is very important. But facts only really work when one, they're embedded in moral narratives. [...]" (Dalton, 2018).

The way public actors talk about climate change and, in particular, the way the discourse frames individual actions, strongly affects a person's thinking and acting with regard to climate change. The narrative about climate change should, as I believe, communicate the seriousness of climate threats and the urgency to act while leaving individual agents with a feeling of empowerment to take effective action. Yet, my expertise is not in public communication – my expertise is in applied ethics. From the ethical perspective, as this book will show, the dominant narrative fails to capture the nature of individual actions with regard to climate change as a collective phenomenon. Moreover, and here the ethical perspective connects to communicating climate science, the narrative conveys a problematic idea of environmental protection. By giving small-scale individual recommendations, it talks individuals into thinking that climate protection is a private affair. In fact, it is exactly the opposite. My hypothesis is that the public discourse uses the wrong moral narrative. Next to elaborating what individual agents ought to do with regard to climate change and exploring the nature of individual climate duties in practice, this book is, more generally, a plea for changing the moral narrative of climate change.

Despite the recent insights from communication science with regard to climate communication, the following section will introduce basic scientific facts about climate change (chapter 1.1). Because these findings provide the empirical background for moral claims and corresponding duties with regard to climate change, I consider it important to give a brief overview right at the beginning of this book. Next, I offer a first insight regarding individual climate duties by presenting the status quo of the discussion (chapter 1.2). The third part of this chapter briefly outlines the idea, argumentation and general structure of this book (chapter 1.3).

1.1 Understanding Climate Change

Assuming that readers have a basic knowledge of processes causing anthropogenic climate change as well as of climate change research, the following is limited to a very brief introduction on basic scientific facts about climate change. The main empirical findings provided by the Intergovernmental Panel on Climate Change (IPCC) in its Fifth Assessment Report (AR5) serves as the main source. The IPCC's Special Report on 1.5 degrees (SR15) will be used to supplement the AR5's findings where needed.[2] This section starts by highlighting the new urgency to act. Subsequently, I connect impacts and risks particularly regarding human well-being and human rights to provide a basis for the following chapters.

The Old and New Urgency to Act

The IPCC declares that

> [i]t is certain that global mean surface temperature (GMST) has increased since the late 19th century [...]. Each of the past three decades has been successively warmer at the Earth's surface than any the previous decades in the instrumental record, and the decade of the 2000's has been the warmest. (Stocker et al., 2013, p. 37)

2 Comments based on the SR15 are limited to content of the accepted Summary for Policy Makers.

Between 1880 and 2012 the GMST has increased by almost 0.9 °C. Considering the period between 1951 and 2012 only, it increased by approximately 0.7 °C. The annual mean surface warming since the 20th century has reversed cooling effects of the past 5000 years in mid-to-high latitudes of the Northern Hemisphere [high confidence].[3] It is *very likely* that the period between 1983–2012 has been the warmest 30-years period in the Northern Hemisphere over the past 800 years [high confidence].[4] Likewise, it is *virtually certain* that the ocean temperature in the upper 700 meters has warmed and is *likely* to have warmed between 700 and 2000 meters (Stocker et al., 2013, pp. 37–39).

Generally, the global climate changes when sunlight absorbed by Earth either exceeds or falls below the amount of warmth emitted by the planet. Briefly stated, radiative forcing (RF) is the difference between sunlight absorbed and energy radiated back to space. For observed warming RF is positive. RF of the global climate is caused by a number of forcing agents including anthropogenic and natural forces (mainly solar radiance) (Stocker et al., 2013, p. 54). Carbon dioxide (CO_2) is the largest single contributor to RF [very high confidence] followed by other greenhouse gases (GHG) such as methane (CH_4) or dinitrogen monoxide (N_2O) all of which are of an anthropogenic origin [very high confidence]. The observed global warming is far outside the range of internal climate variability in estimations from pre-industrial data as well as from model-based simulations: Neither natural forces nor internal climate variability are sufficient to explain the dramatic increase in GMST. Human activities are *very likely* to have caused more than half of the overserved GMST between 1951 and 2010 (Stocker et al., 2013, p. 66). It is therefore safe to say that observed warming is mostly anthropogenic.

The trend of global climate change will further intensify over the coming decades. Due to thermal inertia of the climate system (oceans in particular), further anthropogenic GHG, which are predominantly due to continued burning of fossil fuels, and climate feedbacks[5], warming will continue. It is likely

3 The IPCC uses five different levels to express their confidence in the validity of its findings: very low, low, medium, high and very high. Unless indicated otherwise, these confidence levels will be shown in squared brackets.

4 The IPCC indicates likelihood with seven different values (probability in brackets): virtually certain (99–100%), very likely (90–100%), likely (66–100%), about as likely as not (33–66%), unlikely (0–33%), very unlikely (0–10%) and exceptionally unlikely (0–1%). Throughout this chapter, likelihood will be indicated in italics.

5 Next to the level of water vapour and clouds, the influencing factor behind climate feedback is the so-called albedo. It is a term for the degree to which sunlight is reflected from Earth's surface and depends on brightness. An example is the snow and ice albedo: Since ice is light-coloured and reflective, a large proportion of solar rays that hit the ice is reflected and sent back to space, which limits the amount of warming it causes. As the climate is getting

that GMST will increase by another 0.3 to 0.7 °C until 2035 (Stocker et al., 2013, p. 85).[6] Scenarios for the second half of the 21st century strongly depend on how ambitious present and future mitigation is. The IPCC calculates future scenarios for different GHG concentrations (Representative Concentration Pathways, RCPs) and their corresponding global warming. More specifically, RCPs describe GMST rises for a range of possible RF values in 2100 relative to pre-industrial values (the number indicates the target value for RF). According to all RCPs, GMST is predicted to rise over the 21st century: Until the middle of the 21st century, GMST will increase by 1 °C in a very low emissions scenario (RCP2.6) and by 2 °C in a very high emissions scenario (RCP8.5).[7] Towards the end of the 21st century, GMST is projected to increases by 1 °C (RCP2.6) to 3.7 °C (RCP8.5) (Stocker et al., 2013, p. 90). However, indicated temperatures are mean values. The IPCC indicates the following likely ranges relative to the reference period of 1986–2005:

Table 1 IPCC's predictions regarding GMST increase (Stocker et al., 2013, p. 90)

GMST predictions	Middle of the century (2046–2065)		End of the century (2081–2100)	
Scenario	**Mean value**	**Likely range**	**Mean value**	**Likely range**
RCP2.6	1.0	0.4 to 1.6	1.0	0.3 to 1.7
RCP4.5	1.4	0.9 to 2.0	1.8	1.1 to 2.6
RCP6.0	1.3	0.8 to 1.8	2.2	1.4 to 3.1
RCP8.5	2.0	1.4 to 2.6	3.7	2.6 to 4.8

Note that nations under the UNFCCC agreed to hold global warming to below 2 °C or even 1.5 °C refer to preindustrial levels as a baseline. Predictions using preindustrial levels as a reference are less precise due to major variations, which is why the IPCC limits its findings to the likelihood of exceeding warming of

warmer, ice melts and reveals darker, less reflective surfaces. The result is that a greater part of solar radiation is absorbed, leading to more warming, which again leads to more ice melting – and so on (Allison et al., 2013, pp. 355–356, 359–361; Planton, 2013, p. 1448).

6 Reference to the period 2016–2035 compared to 1986–2005. The temperature increase is consistent with the SR15's findings of current warming rates of 0.2 °C per decade (Allen et al., 2018, p. 4).

7 Note that the reference period is 1986–2005 not pre-industrial levels.

1.5 °C or 2 °C above preindustrial levels (1850–1900). As a consequence, different scholars urge the UNFCCC to use 1986–2005 as a reference when indicating a specific stabilization level (see e.g. Hawkins et al., 2017; Schurer, Mann, Hawkins, Tett, & Hegerl, 2017).[8] Using the 2 °C goal as a reference throughout this book, the important point for now is simply that estimated warming towards the middle and the end of the century is significantly higher with preindustrial levels as a baseline.

Predictions for continued warming are further complicated by possible abrupt and irreversible changes to major elements of the climate system: "Tipping elements are components, which are characterised by critical thresholds (Lenton et al., 2008; Schellnhuber, 2009). A selected number of tipping elements are the West Antarctic Ice Sheet, the Greenland Ice Sheet as well as the Amazon Rainforest, Indian monsoon and the El Niño" (Collins et al., 2013, pp. 1116–1117; Schellnhuber, 2009). Once these elements come close to the threshold, they can be tipped into changes beyond which "abrupt or nonlinear transition to a different state ensues" (Collins et al., 2013, p. 1114). Most of these changes are irreversible, which is why their expected tipping thresholds are also referred to as points of no return (see e.g. Aengenheyster, Feng, van der Ploeg, & Dijkstra, 2018).

Politically and scientifically, the IPCC's scenarios have led to a standard approach of limiting further increase in average global temperature (combining sea and surface temperature) of more than 2 °C compared to pre-industrial temperature. Others, in contrast, advice to limit warming to not more and even less than 1.5 °C (see e.g. Hansen, Kharecha, et al., 2013; Hansen, Sato, Russell, & Kharecha, 2013; Tollefson, 2015). In accordance, nations under the Paris Agreement agreed to limit warming to "well below 2°C above pre-industrial levels and pursu[e] efforts to limit the temperature increase to 1.5 °C above pre-industrial levels, recognizing that this would significantly reduce the risks and impacts of climate change" (UNFCCC, 2015b, article 2a). On invitation of the Paris Agreement, the IPCC recently provided the SR15 on impacts of 1.5 °C warming and related GHG pathways compared to 2 °C warming.[9] Consistent with the AR5's predictions, the SR15 underlines that warming is currently likely to cross 1.5 °C between 2030 and 2052 [high confidence] (Allen et al., 2018, p. 4). The report finds that limiting warming to 1.5 °C is still possible but requires "rapid and far-reaching transitions" (Allen et al., 2018, p. 21).

8 Note further that under the Paris Agreement preindustrial levels were not even defined (Schurer, Mann, Hawkins, Tett, & Hegerl, 2017, p. 563).

9 For invitation reference see UNFCCC (2015a, paragraph 21).

Still, limiting warming to 2 °C temperature rise is so far widely accepted to be politically feasible but likewise, it is understood to be far from being unproblematic (Shue, 2015b, p. 324). Most scholars refer to a cumulative carbon budget for the amount of a 2 °C temperature rise.[10] CO_2 emissions have an extraordinarily long atmospheric life span of several hundreds to thousands of years (Allen et al., 2009).[11] So far, some hundred gigatons (Gt, 1 gigaton equals one billion metric tons) linger in the atmosphere, crossing the level of 400 ppm in 2016 for the first time (Le Quéré et al., 2018, p. 407).[12] With any current amount of carbon emitted, no matter how small, the total amount of atmospheric carbon swells. The single factor for predictions of further warming is therefore the total *cumulative* amount of carbon emissions since 1750 rather than for example annual carbon emissions. The importance of cumulative carbon has been confirmed by the IPCC's AR5 and is now underlined by the SR15 (Allen et al., 2018, p. 6; Collins et al., 2013, p. 1033). Based on the strong correlation between the total cumulative amount of carbon emitted and the increase of average global temperature (both since the beginning of the Industrial Revolution), climate scientists can specify the approximate carbon budget required to reach the goal of limiting climate change to 2 °C. The cumulative carbon budget for 2 °C is approximately one trillion tons of carbon (≙ 3.67 trillion tons of CO_2).[13] Depending on the estimated amount of current atmospheric carbon, predictions for the remaining global carbon budget vary significantly.[14] Right before completion of this book in March 2019, the total amount of cumulative carbon since industrialization was more than

10 For background information on the calculation of the cumulative carbon budget see Allen et al. (2009) and Meinshausen et al. (2009).

11 The biggest source of confusion and error is the difference between carbon and CO_2. Science usually uses carbon to indicate the carbon budget. The central climate component, however, is CO_2. Also, using CO_2 is probably more intuitive because it is what regular individuals all emit. One ton (t) of carbon equals 3.67 t of CO_2. CO_2 is usually indicated in parts per million (ppm) but may also be converted: 1 ppm equals 2.12 Gt carbon or 7.81 Gt CO_2 (see Le Quéré et al. (2018, p. 409) for conversion factors). Use of carbon or CO_2 is equally correct. For reasons of clarity, I choose CO_2 as a measured value and provide conversion where necessary.

12 For the latest update on the Keeling Curve see UC San Diego (2018).

13 Steigleder (2016a) stresses that others estimate a cumulative carbon budget of 700 Gt, which is obviously quite a difference, giving rise to many uncertainties. Further, he highlights uncertainties to the effects of 2 °C warming (Steigleder, 2016a, p. 252). To make my point, I take 2 °C warming and a budget of one trillion tons of carbon as given.

14 For more details on modelling, assessment and uncertainties linked to calculating the global carbon budget see Le Quéré et al. (2018) and Matthews, Zickfeld, Knutti & Allen (2018).

629 Gt of carbon (≙ 2.30 trillion tons of CO_2). So, the global carbon budget is already more than half used up. If carbon emissions continue to rise without significant reductions, the remaining budget will be used up by the year 2035.[15] Time is running out on our ability to limit warming to 2 °C and even more so to 1.5 °C. The urgency to act was evident from earlier IPCC reports. Given the role of cumulative emissions and the importance of limiting warming to anything close to 1.5 °C, there is a new urgency to act. If ambitious measures are not implemented immediately, there are less than 16 years left until the global carbon budget is consumed.[16]

Impacts, Risks and Threats

With substantial and rapid warming throughout the 21st century, the warming-related consequences are manifold. Massive climate effects influence the natural and the human system. As the aim of this book is to justify individual moral duties based on moral rights violations, impacts on the natural system are only briefly mentioned while the focus is on observed and expected impacts that affect the human system with a special focus on human well-being. Again, the IPCC's AR5 serves as the main source and is supplemented by the SR15's findings. Even though the majority of impacts are caused by direct warming, a growing number of impacts is the result of changes in temperature combined with changes in precipitation and extreme weather events.

A selected number of observed impacts to the natural system directly due to warming are: the shrinking of sea ice and glaciers [high to very high confidence] (Stocker et al., 2013, pp. 40–41), the rise of global mean sea level [high confidence], the increased number of extreme events such as heat waves [medium confidence] (Stocker et al., 2013, p. 46), droughts [high confidence], floods [high confidence] and cyclones (Stocker et al., 2013, p. 50)[17], coral reef

15 For data and visualization see University of Oxford (2019).

16 Given that the SR15's findings strongly recommend limiting warming to 1.5 °C, the scientific recommendation for climate policy is to reduce CO_2 emissions by 45% compared to 2010 by 2030, reaching net zero emissions by 2050 (Allen et al., 2018, p. 15).

17 Extreme events vary across regions: Globally, the number of heat-related events is increasing [medium confidence] (Stocker et al., 2013, p. 46). The frequency and intensity of droughts has increased in the Mediterranean region and West Africa while it has decreased in North America and the north-western parts of Australia since 1950 (Stocker et al., 2013, p. 50). Australia and New Zealand have rather been subject to sever floods in recent years (Field et al., 2013, p. 42) while heavy precipitation is likely to have increased over North America and Europe (Stocker et al., 2013, p. 46). The frequency and intensity of cyclones in the North Atlantic is virtually certain to have increased since the 1970s. There is, however, disagreement over the causes of this increase and hence, only low confidence that changes are due to human influence (Stocker et al., 2013, p. 50).

bleaching [very high confidence] (Field et al., 2013, p. 47), shifts in geographic ranges and seasonal activities of terrestrial and freshwater plants and animal species [high confidence] (Field et al., 2013, p. 44) and additional shifts in migration patterns for marine species [very high confidence] (Field et al., 2013, p. 47). Almost all risks to the natural system are higher for 2 °C warming compared to 1.5 °C (Allen et al., 2018, pp. 8–10). Warming of 2 °C could for example trigger the marine ice sheet instability in Antarctica as well as irreversible loss of the Greenland ice sheet [medium confidence] (both tipping elements of the climate system) leading to an increased sea level rise [high confidence] of up to 0.1 meters more by 2100 [medium confidence] (Allen et al., 2018, p. 9).

Observed impacts on the human system are at least as diverse. Warming-related changes range from affected water resources due to the melting of snow and ice [medium confidence] to decreasing wheat and maize yields [high confidence], the decline in fruit-bearing trees [low confidence], the reduction of productivity of coastal fisheries [low to medium confidence, depending on the region], an increased number of malaria infections [low confidence] and negative impacts on the livelihoods of people living in the polar regions [low to medium confidence] (Field et al., 2013, pp. 44–47). Risks to the human system too are also projected to be higher for 2 °C warming compared to 1.5 °C (Allen et al., 2018, pp. 11–12).

Against the potentials of adaptation and risk-reduction measures, warming-related effects may appear to be manageable, but that is clearly an incorrect conclusion. Extreme events, in particular, have direct effects on lives and livelihoods and their intensity and frequency is hardly manageable by adaptation or risk-reduction measures. Extreme events cause displacement of populations, damage to agricultural land, contamination of water supply, damage to infrastructure and settlements, morbidity and mortality (e.g. due to delays in medical treatment or increased transmission of diseases), and psychological distress and disruption of educational service (CIEL, 2011, p. 6; Field et al., 2013, p. 40). Climate change consequences impact a range of internationally guaranteed human rights. Threats to the right to life, water, health, adequate housing, property and education are probably the most visible violations.[18] The variety of non-climate related factors cause differences in adaptive capacity and likewise, differences in vulnerability and exposure (Field et al., 2013,

18 The Centre for International Environmental Law (CIEL) provides an overview of climate-induced human impacts and corresponding internationally guaranteed human rights (CIEL, 2011).

pp. 40, 50, 62).[19] Generally, existing impacts of climate change affect above all people living in poverty and increasingly, indigenous people in the Arctic (Field et al., 2013, p. 51).

The IPCC predicts impacts and risks to increase over the next decades and identifies key risks that directly threaten human well-being (Field et al., 2013, pp. 59–60). Key risks [all with high confidence] reflect on the variety of non-climate related factors as key vulnerabilities.[20] Low-lying coastal zones and small island states in particular are vulnerable to sea level rise, coastal flooding and coastal erosion due to their exposure. The key risks are death, injury and disruption of livelihoods as well as decreasing food supply and drinking water. Social vulnerability with respect to sea level rise and coastal flooding is amplified for the widely unprotected urban populations (substandard housing, lack of essential infrastructure, lack of insurance) leading to loss of resources and sense of place and identity. For the same reasons (insufficient infrastructure and poor-standard housing), risks of extreme precipitation and inland flooding are concentrated in urban areas. Related collapse of health, emergency and electric power systems further increase the systemic risks to cope with climate change impacts. Warming, droughts and precipitation changes threaten food security and water supply in both urban and rural areas but with higher risk in poor regions. The poor endowment of farmers in dry areas, in conjunction with limited ability to compensate for losses in water dependent farming, leads to loss of agricultural productivity and livelihoods, which further accelerates the risk of insufficient food supply (Field et al., 2013, pp. 59–60, 68, 70–71). Existing health problems are predicted to intensify until the middle of the 21st century [very high confidence]. Ill-health is expected to increase for low-income developing countries in particular [high confidence] resulting from undernutrition [high confidence], heat waves, food- and water-borne diseases [all with very high confidence] as well as vector-borne diseases [medium confidence] (Field et al., 2013, p. 71).

Poor countries and marginalized poor people within countries are particularly vulnerable to present impacts and future risks as a result of their lower adaptive capacity and greater exposure to climate change. Negative climate impacts, in turn, make it more difficult to escape poverty. Present climate consequences already pose a threat to sustainable development. With predictions

19 Non-climate related factors include but are not limited to social, economic, cultural, political and institutional factors. For an overview see Field et al. (2013, p. 50).

20 The following description is of an exemplifying nature. It is limited to exposure and social vulnerability to climate hazards, leaving aside economic, environmental and institutional vulnerability.

of risks in mind, future effects of climate change will make sustainable development even more difficult to achieve. For the second half of the 21st century in particular, risks are substantially lower in a low-emission scenario (RCP2.6) in comparison to a high emission scenario (RCP8.5) (Field et al., 2013, p. 62). Mitigation reduces warming directly, decreases future risks and increases the time available for adaptation. But mitigation is more than that. Mitigation is the key premise for many regions to grow out of poverty. (Field et al., 2013, pp. 73, 87).

Though underdeveloped regions are certainly most vulnerable, current impacts and future risks are not limited to developing countries. The U.S. Global Change Research Program, for example, predicts various risks across the United States (US), ranging from unknown but likely severe impacts on groundwater, increased "diseases transmitted by food, water, and insects", difficulties to meet "air quality standards necessary to protect public health" and increased risks of illness and death through extreme weather events to (existing and expected) conflicts over water supply (Karl, Melillo, & Paterson, 2009, pp. 47–48, 89). The Stern Review predicts that "heat waves like 2003 in Europe, when 35,000 people died and agricultural losses reached $15 billion, will be commonplace by the middle of the century" (Stern, 2007, pp. 10–11, 112). And the IPCC in its Fourth Assessment Report (AR4) predicts water security to be an increased challenge to many regions in Australia and New Zealand as early as 2030 (Hennessy et al., 2007, p. 509).

1.2 Individuals and Climate Change: Current State of the Debate

Facing the extreme current and expected impacts, there is little disagreement that harm due to climate change is morally unacceptable. Likewise, there is little disagreement that climate change is caused by a variety of actions performed by a vast number of agents. Even at the level of individual agents, as the main research object of this book, climate change is caused only collectively by an almost infinite number of individual contributions. On the one hand, individual contributions only represent a small share of global emissions only and individual efforts to tackle climate change are too tiny compared to the global impact on the climate system. On the other hand, climate change is not a natural disaster but seems to be collectively avoidable if only every individual agent was to adopt a green lifestyle. From a moral perspective, do individuals have a duty to refrain from emission-intensive actions?

On the account of traditional theories of individual responsibility, which focus on a clear and direct causal connection between individual actions and climate change, it is hard to justify such a duty.[21] After all, it is not the individual action in isolation that causes or prevents global climate change. Early contributions to the debate on individual duties with regard to climate change, have therefore defended a position that regards individual actions to be negligible and gives way to a purely institutional account. Walter Sinnott-Armstrong (2005) is probably the most prominent representative. According to institutional accounts, fighting climate change is the job of governments and politicians but not of individuals.[22] In response, many different objections have been proposed. Augustin Fragnière (2016) has reviewed the incredible number of approaches made on individual climate duties. He summarises that most writers agree that individuals face at least some kind of duty but disagree about nature and extend (Fragnière, 2016, p. 809).

The challenge that individual actions make (or appear to make) no difference to collective harms like climate change has been referred to as "causal inefficiency" (Garvey, 2011; Hiller, 2011) or "inconsequentialism" (Sandberg, 2011; Sandler, 2010). Fragnière (2016) breaks the problem down to four key arguments discussed in philosophical literature:

(1) Individual actions are too small to be significant or perceptible,
(2) given a climatic threshold, individual emissions cause no marginal harm,
(3) there is no clear, direct causal pathway between particular individual actions and climate-induced harm and
(4) there are settings (e.g. flying) where the same amount of emissions will be emitted anyway regardless of the individual action.

In the past years, philosophers have put in great effort to challenge these arguments. Broadly, writers work in one of the following two directions: Either they modify existing theories, or they develop new theories that define duties without relying on the traditional conditions of individual moral responsibility. Each account is complex and is followed by an even more complex debate that cannot be fully reproduced here. The description below serves as a rough overview. It largely follows the review provided by Fragnière (2016). Because Fragnière occasionally expresses his opinion on parts of the discussion too, his and also the most recent contributions are added where necessary or useful.

21 The following and chapter 2 will elaborate on this problem in more detail but it has been repeatedly discussed before (see e.g. Jamieson, 2010; Lichtenberg, 2010; Scheffler, 1995).

22 The general idea has been raised earlier by B. Johnson (2003) with reference to the tragedy of the commons. Chapter 2 will examine Sinnott-Armstrong's argument.

Insignificance and Imperceptibility

Regarding the claim that individual actions are insignificant or imperceptible (Sinnott-Armstrong, 2005), several philosophers have tried to prove the general idea wrong that individual contributions can be without an effect at all. Their key argument is that imperceptibility must not be confused with non-existence (Hiller, 2011; McKinnon, 2014; Otsuka, 1991; Sandberg, 2011). Indeed, the judgement that individual actions make no difference to climate change creates what philosophers refer to as a *sorites paradox*: if no individual action made a difference, anthropogenic climate change would not exist (McKinnon, 2014, pp. 37–38).[23] Opponents may stress that effects must be more than trivial to be morally relevant (Feinberg, 1984, p. 269). By following set-theoretic thinking as defended by Derek Parfit (1984), many philosophers point out the risk of overlooking the effects of sets of acts and ignoring small or imperceptible effects (Almassi, 2012; Attfield, 2009; Cuomo, 2011; Hiller, 2011; W. Peeters, De Smert, Diependaele, & Sterckx, 2015; Schwenkenbecher, 2014; Vanderheiden, 2008).[24] Again, others disagree with Parfit's thinking (Jamieson, 2014, p. 288; Maltais, 2013; Sandberg, 2011). Some scholars have tried to empirically invalidate the claim that individual actions cause nothing but insignificant effects: John Nolt estimates that the amount of harm the emissions of an "average American" citizen produces over a lifetime, affects "the suffering and/or death of [up to] two future people" over the next millennium (Nolt, 2011, p. 9). John Broome calculates the effect of lifetime emissions to cause six months of healthy human life or monetary costs of up to 65,000 US-Dollar (Broome, 2012, p. 224) and Avram Hiller compares an individual action's effect with "ruining someone's afternoon" (Hiller, 2011, p. 357). As a reply, some philosophers have stressed the relevance of marginal effects instead of total individual emissions (Kawall, 2011) and again others have challenged the moral significance of statistical deaths (Seager, Selinger, & Spierre, 2011).

---

23 The Oxford Dictionary of Philosophy defines a sorites paradox as follows: "One grain of sand is not a heap. And for any number n, if n grains of sand are not a heap, then the addition of just one more grain does not make them a heap. But in that case you can never get a heap, for each grain you add leaves you just as much without a heap as before. […]" (Blackburn, 2008, p. 344).

24 Facing globalization and interconnectedness, Parfit (1984) stresses that we must change the way we think about the effects our actions have. He defines five "mistakes in moral mathematics" (Parfit, 1984, pp. 67–86). At least two of them (ignoring the effects of sets of acts as well as of small and imperceptible effects) may apply to climate change.

Thresholds and No Marginal Harm

Thresholds are essential to the climate debate. Given the natural absorbing capacity of the Earth, there is some level of emissions that, were everyone emitting below that level, there would be no harm. Some philosophers clarify that individual annual contributions before overshooting the planet's capacities each year should still be considered to cause relevant harm. They stress that annual emissions are being emitted in the context of already too high emissions (Almassi, 2012; Fragnière, 2016). Given the IPCC's findings presented in the first section of this chapter regarding cumulative carbon emissions in particular, an argument trying to make some amount morally legitimate each year can safely be rejected. However, building on the planet's natural threshold, another argument goes that individual emissions above the threshold do not make a difference. Each contribution is sufficient but not necessary to cause climate change. These cases are referred to as cases of overdetermination. Many philosophers have argued that climate change is such a case (Barry & Øverland, 2015; Cripps, 2013; Hohl, 2017; Jamieson, 2014; B. Johnson, 2011; Sandberg, 2011; Sinnott-Armstrong, 2005). Fragnière points to the discussion about the moral status of overdetermined cases (Lawford-Smith, 2016; W. Peeters et al., 2015; Vanderheiden, 2008) and concludes: "Climate change is actually not a case of overdetermination." (Fragnière, 2016, p. 801). He compares climate change with a typical overdetermined case of five soldiers firing at a prisoner. Each shot would have been sufficient to kill the prisoner, but none is necessary. In his opinion, climate change is different. Beyond some climatic threshold "each supplemental amount of GHG makes the problem worse" and therefore causes marginal harm (Fragnière, 2016, p. 801). That, however, sends the discussion back to insignificance and imperceptibility.

Another line of thought builds on micro-thresholds or the existence of several climatic tipping points. It argues that individual actions risk triggering the next point or threshold. Scholars who defend such a line of thought suggest focusing on expected marginal harm instead of actual marginal harm (Almassi, 2012; Cullity, 2015; Lawford-Smith, 2016). Critics point to the fact that there is no evidence that it is possible for individuals to trigger a threshold by individual actions (Cripps, 2013; Maltais, 2013).

No Direct Causality

Given that climate change is caused only by the accumulation of tiny amounts of emissions, individual actions are connected to the effects of climate change only indirectly. Individual polluting actions in isolation therefore fail to fulfil the relevant criterion for causality (Adler, 2007; Fragnière, 2014; Jamieson, 2010;

Posner & Weisbach, 2010, p. 232; Sinnott-Armstrong, 2005; van de Poel, Nihlén Fahlquist, Doorn, Zwart, & Royakkers, 2012). Against this line of thought, some philosophers have claimed that a globalized world with "new harms" (Lichtenberg, 2010, p. 557) requires a fundamental revision of the traditional causal understanding towards an understanding that is able to fully consider indirect effects (Attfield, 2009; Kernohan, 2000; Lichtenberg, 2010). Others suggest that, instead of direct causality, the assessment of effects should focus on individual contribution (Braham & van Hees, 2012; Hohl, 2017; Nefsky, 2017; Pellegrino, 2018; Shockley, 2017; Wright, 1985).

Same Amount of Emissions

Regarding the effects of air travel, some writers have claimed that the number of flights will remain unchanged and so do emissions regardless of individuals booking a seat or not (Garvey, 2011; Sandberg, 2011). In response others evaluate the *expected* harm of individual actions (Almassi, 2012; Kagan, 2011; Morgan-Knapp & Goodman, 2015; Pellegrino, 2018). The key argument is that expected harm from individual action, such as buying a plane ticket, is positive. Assuming that markets work according to the paradigm of supply and demand, buying a plane ticket is expected to increase the supply of flights in the long-run and with it the amount of emissions from air travel.[25]

Facing the complex discussion on causal inefficiency, other philosophers have defended accounts that try to fully forgo proving causal efficiency by referring to virtue, universalization, integrity and participation or fairness.

Virtue

On virtue-based accounts, the focus is on character traits that individuals ought to have by changing their behaviour in a certain way (Jamieson, 2007b; Sandler, 2010). With regard to climate-related issues, virtues, such as humility, temperance, mindfulness or respect for nature, are usually designed to bridge the gap between individual acts and collective outcomes (Jamieson, 2007b, 2010, 2014). Given the high reliance on fossil fuels and a society that is not yet prepared to act collectively, critics argue it may be hard to live up to these character traits in modern societies (Cripps, 2013). Claiming that virtuous behaviour is "morally ideal" but not "morally required", others argue that virtues simply fail to qualify to be moral requirements at all (Kingston & Sinnott-Armstrong, 2018, p. 171).

25 For further clarification see Fragnière (2016, p. 803).

Universalization

Accounts of universalization focus on individual duties to reduce personal carbon emissions by referring to hypothetical universalization. Adopting principles of Kantian ethics, they ask what the individual duty would be, if everyone acted that way (Garvey, 2011; Rentmeester, 2010; Wallace, forthcoming). According to Kant's Categorical Imperative, the decisive factor for an act to be morally permissible is that individuals act on a maxim that may be universalized without contradiction. Analogous to Kant's testing of the maxim 'make false promises', Casey Rentmeester (2010), for example, tests the maxim 'commute to work' and finds that if everyone commuted to work, climate change would increase dramatically, leading to civilization collapse, which would in turn undermine the mere possibility of commuting to work. Others have argued that even though the consequences are very bad, maxims that focus on polluting actions do not cause contradiction (Sinnott-Armstrong, 2005). Others clarify that the discussion is with regard to the logical and practical contradiction, but many philosophers agree that such testing of maxims depends on empirical knowledge (Cripps, 2013; Fragnière, 2016; Hohl, 2017; Kutz, 2000). Indeed, polluting actions are not in themselves contradicting if universalized. If the planet was, for example, much less populated or if the natural absorptive capacity of the planet was infinite, polluting actions would not be a problem. On some readings, maxims are indeed ruled out if the contradiction is practical. Some object that such readings have to include information about the context and the expected acting of others and would thereby move too far away from Kant's initial idea of internal testing (Cripps, 2013). In contrast to testing certain maxims with regard to particular polluting actions, testing rather general maxims such as 'live unsustainably' is another option (Wallace, forthcoming). However, others object that such a general maxim is a poor guideline for particular individual actions (Fragnière, 2016). Again, some scholars reply that Kantian ethics is generally too far apart from considering environmental consequences as side-effects of individual everyday actions (Jamieson, 2007b; Sandler, 2010).

Integrity and Participation

Writers relying on integrity or participation argue that an individual should refrain from polluting actions either because it brings harm through the individual's agency (together with others) (Broome, 2012; Lichtenberg, 2010) or because the individual participates or is complicit in a system of harmful global practices (Hale, 2011; Kutz, 2000; Lichtenberg, 2010). Individuals' actions express our values and beliefs to others, which makes complicity bad. Others reject the complicity-argument by saying that polluting actions are not intrinsically wrong, which makes individual actions relevantly different from complicity in other collective wrongs such as racism or child labour (Maltais, 2013; Sinnott-Armstrong, 2005).[26] An integrity-argument has been pointed out by Marion Hourdequin (2010). She would rather refer to consistency as an internal integrity to bring one's values, believes and actions in line with each other.

Fairness

On fairness-accounts, every individual agent ought to do their part of the collective action to tackle climate change (Schwenkenbecher, 2014). One way is to reason that individuals are obliged to do their part if their personal emissions exceed their fair share because they not only contribute to harm but because they unfairly emit someone else's share. Given that most people living in the industrialized countries emit far beyond their share, Christian Baatz argues that those individuals who emit beyond their fair share have an imperfect Kantian duty to reduce their emissions "as far as can reasonably be demanded of them" (Baatz, 2014, p. 15). Baatz explains that limiting one's own emissions to a fair share may be unreasonable if limiting requires giving up on aspects of a decent life. Due to differences regarding personal circumstances, he argues that reasonability requires assessment "on a case by case basis" (Baatz, 2014, p. 15). Baatz has been criticized for failing to provide a convincing justification of the argument of having a fair share (Bowman, 2014), for applying a subjective account of what can be demanded (Duus-Otterström, 2014) and for leaving aside options for offsetting (Neuteleers, 2014).[27]

The above is only a brief overview of almost fifteen years of discussion regarding individual actions and climate change. In view of the closing window that is left to effectively fight climate change, I believe, it is time to move away from accounts that try to prove causality of individual actions.[28] Yet, I also believe that there is no relevant reason to forgo causality altogether. Climate change is caused collectively. Thus, climate change is avoidable on the collective level too and that provides sufficient reason, I believe, to stick with morally criticizing action – collective action. However, as I will show, focusing on collective action does not mean to leave individuals without a duty either.

Though connected to several elements of the debate, a different approach that is not conducted towards individual lifestyle changes directly is the

26 The examples for collective wrongs are taken from Fragniére (2016).

27 For a further reply see Baatz (2016).

28 Nonetheless, I will discuss a recent and probably most coherent approach focussing on contribution defended by Hohl (2017) in the course of this book.

theory defended by Elizabeth Cripps (2013). Facing climate change as a collective problem, she argues for a derivative individual duty to promote collective level action. Building on the extensive literature review outlined above, Fragnière claims that "[v]irtually every author (at least among those taking a stance) acknowledges that individuals have some duty to promote collective action" (Fragnière, 2016, p. 807). The idea to focus on collective action rather than individual lifestyle changes is part of early institutional accounts (such as the one defended by Sinnott-Armstrong, 2005). Cripps defends the primacy of promotional duties over duties to change individual behaviour. Nonetheless, accounts in defence of isolated individual duties usually account for promotion too. They defend a promotional duty that builds subsequently on reducing one's personal carbon footprint directly. These individual promotional accounts rely on the communicative potential of changing one's own behaviour towards a green lifestyle. The communication potential usually refers to setting an example and creating domino or amplifying effects. The argument goes: Regardless of a justification of individual behavioural changes, my riding a bike instead of taking a car or my travelling by train instead of taking a plane will encourage others to act alike. In contrast, institutional accounts of promotional duties argue that tackling climate change requires changes in the regulations and public institutions that shape our everyday behaviour. In order to locate Cripps' account, it is helpful to introduce collective accounts of promotional duties in contrast to individual and purely institutional promotional accounts. Collective promotional accounts aim to justify the moral duty to promote the acting of sets of individuals.[29] On promotional accounts, Fragnière further reviews: "But when it comes to a more precise justification and description of promotional duties, the debate is still at an early stage [...]." (Fragnière, 2016, p. 807). Following Cripps' approach, I am prepared to justify, modify and advance the normative foundation, the idea of weakly collective duties as well as the nature of promotional duties.

1.3 Outline and Structure of the Book

This book deals with individual duties to fight climate change. Speaking about individual duties, I simply ask what regular individual agents ought to do with

29 Obviously, institutional accounts are collective accounts. They narrow collectivities to institutionalized collectivities (in the sense of structured, organized and corporate) that are able to act intentionally. In contrast, Cripps develops an account that captures individuals as collectivities yet unable of intentional action.

regard to climate change. Assessing individual duties is certainly a complex and challenging task. These duties, it seems, are not grounded on the standard thinking of individual responsibility focussing on the direct causal effects of either individual action or individual failure to act. The duties that I argue for are grounded in the moral viciousness of peoples' severe suffering.

There seems to be great disagreement among philosophers regarding the relationship between responsibility and duties. In the climate ethics discourse, some philosophers understand responsibility in a backward-looking sense of guilt, blame, and even punishment. Some others apply causal responsibility and focus only on those who cause climate change and again some others seem to complement retrospective responsibility with moral duties that refer to forward-looking responsibility. To provide some clarification, I understand responsibility in a forward-looking sense and apply a broad understanding. In view of something that is morally wrong (in this book: climate-induced, severe harm), responsibility addresses all those moral agents who are required to respond in some way. Thus, all those who are responsible, have a duty to act accordingly.

Applying one of the two main deontological theories, I defend a rights-based moral approach to climate duties. Against the commonly held misunderstanding that deontological theories are indifferent to consequences from certain actions, Klaus Steigleder (2016b), following Alan Gewirth (1978), has argued that rights-based moral theories cover consequences – consequences of a special kind. In contrast to consequentialist theories targeting cumulative effects or overall outcomes, rights-based theories consider distributive consequences to individual agents. An individual agent may be obliged to refrain from acting in a certain way if her action or its consequences interfere with the rights of other agents (Steigleder, 2016b, p. 178).

Focusing on the prevention of further climate-induced harm, it is safe to say that no individual agent could tackle climate change on her own. In view of the collective nature of climate change, it seems reasonable to apply duties to collectivities as a result of their collective responsibility. However, standard theories of collective responsibility are just as poorly equipped with regard to climate change as our standard thinking of individual responsibility. Following Cripps (2013), I will argue that there is neither a collectivity that intentionally causes climate change nor a number of individuals who intentionally contribute to climate-induced harm. Given that individual and collective approaches are unable to fully capture the climate challenge, Cripps' theory is strongest for rethinking collectivities. As will be shown, the new understanding of collectivities addresses sets of individual agents that are weakly connected by mutual dependence. These weak collectivities face weakly collective duties to organize themselves as necessary to enable climate action.[30] The relevant weak collectivities with regard to climate change are the Young, the Able and the Polluters. Individual duties derive from weakly collective duties: In the absence of collective-level action, individuals as members of the relevant weak collectivities primarily ought to promote collective climate action.

The next chapter accelerates readers' problem awareness and dives into the details of identifying individual moral duties for climate change as a collective problem (chapter 2). It explains why existing theories fail to capture moral duties of individual agents in the case of climate change and sets the stage for Cripps' new approach.

Chapter 3 introduces the idea of individuals as collectivities. The first part examines Cripps' theory in detail, presents the moral foundation for weakly collective duties, the concept of non-intentional collectivities and the normative principles guiding duties for each collectivity. Further, I explain Cripps' reasoning by answering why it is not enough to do one's part and why promotional duties are primary to direct individual duties (chapter 3.1). In chapter 3.2, I introduce three objections raised against individual moral duties in general and promotional duties in particular: the no-competence objection (chapter 3.2.1), the fairness-objection (chapter 3.2.2) and the demandingness-objection (chapter 3.2.3). Chapter 3.3, introduces the three major areas that I wish to advance: the normative foundation (chapter 3.3.1), the argumentation for weakly collective duties (chapter 3.3.2) and the nature of promotional duties (chapter 3.3.3).

According to the areas that I wish to advance, the following chapters are designed to offer an alternative normative foundation (chapter 4), to advance the concept of weakly collective duties (chapter 5) and to describe the nature of promotional duties in more detail (chapter 6).

The first part of chapter 4 introduces the rights-based moral theory defended by Gewirth (chapter 4.1). Starting off with some preliminary remarks on basic concepts, considerations and the method used (chapter 4.1.1), I subsequently elaborate Gewirth's foundational argument, which leads to a supreme moral principle (chapter 4.1.2). The first section closes my two additional remarks (chapter 4.1.3). The second part of chapter 4 examines the implications of the foundational argument in practice (chapter 4.2). I examine the direct applications (chapter 4.2.1) as well as the indirect applications (chapter 4.2.2) before demonstrating how Gewith's theory is able to solve cases of conflicting rights (chapter 4.2.3). The third part of chapter 4 examines Cripps' ideas of weakly

30 Note that the term 'weakly' refers to the sense in which the harm is collective, not to the force of the duty (Cripps, 2013, p. 3).

collective and individual promotional duties from a rights-based perspective (chapter 4.3). Following Gewirth, I attempt to provide guidance on demandingness and reasonable costs in the fourth part (chapter 4.4).

Chapter 5 advances Cripps' idea of weakly collective duties. I start by defending weakly collective duties (chapter 5.1) before examining the case of the Polluters (chapter 5.2). With regard to the Polluters, I consider predictability and avoidance as the key conditions for assigning weakly collective responsibility to the Polluters (chapter 5.2.1). The following section discusses Cripps theory in distinction to an approach of direct individual duties defended by Sabine Hohl (2017) by asking whether climate change is yet a matter of individual contribution (chapter 5.2.2). Subsequently, I elaborate whether individuals have a limited duty to personally refrain from emitting (chapter 5.2.3) before I defend the need for a threshold-level of emissions (chapter 5.2.4). The second part of chapter 5 closes by an important remark regarding the Able. Yet, this remark results from the discussion regarding the Polluters. Because the Able have received wide-spread attention in the philosophical debate already, I focus primarily on the case of the Polluters and the Young. With regard to the Young, I wish to focus on a particular sub-group of individuals in the third part: young parents (chapter 5.3). This chapter asks whether young parents have a duty to mitigate as young parents (chapter 5.3.1) and whether young parents have a duty to ensure mitigation in the long-run by raising children to be climate-aware and responsible agents (chapter 5.3.2).

Chapter 6 elaborates the nature of promotional duties in greater detail. The first section suggests two channels of promotional action (chapter 6.1). The second section attempts to conceptualise the transformation from individual agents who have a promotional duty to required collective climate action (chapter 6.2). Next, I elaborate different factettes of the promotional duties of the Young (chapter 6.3), the Able (chapter 6.4) and the Polluters (chapter 6.5). Regarding the Polluters, I primarily focus on promotional duties (chapter 6.5.1) but also on what it means to live up to the promotional duties personally (chapter 6.5.2).

CHAPTER 2

The Challenge: Identifying Individual Duties for Collective Harm

In some situations, it is easy to identify the moral duty of individual agents to act in a certain way: A has a duty not to kill B even if she were to receive one billion dollars. C has a duty not to attack D even if D is barring C's way. Assuming that E is not in dire need, E has a duty not to steal from F even if F is a very rich person and would not even notice her loss. Our standard thinking of individual responsibility that focuses on the direct effects of individual actions or individual failures to act tells us that A, C and E are morally responsible, which imposes a correlative duty not to act in these ways. These are "paradigm moral problems" as Dale Jamieson (2007a) puts it, where "an individual acting intentionally harms another individual; both the individuals and the harm are identifiable; and the individuals and the harm are closely related in time and space" (Jamieson, 2007a, p. 475). Unfortunately, climate change is none of these problems. Climate change is a large-scale problem caused by a vast number of actions performed on a daily basis by a similarly vast number of agents. Climate-induced impacts severely harm other agents – agents who are sometimes distant in time and space. And yet, my driving my car or taking a plane does not have the same moral status as A killing B, C attacking D or E stealing from F. Driving or taking a plane in isolation neither causes climate change nor does its omission make a difference to threats of climate change. Individual polluting actions are – as Cripps puts it – "harm ingredients" to an overall outcome that take effect only if they are put in combination (Cripps, 2013, p. 120).

Taking up on this line, Sinnott-Armstrong (2005) has famously argued that there is no moral duty for individual agents not to perform emission-intensive actions such as taking a fun-drive in a gas guzzling Sport Utility Vehicle (SUV) on a Sunday afternoon.[31] After analysing several principles that might provide grounds for an individual moral duty to refrain from fun-driving, he finds that fun-driving does not cause harm in any relevant sense even though it unnecessarily emits carbon (Sinnott-Armstrong, 2005, p. 311). His key argument is that the individual act of driving for fun is neither sufficient nor necessary

31 Wasteful fun-driving is an example that has been often used by proponents and opponents of direct individual moral duties to refrain from emission-intensive actions (see e.g. Baatz, 2014; Hiller, 2011; Jamieson, 2014; Sinnott-Armstrong, 2005).

 | DOI:10.30965/9783957437426_003

for climate change (Sinnott-Armstrong, 2005, p. 297). Small quantities of carbon could "help plants grow", thus carbon in itself is not problematic either (Sinnott-Armstrong, 2005, p. 298). Climate change occurs due to massive quantities being released into the atmosphere, while slowly exceeding the threshold. As climate change is a massive phenomenon, he argues, the individual act does not make a difference and is therefore negligible (Sinnott-Armstrong, 2005, p. 298). Given that individuals in isolation are unable to solve climate change, Sinnott-Armstrong stresses that governments have a moral obligation to fight climate change. In contrast to individuals, governments can make a difference. Instead of refraining from emission-intensive actions, individuals should rather enjoy fun-driving while making governments fulfil their duty by working to make fun-driving illegal by law (Sinnott-Armstrong, 2005, pp. 295, 300, 305).

Numerous philosophers have challenged Sinnott-Armstrong's view. As an adequate reconstruction of every reply goes beyond the scope of this book, I have limited myself to presenting a rough overview (chapter 1.2). After Aaron Maltais (2013) has reviewed many of the proposed objections against Sinnott-Armstrong, Ewan Kingston and Sinnott-Armstrong (2018) have recently replied to key objections themselves. With regard to group causation they write:

> [W]hen millions of people joyguzzle during a year, they do not coordinate, intend to cause harm, or directly transfer energy to any particular target. We do not say that these factors are necessary to make single acts in a harmful group of action wrong, but we strongly suspect these factors create the impression of causation […]. (Kingston & Sinnott-Armstrong, 2018, p. 172)

This is exactly the starting point for Cripps to develop a theory for weak collectivities and to identify their duties. Kingston and Sinnott-Armstrong (2018) move on with the following:

> If proponents of the argument from group causation believe that they have example [*sic*] to support the principle without the element of intention, conspiracy, or direct energy transfer, then we would be happy to consider them. But so far we are not convinced that such examples exist. (Kingston & Sinnott-Armstrong, 2018, p. 173)

The following section, in connection with chapter 3.1, will show that Cripps' account varies significantly from traditional group-based accounts. Cripps has presented an account that does not rely on intention and is still able to assign moral duties to collectivities – but collectivities in a weak sense. Interestingly,

Kingston and Sinnott-Armstrong (2018) refer to Cripps (2013) with regard to the question, whether a moral duty to get governments to do their job creates a duty to refrain from emission-intensive actions oneself (Kingston & Sinnott-Armstrong, 2018, p. 183). They limit Cripps' approach to the applied part of her theory. The theoretical part, however, is in fact able to provide what Kingston and Sinnott-Armstrong – in their own words – "would be happy to consider" (Kingston & Sinnott-Armstrong, 2018, p. 173). Their view on Cripps' theory is pretty much in accordance with many others who consider her attempt as just another defence of political duties. This creates the impression that the strength of Cripps' theoretical part has been overlooked since its publication.[32]

Cripps has criticized Sinnott-Armstrong for focussing only on state-level changes (Cripps, 2013, p. 141).[33] Indeed, Sinnott-Armstrong neglects that there is something in between the personal and the state-level. He overlooks the difference between individuals in isolation and individuals in combination as collectivities. As members of collectivities individuals have a duty to act against climate change. That is not, as chapter 3.1 will show, in the sense of blameworthiness or guilt but in a sense that individuals as members of the respective collectivity cannot claim themselves to be fully apart from climate change.

The number of cases of collective harm during transatlantic slave trade, colonial rule, apartheid or genocide during World War II lead to an increasing dissatisfaction with standard individual moral theories back in the 1950s (May & Hoffman, 1991; May & Tuomela, 2007). Philosophers initiated a discussion about collective responsibility – the responsibility of any collectivity taken as a whole. It has commonly been applied to organized institutional collectivities such as states or corporations. Sinnott-Armstrong's focus on state-level action must probably be seen in par with this line of thought. Likewise, collective responsibility has been attributed to sets of individuals: A group of teenage friends pushing someone's car over a cliff, a family that commits murder to restore family honour and maybe even a set of bystanders who tolerate or ignore someone being harmed and abused.[34] Over decades, the dominant view was that sets of individuals form collectivities either because members think

32 Fragnière (2015) is an exception.

33 Others even claim that contrasting the state-level solution on the one hand with a purely individualistic solution on the other is inaccurate (Hohl, 2017, p. 155; Schwenkenbecher, 2014).

34 The example of people jointly pushing a car over a cliff has repeatedly been used (see e.g. Braham & Holler, 2009, pp. 149–150; Mellema, 1997, pp. 40–41; Sverdlik, 1987). For a similar version see Hardin (1988, pp. 156–157).

of themselves as being a member of the collectivity or because the collectivity jointly intends or intended to act in a certain way.[35] Cripps refers to this view as the intentionalist view (Cripps, 2013, p. 28).

One of the most influential intentionalist accounts has been extensively defended by Margaret Gilbert. In early contributions Gilbert has focused on peoples' actions together and jointly being able to act (Gilbert, 1989). Later, she focuses rather on collective intention (Gilbert, 1997, 2006, 2009). Starting off from a case of two individuals goings for a walk, Gilbert claims that such an activity requires so-called "joint commitment" as "a kind of commitment of the will" (Gilbert, 2006, p. 134). Gilbert explains: "In this case [the case of the joint commitment as a commitment of the will], the wills of two or more people create it, and two or more people are committed by it." (Gilbert, 2006, p. 134). Joint commitment is brought about only by individuals jointly together. Bringing something about jointly does not require a "single centre of consciousness" or "a distinct form of 'subjectivity'" (Gilbert, 2006, p. 134). It is rather a commitment to do something "as a body" (Gilbert, 2006, p. 137, 2009, pp. 179, 180–182). Doing something as a body may include having an intention, belief, acceptance, and so on. To intend something as a body, for example, is the commitment of two or more individuals to constitute themselves to intend something as uniform as possible. Individuals jointly commit to establish a "single body" (Gilbert, 2006, p. 137), which is probably best understood as a kind of unified entity. Joint commitment implies a normative relation between agents: Each agent is committed in some way, has the duty to act according to the joint commitment and is entitled to claim other agents to conform their actions accordingly.[36] Agents mutually owe these actions to each other (Gilbert, 2009, pp. 175–176).

Taking up upon this structure of joint commitments, Gilbert introduces a type of collectivity, which she calls plural subject: "A and B constitute a *plural subject* (by definition) if and only if they are jointly committed to doing something as a body – in a broad sense of 'do'" (Gilbert, 2006, p. 145). As plural subjects, individuals may have the mental states mentioned above: intentions, beliefs, acceptance, and so on. Everyday speech covers these states of plural subjects for example by saying 'we believe that X is a good political strategy', 'we accept that law', 'we intend to climb the mountain'.[37] In this particular

35 Shockley (2013) provides further classification.

36 Note that Gilbert does not talk about *moral* obligation even though she compares the nature of mutually owing something to the nature of claim-rights. It is a "direct obligation" that exists because of the social relation between agents in jointly committing to act (Gilbert, 2009, p. 175).

37 For more examples see Gilbert (2006, p. 145).

sense, Gilbert regards individuals as social groups to have – what I refer to as – collective intention. Gilbert names it "shared intention" but highlights that joint intention or collective intention are equally appropriate (Gilbert, 2009, p. 168, footnote 6).

Collective or shared intention may be broadened to cope with large scale cases (Gilbert, 2006, chapter 8.3, 2009, p. 168, footnote 5). More generally, Gilbert writes: "Members of some population P share an intention to do A if and only if they are jointly committed to intend as a body to do A." (Gilbert, 2009, p. 179). That does not mean that shared or collective intention necessarily requires correlative personal intentions of the individual parts (Gilbert, 2009, p. 172). The two individuals from the basic case above might agree to go for a walk along the river tomorrow. One of them might change her mind over night, still join for the walk because they jointly agreed to meet but tells the other individual about her lack of intention half way down the river. Gilbert believes that having correlative personal intentions can be a sufficient condition for collective intention but having correlative personal intentions is not a necessary condition for collective intention (Gilbert, 2009, p. 173). Collective intention cannot be changed or rescinded unilaterally by a single agent. It requires the concurrence of all agents to initiate change or to even release one agent from participating in collective intention (Gilbert, 2009, p. 173).

Moreover, Gilbert applies her theory to societies. For societies, there is a high degree of impersonality and anonymity: Members might only know a fraction of the group personally and they might not even know that particular members of the group exist (Gilbert, 2006, p. 174). And still, according to Gilbert, impersonal and anonym societies qualify as plural subjects based on two conditions: First, all members of the population must have expressed their readiness to participate in the joint commitment and, second, this readiness must be "population common knowledge" (Gilbert, 2006, p. 175). One way is to refer to existing institutions such as citizens of the European Union (EU), people living in the United Kingdom (UK), the Americans, and so on. However, Gilbert further argues that no such institutional reference is in fact necessary. Sharing any concept of being a population is sufficient (Gilbert, 2006, p. 175). Examples may be 'those who grow wine along the Rhine', 'indigenes people living above the polar circle', or 'the professional hunters in the South'. None of the examples refers to existing institutional settings and still, they qualify as plural subjects united by a joint commitment.[38]

Gilbert's conceptions of joint commitment, plural subjects and shared intention have been heavily criticized. Some critics claim to identify circularity

38 For more examples see Gilbert (2006, p. 175).

(Tollefsen, 2002; Tuomela, 1992). They argue that the result of a joint commitment to form a plural subject, in fact, presupposes a joint commitment to form such a subject. The argument could go as follows: The joint commitment of two agents to go for a walk as a plural subject presupposes that both are jointly committing to jointly commit to go for a walk. As mentioned above, regarding the expression "a kind of commitment of the will" (Gilbert, 2006, p. 134), Gilbert holds that before the two are jointly acting as a single body, they must be willing to do so in the sense of openly showing to be ready to act and the readiness to jointly commit must be common knowledge to all parties (Gilbert, 2006, p. 138). Nevertheless, the risk of circularity or at least of infinite regress is probably not to be dismissed fully. Further, owing each other to act in a certain way may not be regarded as a mutual obligation that *constitutes* joint action. Michael Bratman (1999, chapter 7) for example regards it rather as a *consequence* of joint action by creating expectations and fulfilling them. In Bratman's view, commitments are essentially part of the structure of joint action or joint intention. Another criticism targets Gilbert's claim of being able to capture large groups, communities or societies: Starting off with the case of two individuals jointly agreeing to go for a walk, there might be difficulties that arise from directly transferring joint commitments from two agents to numerous, anonymous agents given that collective intention can be changed or rescinded only concurrently (Baltzer, 2002). Still others criticize Gilbert's assumption that collective intention may be strictly independent of personal intentions (see e.g. Schlothfeldt, 2009; Tuomela, 1995).

The remarks above introduced a very comprehensive theory that has been discussed extensively throughout the years. Also, my remarks on Gilbert are only representative of a number of intentionalist theories.[39] The following should have become clear: Gilbert's contributions are valuable and elaborate works regarding collective intention, and they provide a very profound starting point when it comes to intentional collective action and corresponding collective responsibility. In order to count as a collectivity in the strong, intentional sense, individuals must either jointly commit to do something, or in large-scale cases, members of a set must be organized or connected in a way that forms a number of individual agents into a unified entity.

Returning to the case of climate change, it is hard to see where individual agents jointly commit to pollute. Also, individual agents are neither connected in a way that constitute a unified entity, nor are they organized to cause climate change. Our using the car in order to drive the kids to school, fetch

39 Further intentionalist accounts are for example defended by Caws (2005), Gould (1988) or Tuomela (2002).

roles for breakfast or be in a business meeting with the boss on time and even, taking up Sinnott-Armstrong's example, fun-driving with a powerful SUV is no joint commitment to act as a body to cause climate change. Individuals try to achieve their personal goals: getting the kids to school safely and on time, getting a good breakfast, satisfying the boss, enjoying the feeling of power, superiority and recklessness of hundreds of horsepower, and so on. By achieving their goals, individuals are situated in relation to each other so that climate change happens to occur. Individuals no way jointly commit to act in these ways. As a consequence, individuals can neither be regarded as a unified entity nor to have collective intention.

Yet, it may be objected that individuals, as human beings with social ties living in a globalized and connected world, do not act in bubbles either. Indeed, individuals can at least be expected to foresee that there are enough others acting in similar ways. Also, they can reasonably be expected to anticipate that their actions cause harm in aggregation with others. Both conditions will be examined in more detail later in this book (chapters 3.1 and 5.2). Nevertheless, it is one thing to expect individuals to anticipate their relation to other peoples' actions and it is quite another to hold a set of individuals responsible for intentionally causing climate change. Holding individuals collectively responsible requires them to express their membership or their commitment to participate. I consider that an improper imputation.

A different claim could be that people are united by criteria that set them up as a collective entity. In that case, the criterion would have to clearly distinguish the collectivity from others as in the above cases of 'those living in the UK' or 'the professional hunters'. It must be common knowledge and members must have expressed their readiness to participate. With regard to climate change I can think of two criteria that might provide ground for a collectivity. The question is, however, if they are sufficient to establish a united entity. The first is 'people living in the industrialized countries' and the second is 'humanity'. After all, citizens of the industrialized countries cause major emissions and any human being contributes to climate change. Then, collective responsibility could be assigned in accordance to the criteria 'industrialization' or 'common humanity'.

Regarding 'industrialization', my clarification above applies. Individual agents who live in the industrialized world hardly intend to cause climate change. I assume that almost no individual agent boards the plain an tells herself: "Today, I, as a member of the industrialized countries, intend to do my part for heating up the planet." Yet, there is something about contributing to climate change and I will address this issue later – but in any case, it is not intentional. With regard to effectiveness, climate mitigation requires a prospective view and, more importantly, it requires a global perspective. People living in countries such as China or India who just started to emit are a necessary part of the solution to climate change. Also, people from poor countries will not develop themselves out of poverty without emissions. Taking it seriously that duties are grounded in the moral viciousness of peoples' suffering, the criterion that establishes a collective entity must be a different one than the criterion 'industrialization'. It requires a broader view taking into account all those agents who can legitimately be expected to do something about peoples' suffering. In my view, the first criterion is too narrow.

Reference to 'humanity' as a whole does not provide a solution either. In fact, if everyone is a member, no one is a member. Humanity is not a distinct, organized, collective body. Moreover, as I will argue in the course of this book, people living in completely underdeveloped regions like the tropics of Papua New Guinea, Nepal or South Sudan who literally light a fire during the night can hardly count as members of a collectivity who can do something about climate change. Humanity is simply a too broad criterion to establish a collective responsibility.

With regard to climate change, neither concepts of individual responsibility, nor standard concepts of collective responsibility are sufficient to justify either individual or collective duties. Yet, climate change requires action – collective action. The challenge is the following: In what conditions may a set of agents that does not yet constitute a collectivity in the conventional sense have a duty to act against climate change? And following, what are the duties of individuals given that collective action is urgently needed, but no set of individuals is organized to perform collective action? Cripps' theory has taken up upon these challenges. The following chapter starts by introducing her account. Further, it faces possible criticism and suggests areas of modification and advancement.

CHAPTER 3

Individuals as Collectivities

Chapter 2 argued that neither accounts of individual responsibility nor standard accounts of collective responsibility, such as the theory defended by Gilbert, are able to capture the significance of individual actions within the context of climate change. Cripps has developed a theory that is able to cope with the challenge of individual actions regarding collective harm. The first section introduces her theory (chapter 3.1). The second section examines the most serious objections (chapter 3.2) before I elaborate limitations, necessary modifications and possible advancements in section three (chapter 3.3). Chapter 3.3 provides the 'roadmap' for chapters 4, 5 and 6.

3.1 Elizabeth Cripps' Approach: Weakly Collective Duties and Individual Promotional Duties

Within the ethical discussion on individual duties with regard to climate change, the theory defended by Cripps (2013) constitutes one of the most comprehensive and thorough attempts to develop a coherent theory towards individual moral duties. Cripps' theory is the foundation of my book. Chapter 1 and 2 already touched upon some elements. The aim of this chapter is to substantiate these remarks. The following elaborates the normative foundation of Cripps' approach (chapter 3.1.1) and introduces her idea to rethink collectivities as weak collectivities to capture non-intentional collective action (chapter 3.1.2). Next, with regard to climate change, weakly collective duties to the relevant non-intentional collectivities are presented (chapter 3.1.3) before the case for derivative individual duties and the primacy of promotional duties with regard to effectiveness, fairness and efficiency is demonstrated (chapter 3.1.4).

3.1.1 *The Interest-Based Normative Foundation*

Facing the serious suffering of humans due to climate change, Cripps assumes that some "human functionings" are centrally important to live a flourishing human life (Cripps, 2013, p. 7). By taking the capability approach as developed by Amartya Sen and Martha C. Nussbaum as a starting point, she defends that we, as humans, have a fundamental interest to build and maintain human functionings (Cripps, 2013, pp. 7–10). Cripps commits her foundation to "only

 | DOI:10.30965/9783957437426_004

those elements which it is hardest to deny are central to human life" (Cripps, 2013, p. 8) and suggests choosing five capabilities: life, bodily health, bodily integrity, affiliation and practical reason (Cripps, 2013, p. 7).

Assuming that "it is a bad thing if individual humans are seriously harmed" (Cripps, 2013, p. 11), she assumes two basic moral principles: The no-harm principle and the principle of beneficence. Cripps defines the no-harm principle as follows:

> **No-harm principle**
> An individual (moral agent) has a moral duty to avoid inflicting serious suffering (deprivation of a fundamental interest) on another human being or human beings (moral subject(s)) *at least* if she can avoid so doing without suffering comparable harm herself. (Cripps, 2013, p. 11)

The no-harm principle not only requires avoidance of harm but also compensation for imposed harm (Cripps, 2013, p. 12). However, following a point stressed by Simon Caney (2010), Cripps points out that avoiding harm is always preferable to compensation (Cripps, 2013, p. 12). Caney has emphasised the primacy of avoidance over compensation with regard to the transnational solution to climate change. He writes: "On one way of thinking about harms, if one imposes one cost on people but also bestows on them a benefit, then the two may cancel each other out, and the affected person has no cause for complaint." (Caney, 2010, p. 172). Mitigation prevents harm from being increased any further, adaptation enables the protection of human interests and compensation neutralizes peoples' loss. Compensation is the necessary third option: First, because collective mitigation and adaptation will not be sufficient to avoid all harm even if collective action was initiated right now and, second, because failure to act collectively is an all "too plausible possibility" (Cripps, 2013, p. 12). For fundamental entitlements, it is wrong to counterbalance failure to successfully mitigate and adapt to climate change by simply paying a sum of money. Caney exemplifies the matter in the following way:

> [I]t is obviously impermissible for one person to assault someone else with a view to giving them a large benefit in order somehow to cancel out the harm. Similarly, one cannot destroy someone's property and then simply write a check and then think that the victim has no cause for complaint. (Caney, 2010, p. 172)

The point here, for Cripps and Caney, is that if there is a fundamental entitlement that generates a corresponding duty, interference may never be justified through compensation. Or, as Cripps puts it: "[M]aking it up is always second-best" (Cripps, 2013, p. 12).

The second basic moral principle is the principle of beneficence, which Cripps presents in two versions:

> **Weak principle of beneficence**
> An individual (moral agent) has a moral duty to prevent the serious suffering (deprivation of fundamental interest) of some human being or human beings (moral subject(s)) if she can do so *at minimal cost* to herself. (Cripps, 2013, p. 13, emphasis added)
>
> **Moderate principle of beneficence**
> An individual (moral agent) has a moral duty to prevent the serious suffering (deprivation of fundamental interest) of some human being or human beings (moral subject(s)) if she can do so *at less than significant cost* to herself. (Cripps, 2013, p. 13, emphasis added)

The weak principle of beneficence requires individual agents to bear only minimal cost. Suppose agent A can save agent B's life by pressing a button. Pressing a button is certainly a minimal cost and A has a duty to save B. In contrast to the weak version, there is a moderate and a strong version as defended by Peter Singer (1972). Singer argues "[I]f it is in our power to prevent something very bad from happening, without thereby sacrificing anything morally significant, we ought, morally, to do it" (Singer, 1972, p. 231). Regarding the strong version of the principle of beneficence, a moral agent is obliged to prevent serious harm until she has to give up something of comparable cost. According to the strong version, the only case where A does not have a moral duty to save B's life if A can only do so by risking her own life. Cripps locates the moderate version between the two and writes: "The moderate principle is more demanding [than the weak version], but does still impose limits." (Cripps, 2013, p. 13).

The moderate principle limits costs to "less than [morally] significant costs" (Cripps, 2013, p. 13). Cripps admits that there is a broad range starting "from comparable cost to the agent, at one end, to trivial costs which even the weak principle would exclude, at the other" (Cripps, 2013, p. 13). The disagreement here is linked to over-demandingness. What amount of cost is bearable for individual agents? The morally relevant cut-off point for A's duty to save B's life may range from pushing a button over ruining one's clothes to risking a broken arm. Cripps believes it to be impossible to specify "an exact line" (Cripps, 2013, p. 13). Cripps provides a rather crude measure that limits morally significant costs to anything that "seriously interferes, even in a temporary basis, with the exercise of central human functions" (Cripps, 2013, p. 14). The premise seems to be to protect the fundamental interests of those facing climate-induced deprivation while satisfying the fundamental interests of those under the moral duty.

3.1.2 *Revising Our Understanding of Collectivities*

Assuming that climate change is not a "paradigm moral problem" (Jamieson, 2007a, p. 475)[40], Cripps takes the collective character of climate change as given (Cripps, 2013, p. 2). She denies both an understanding that places individual agents under the duty to refrain from emission-intensive actions grounded in their personal footprint but also the standard understandings of collectivities, which grounds collective duties on intentionality or common knowledge (Cripps, 2013, pp. 1–2). With regard to climate change, the difficulties the standard understanding of collectivities is unable to deal with are twofold: Neither is there any collectivity that intentionally caused climate change, nor did individuals together act intentionally or knowingly contribute to collective action. Yet unable to solve climate change collectively, there is no prima facie collective duty in the standard thinking (Cripps, 2013, pp. 2–3).

Considering contributions by Gilbert (1989, 2006) to demonstrate traditional, intentional understanding of collectivities, Cripps suggests to rethink collectivities (Cripps, 2013, pp. 28–32).[41] She argues that in a globalized world

> [w]e can influence one another's lives at previously unthought-of levels, across both distance and time. As a result we find ourselves frequently in situations calling for collective action, even if we are neither 'set up' for such action nor consider ourselves in any way bound together. (Cripps, 2013, p. 28)

To illustrate her point, Cripps considers two examples, which will be picked up throughout this book: The "beach rescue" and the "swimming teenagers" (Cripps, 2013, pp. 60–61).[42]

In the first example, a number of unconnected tourists are on the beach while a child starts to drown. Saving the child's life requires the tourists to organize themselves to, for example, put a boat to the water, boat up to the child and pull her out of the water. Even though the tourists are not organized in any way and they would probably not even consider themselves to be connected in any way, collective action is required (Cripps, 2013, p. 60).

---

40 Recalling from chapter 2, a "paradigm moral problem" is where "an individual acting intentionally harms another individual; both the individuals and the harm are identifiable; and the individuals and the harm are closely related in time and space" (Jamieson, 2007a, p. 475).

41 With reference to personal correspondence Cripps stresses Gilbert's willingness to discuss alternative interpretations of collectivities and the way collectivities are formed (Cripps, 2013, p. 215, footnote 5). To contrast Cripps' approach against existing theories, I followed Cripps in using Gilbert's theory. Yet, criticism applies to other intentionalist accounts too.

42 Originally, the example of the beach rescue is from May (1992, pp. 110–111).

In the second example a number of teenagers, who do not personally know each other, happen to pass a lake. There is a child swimming in the middle of the lake – also independent of others. All teenagers, again independently, jump into the water and start swimming and splashing. Their swimming and splashing causes major turbulences, which threatens to drown the child. There is no collective intention to harm the child – in fact, there is no intention at all. And yet, cooperation of all teenagers is required not only to save the child, but also to sustainably protect other children from drowning by limiting turbulences in the long run (e.g. by limiting the number of people swimming) (Cripps, 2013, p. 61).

Though the application of intentional accounts might be promising in many cases (see chapter 2), intentional accounts seem to be unable to capture either the tourists or the swimming teenagers as collectivities, which have a moral duty to act. The tourists are unconnected and not organized. Yet, the child's rescue is required. The rescue cannot be secured without the tourists' cooperation. The teenagers have no intention to harm the child and yet, the situation requires cooperation between them to save the child.

On the conceptual level, Cripps identifies a further shortage of intentional theories. She believes that the traditional, intentional understanding of collectivities fails to identify three "special case individuals" as part of a collectivity: infants, individuals with intellectual disabilities, and rebellious teenagers (Cripps, 2013, pp. 29, 29–31). She holds: "Although often intuitively regarded as members, these individuals are not or cannot be aware of a shared purpose or goal, do not or cannot consider themselves as being part of the group, and do not or cannot jointly intend to do anything." (Cripps, 2013, p. 28). Cripps considers families, which are standardly taken as collectivities: Even though infants, rebellious teenager and people with severe mental disabilities standardly count as family members, they are aware neither of the shared purpose of being a family nor of being a member of the family. Also, they cannot participate intentionally. Even stronger, the teenage rebel would probably deny being part of the family or needing basic supplies provided by the family such as accommodation, protection or food. And yet, babies, teenage rebels and mentally disabled must be recognized as integral parts of the family (Cripps, 2013, pp. 29–30). Cripps admits that a core set of individuals that assigns membership to others might be an option (Cripps, 2013, p. 32). With special emphasis to people with intellectual disabilities, Cripps concludes that such as an "unqualified rule-setting authority" is "extremely unappealing" (Cripps, 2013, pp. 31, 32). Likewise, infants and teenage rebels should not depend on others to recognize them as family members. What is needed is an understanding that fully covers those individuals who are actually entitled to be a member of the collectivity (Cripps, 2013, p. 31).[43] Cripps' way out is to develop a collective account that does not rely on intention.[44] This account is the "non-intentionalist model" (Cripps, 2013, p. 28):[45]

> A set of individuals constitutes a collectivity if and only if those individuals *are mutually dependent* for the achievement or satisfaction of some common or shared purpose, goal or fundamental interest, *whether or not they acknowledge it themselves*. (Cripps, 2013, pp. 29–30, emphasis added)

Instead of intentionality, agents' interdependence identifies a set of individuals as a collectivity. Cripps defines three "conditions of collectivityhood" (Cripps, 2013, p. 34):

(1) "[T]he significance of certain actions by individual members cannot be captured without reference to the collectivity as a whole." (Cripps, 2013, p. 34)
(2) Collective action and collective achievements are distinct from individual action and individual achievements and
(3) even if individual members leave or join, the collectivity keeps being a collectivity (Cripps, 2013, pp. 34–35).

In response to the problem of causal inefficiency of individual actions, set-theoretic thinking following Parfit (1984) is an option (see chapter 1.2). However, expressing the relation between individuals or individual actions as sets is only the first step.[46] Condition (1) claims an additional and "ineliminable" reference to the collectivity as a whole (Cripps, 2013, p. 34). Cripps writes:

> [B]ecause of the relations those individuals are in, there are indefinitely many combinations of individual acts that could bring about any given outcome at the collective level. Thus, it is necessary to refer to that collective level – to the set of individuals in those relations – to capture all of these possibilities in one statement. (Cripps, 2013, p. 35)

43 Cripps highlights that these conceptual remarks may be applied to tribes and states too (Cripps, 2013, p. 29). For the application to states see Cripps (2011b).

44 Note that Cripps does not deny the existence of intentional or partly intentional collectivities (Cripps, 2013, p. 33). With regard to certain examples including climate change, however, these modes require supplementation.

45 Cripps also works on expanding the non-intentionalist model to non-human animals as members of collectivities (Cripps, 2013, pp. 85–111). Though this is a valuable addition, I limit my analysis to human beings.

46 The underlying reasoning will be substantiated in chapter 5.2.2. Mainly in distinction to a recent contribution by Hohl (2017) who defends set-theoretic thinking.

With regard to tackling climate change, an individual's actions in isolation are not capable of causing what the collectivity causes. Likewise, it is impossible for an individual to achieve what the collectivity can achieve (condition 2). The collectivity remains a collectivity grounded in its members' mutual interdependence even if members join or leave (condition 3).

If mutually dependent sets of individuals constitute collectivities whether or not members acknowledge it themselves, then how does an individual become a member? According to Cripps, there are two cases of mutual dependency where an individual becomes a member of the collectivity: The first case applies if an individual acknowledges a *personal goal or purpose* whose satisfaction can only be achieved together with others, whether or not she denies the latter. The second case is if her *fundamental interests* can only be secured together with others whether or not she is aware of it. In these cases of mutual dependency, the collectivity exists regardless of whether or not the collectivity recognizes itself as such or individuals consider themselves to be a part of it (Cripps, 2013, p. 33). Collectivities exist in this very weak sense of collectivity-hood. With regard to climate change, I will simply refer to the relevant collectivities as weak collectivities.

3.1.3 *Weak Collectivities and Weakly Collective Duties*

With regard to climate change there are three relevant weak collectivities that have a moral duty to act. Cripps labels them as "The Young" (Cripps, 2013, p. 27), "The Able" (Cripps, 2013, p. 61) and the "Polluters" (Cripps, 2013, p. 68). To simplify the reading flow, I will leave out any special emphasis and refer to these three weak collectivities as the Young, the Able and the Polluters throughout this book.

The Young

The first relevant set of individuals that complies with the "non-intentionalist model", is the weak collectivity of the Young. The Young include "young generations of humans across the globe" (Cripps, 2013, p. 27). Current and predicted impacts of climate change are significant, ongoing risks to human functionings. It is often believed that climate induced threats will only hit people in underdeveloped regions. Underdeveloped countries are indeed especially vulnerable to climate change, but climate change will impact countries globally (see chapter 1.1). It is impossible to specify how high the risks are and to evaluate "who will be the unlucky ones" (Cripps, 2013, p. 40). For young people globally, climate change will become a bitter reality within their lifetime if further warming is not mitigated successfully. Since they cannot avoid facing climate risks, the Young have a fundamental self-interest in avoiding them in the first place. The Young are not yet organized but mutually dependent on each other to secure their fundamental interest in mitigation. Once members of the Young can reasonably be expected to *know* that they are mutually dependent[47], there is a "prudential incentive" and a "weakly collective duty" to organize themselves to mitigate climate change (Cripps, 2013, pp. 50–51). The duty of the Young is not based on individual self-interest – it is a duty of all to each (Cripps, 2013, p. 34).

Recall from the first section that the principle of beneficence requires the prevention of serious suffering if agents can do so at either minimal cost (weak version) or at less than significant cost (moderate version). Weakly collective duties based on the principle of beneficence should neither impose too high costs on each individual member ("cost of action condition"), nor on the collectivity to get organized in order to act collectively ("cost of coordination condition") (Cripps, 2013, p. 51). Given that the Young organize themselves to secure their own fundamental interests, Cripps argues

> it cannot be too costly for you to do something, in a morally relevant sense, if it actually benefits you to do it. The only proviso – and this is the important one – is that it must be possible to organize to secure that fundamental interest for each without sacrificing another such interest. (Cripps, 2013, p. 52).

In the case of the Young, she therefore applies the weak version, which requires moral agents to prevent serious suffering at minimal cost. The normative basis that justifies the Young's moral duty is the "principle of moralized collective self-interest" (Cripps, 2013, p. 48):

> A set of human beings (moral agents) who are mutually dependent through a common fundamental interest have a weakly collective duty to cooperate to secure that interest, so long as this is possible without those individuals having to sacrifice some other fundamental interest. (Cripps, 2013, p. 48)

Not only not-yet organized collectivities but also so-called "potential collectivities" may face a weakly collective duty (Cripps, 2013, p. 50).[48] A potential

47 Cripps notes that this formulation includes cases where individuals do not consider themselves a member of the collectivity or do not accept protecting fundamental interests as a personal goal. In contrast, assigning moral obligations while the individual cannot *know* that there is mutual dependence would be implausible (Cripps, 2013, pp. 51–52).

48 Different sections point to the fact that there is a difference between a not-yet organized collectivity and a potential collectivity: In the introduction Cripps writes "[...] I argue that sets of individuals – as *not-yet-organized collectivities or potential collectivities* – can acquire weakly collective duties." (Cripps, 2013, p. 3, emphasis added). Next, the idea of potential collectivities is introduced only after the Young are defended to face a weakly

collectivity is any set of individuals, which would constitute a collectivity if they had a common goal, for the achievement of which they were mutually dependent (Cripps, 2013, p. 59). The number of potential collectivities is basically unlimited. With regard to climate change, the set of potential collectivities is way smaller: climate change requires only those potential collectivities "whose members ought (morally) to cooperate to pursue those ends through which they would be mutually dependent" (Cripps, 2013, p. 60). "Should-be collectivities" as subsets of potential collectivities are potential collectivities with a moral duty to organize to act collectively in pursuit of a particular goal (Cripps, 2013, pp. 21, 59). More formally, a should-be collectivity is "[a] set of individuals who: (i) would, were they to espouse some goal or goals, constitute a collectivity and (ii) have a moral duty to espouse that goal." (Cripps, 2013, p. 60). Recall Cripps' examples of the tourists at the beach as well as the swimming teenagers who are mutually dependent on each other to save the drowning child. Both collectivities are neither intentional nor organized or bound together in any way. And yet, facing the child's harm, they ought to cooperate to achieve their goal of saving the child. The tourists at the beach and the swimming teenagers are typical should-be collectivities.

In light of these examples, but with regard to climate change, there are two should-be collectivities with a weakly collective duty to tackle climate change: the Able (just like the tourists at the beach) and the Polluters (just like the swimming teenagers). Each should-be collectivity faces a weakly collective duty to act in response to climate change, which is based on collectivized versions of either the principle of beneficence or the no-harm principle. Cripps believes that the collectivized version "follows naturally" from the basic version introduced earlier (chapter 3.1.1) (Cripps, 2013, p. 61). Having outlined Cripps' terminology of potential and should-be collectivities, I suggest sticking to the term 'weak collectivities' to refer to the Young, the Able and the Polluters.

collective duty. Lastly, according to the definitions provided in the annex, Cripps defines the Able and the Polluters as *potential collectivities* while naming the Young to be a *collectivity* only (Cripps, 2013, p. 203). I understand Cripps this way: every potential collectivity is a not-yet organized collectivity, not the other way around. The Young have a weakly collective duty based on their mutual dependence to secure their self-interest even though they are not organized yet. The Able and the Polluters are not organized either, but in addition, they are not *actually* mutually dependent but only *potentially* mutually dependent on each other. Only in this very weak subjunctive form, would they potentially form a collectivity if they had a common goal. Facing that they, in fact, have a common goal with regard to climate change, they are *should-be collectivities*. The concept of should-be collectivities will be introduced in the following. All three collectivities comply with Cripps "non-intentional model" (Cripps, 2013, p. 28), which is why I chose to refer to all three of them as weak collectivities.

The Able

The "global affluent" face a weakly collective duty grounded in the "collectivized moderate principle of beneficence" (Cripps, 2013, pp. 59, 61):

> A set of human beings have a duty to cooperate to prevent the serious suffering (deprivation of fundamental interest) of another human being or human beings if they can do so at less than significant cost to each. (Cripps, 2013, p. 61)

The Able have a weakly collective duty to reach their common goal of preventing climate-induced harm. Preventing harm requires cooperation. Their ability to aid justifies their moral duty regardless of whether or not they caused harm themselves. Moral duties of the Able are not limited to mitigation but also cover enabling adaptation (Cripps, 2013, p. 77).

Recall that the principle of beneficence requires that the "cost of action condition" and the "cost of coordination condition" are both fulfilled (Cripps, 2013, p. 51). Claiming that collective action of the Able "does not impose too high cost on each individual" (cost of action condition fulfilled) and neither does organising and coordinating to act collectively so (cost of coordination condition fulfilled), Cripps applies the moderate version (Cripps, 2013, p. 51). With regard to cost of action, the Stern Review (2006) claims that the required financial effort for cost of mitigation is about 1% of the annual Gross domestic product (GDP) by 2050 (Cripps, 2013, p. 64). While admitting that costs of adaptation may be hard to estimate and are thought to be tens of billions of dollars, Cripps highlights the central characteristic of the Able: their (financial) ability to aid in cooperation (Cripps, 2013, p. 64). Who else could be able to manage the burden, if they are not? For distributing costs on state-level, Henry Shue (1993) has made the similar point that all the global affluent would have to give up in order to manage the costs of mitigation and adaptation are luxury goods (Cripps, 2013, p. 65). If that holds true for the Able on the non-state level too, it would indeed be less than a significant cost. Considering the costs of coordination, Cripps argues that coordination will certainly be costlier to some members than to others. Referring to existing interaction between individuals on the global level, she assumes that costs are bearable (Cripps, 2013, pp. 65–66).

The Polluters

The Polluters are "[t]hose emitting greenhouse gases above the level at which, were all emitting at that level, climate change would not be worsened, and whose individual emissions over that level are avoidable without the loss to them of some fundamental interests." (Cripps, 2013, p. 59).[49] None of the Polluters' members individually harms another agent or agents, none intends to contribute to collective harm, and none intends the harm to happen at all. This rules out individual responsibility and collective responsibility in a strong, intentional sense. Nevertheless, harm occurs due to the combination of single actions (Cripps, 2013, p. 68). According to Cripps, members of the Polluters are therefore "weakly collectively responsible" (Cripps, 2013, p. 69):

> A number of individuals who do not yet constitute a collectivity (either formally, with an acknowledged decision-making structure, or informally, with some vague defined common interest or goal) can be held collectively morally responsible for serious harm (fundamental interest deprivation), which has been caused by the predictable aggregation of avoidable individual actions. (Cripps, 2013, p. 69)

The Polluters' bear weakly collective responsibility for climate change if their actions' impact is predictable and their actions are avoidable.[50] That is true even though each member intended only to perform her own single action. Predictability covers two meanings: First, individual agents could reasonably be expected to have foreseen that their action in aggregate causes harm. Second, the individual agents could reasonably be expected to have foreseen that enough other individual agents are acting in a similar way. The point is that harm caused in aggregation is not pure coincidence but is predictable for each individual. Further, if there are alternative actions that would not violate any of the individual agent's fundamental interests, the individual action in question is avoidable. Taking predictability in the two meanings and avoidance together, Cripps defines the "three-part sufficient condition for weakly collective responsibility" (Cripps, 2013, p. 69). The normative justification of the Polluters' duty to act is probably the most critical issue. Chapter 5.2 will therefore elaborate predictability and avoidance in greater detail.

---

49 Cripps limits the Polluters' harm to the one caused by the current generation.

50 Cripps' understanding of the relationship between responsibility and duties remains rather blurry. A note highlights that she applies a forward-looking understanding (Cripps, 2013, p. 255, note 4). She refers to responsibility only with regard to the Polluters as if she applied a narrower understanding of responsibility that is complemented by duties assigned to the Young and Able. Yet, she defines 'moral agents' as "[m]oral duty-bearers; those who can be held morally responsible for their actions" (Cripps, 2013, p. 208). I therefore assume that Cripps applies the same broad understanding of responsibility as I do that holds all those responsible have reason to respond to moral wrongs and thus a duty to act accordingly (see chapter 1.3).

The no-harm principle assigns a moral duty not to inflict predictable and avoidable harm on others (Cripps, 2013, p. 74). Given that the Polluters are weakly collectively responsible, the no-harm principle may be collectivized in the following way:

> A set of human beings have a *moral duty to organize themselves* as necessary to prevent serious harm (deprivation of fundamental interests) to another human being or human beings *resulting from the predictable aggregation of their avoidable individual acts.* (Cripps, 2013, p. 74, emphasis added)

Connecting to the first section of this chapter, there is not only this kind of negative duty but also a secondary duty to compensate in cases where the negative duty not to harm others was violated (Cripps, 2013, p. 74). The Polluters weakly collective duty is to mitigate, enable adaptation and to compensate the victims of climate change (Cripps, 2013, p. 77).

The essential findings of Cripps' threefold argumentation for weakly collective duties to tackle climate change are the following:

(1) The Young have a moral duty to <u>mitigate</u> due to their *collective self-interest.*
(2) The Able have a moral duty to <u>mitigate</u> and enable <u>adaptation</u> based on their (financial) *ability to act.*
(3) The Polluters have a moral duty to <u>mitigate</u>, enable <u>adaptation</u> and to <u>compensate</u> based on their *weakly collective responsibility.*

Cripps presents her approach as flexibly as possible. She highlights the possibility of accepting one argument but rejecting another. Arguments for the Young, the Able and the Polluters are complementary but the weakly collective duty to each collectivity remains independent (Cripps, 2013, p. 78).

With regard to climate change, Cripps does not tie her account down to any particular kind of collective action. She writes:

> I have left it open whether the weakly collective duty would best be fulfilled via an effective agreement between states, by giving greater authority to some existing international body, by establishing some stronger global-level institution, or even – at the other extreme – by some global-level agreement between individuals or sub-state collectivities, perhaps coordinated by NGOs or some other voluntary body. (Cripps, 2013, p. 143)

With reference to Stephen M. Gardiner (2011), Cripps further stresses that "we should not automatically assume that the state system is an inevitable or necessary part of a future in which we fulfil our weakly collective duties" (Cripps, 2013, p. 228). Though Cripps certainly has a point in that collective action may take various forms and that existing institutions are not a necessary requirement for collective action, she thereby decreases the strength of her argument. Effective climate action is not a matter of individuals *or* institutions. Rather, climate action is about achieving synergy effects in collaboration of both. In order to impact existing institutions, regular individuals as members of weak collectivities are required to organize themselves accordingly. At the same time, this process may be encouraged by existing institutions. The ethical discussion with regard to climate change has extensively elaborated normative questions in connection to state-level and institutional climate action. It is about time to focus on the lowest level of a bottom-up approach to climate change. This book elaborates what regular citizens as members of the relevant collectivities ought to do together.[51]

3.1.4 *Derivative Individual Moral Duties and the Primacy of Promotion*

Faced with the fact that weak collectivities are still unprepared to act collectively, Cripps proceeds by asking what individual agents as members of the relevant weak collectivities ought to do in the absence of collective action. Cripps identifies three kinds of individual duties derived from the weakly collective duty:

(1) "Mimicking duties" require individual members "to do what would be required of one as part of a fair collective scheme to fulfil the duty" (Cripps, 2013, p. 116).
(2) "Promotional duties" require individual members "to attempt to bring about the necessary collective action" (Cripps, 2013, p. 116).
(3) "Direct duties" require individual members (i) "to protect the interests at stake" and (ii) "to mitigate the harm directly" either in isolation or together with a "likeminded subset" (Cripps, 2013, p. 116).

Cripps' provides a thought experiment called the "weak bridge", which she refers to throughout her book and which might help to illustrate the differences (Cripps, 2013, p. 133). Consider a bridge that is unsafe to be crossed for more than 50 people at a time. There are children swimming underneath the bridge. If the bridge collapsed, the children as well as the pedestrians crossing the bridge would be severely harmed. Now, hundreds of people are pushing onto the bridge. It happens that I am part of the crowd, so I have three options: I could either slip out of the crowd, walk away, do my part and see whether my mimicking will turn out to be fruitless (Cripps, 2013, p. 148). Or, I could directly

51 For readers familiar with the discussion, I wish to clarify that the kind of collective action that members of the relevant collectivities ought to do together neither exclusively aims for a civil movement driven by virtuous citizens, as the one defended by Jamieson (2011), even though Cripps includes this option nor does it aim for developing 'appropriate' collective intentions in the sense of compatible with climate change that would be in line with Kutz's (2000) idea of complicity.

help by climbing down to the riverbank and trying to pull the children out of the water (Cripps, 2013, p. 151). Maybe, some others will even agree to assist. Nevertheless, if more than 50 people rush onto the bridge, the bridge will collapse, which results in great harm. My third option is to climb onto a tree, a wall or a hill, shout out to the crowd in order to raise awareness and call for others not to cross the bridge. I could also try to organize sets of 50 people and control their crossing (Cripps, 2013, p. 143). After we collectively and successfully prevented the serious harm, I might write letters to local politicians asking them to install a barrier to regulate the number of persons crossing the bridge or I might call the local building authority and ask them to replace the bridge with a more stable one. What the collectivity can achieve is different from what individuals in isolation can achieve and – more importantly – it is sustainable.

With regard to climate change, Cripps considers an individual agent from the UK and provides a list of promotional duties that is "by no means exhaustive" (Cripps, 2013, p. 143):

> She could campaign and vote for candidates for local and national office who are committed to promoting international and national mitigation and adaptation efforts. [...] In the absence of such candidate, she could stand herself or encourage a suitable candidate to do so. She could petition her local MPs or MEPs [members of the (European) parliament] to change their policies. She could send emails to the UK Prime Minister, the US president, the UN Secretary General, the President of the European Commission, or the President of China. She could write them open letters in national or international newspapers or on social media sites. She could sign and circulate petitions, in person or online. She could join or organize marches at international summits. She could write articles, set up websites, and otherwise call for adherents to the kind of norm-changing 'global citizenship movement' suggested by Dale Jamieson. (Cripps, 2013, p. 143)

In cases of collective harm where there is no set of people prepared to act collectively, and climate change is such a case, the primary derivative individual duty is promotional (Cripps, 2013, p. 138). Direct duties are second-best if collective action is impossible[52] or promotional action is too demanding (e.g. in totalitarian regimes where promoters risk their life), exhausting (promoters might have capacity exceeding their promotional potential) or if acting directly can better advance promotional duties (e.g. celebrities who attract public interest by helping victims) (Cripps, 2013, p. 151). According to Cripps, mimicking may only be required as part of promotional or direct duties for example

52 In fact, in these cases there is no weakly collective duty (Cripps, 2013, p. 150).

if it increases the authenticity, reliability and the reputation of those agents being concerned with promoting collective action (Cripps, 2013, pp. 139, 152). With regard to climate change Cripps stresses: "In making unilateral emissions cuts I cannot consider myself to be doing all that my climate duties demand of me. Moreover, in cases of conflict between mimicking and promotional or direct duties, there is no plausible philosophical case for prioritizing the former." (Cripps, 2013, p. 139).

The primacy of promotion is complemented by quite an extensive reasoning. According to Cripps the three key arguments are: "Effectiveness, fairness, [and] efficiency" (Cripps, 2013, p. 143). The following breaks down the reasoning to these three key-arguments and adds clarification where needed. Chapter 3.2 will take up her reasoning with regard to corresponding objections in more detail.

Effectiveness

Reducing personal emissions in isolation or helping victims of climate change directly is ineffective to mitigate climate change. This is in line with various philosophers working on climate change. Mimicking is suitable only if it is sustainably secured by a collective scheme where every member does her part. Assuming theoretical full-compliance does not help either. Based on the current status, widespread non-compliance is very likely. "Without individual duties to bring about collective action, the collective level-duty will remain unfulfilled." (Cripps, 2013, p. 144). Unlike mimicking, promotional actions are not "throw-away acts" but the combination of promotional actions becomes "greater than the sum of its parts" (Cripps, 2013, p. 148).

Given that individual agents' actions amount to a substantial but not to a major part of global emissions, effective mitigation requires large-scale technological and structural changes across sectors.[53] These changes require promotional action but not reducing one's own emissions.

Fairness

With regard to Liam B. Murphy (1993, 2000), critics may ask why individuals should do more than under a fair collective scheme. Why should individuals be "taking up the slack" for others (D. Miller, 2011)? On this account, anything but mimicking is assumed to be unfair to the compliers (Cripps, 2013, p. 157). Cripps responds by rejecting Murphy's use of fairness: Fairness to the agent bearing the duty has to be seen in reasonable relation to the grounds of weakly

53 With reference to Aufrecht (2011), Booth (2012), Cuomo (2011) and Tan (2015), Fragnière (2016) also points to structural changes.

collective duties. Namely, in relation to other agents' severe suffering due to climate change. As long as duties are grounded in the moral viciousness of other agents' suffering, it is hard to explain why individual agents should not be expected to do more only because it is more than what would be required under full compliance (Cripps, 2013, p. 159). As long as demanding more does not violate any of the individual agents' fundamental interests, the primary focus of fairness remains the victim's suffering. Therefore, individual agents cannot expect that duties are limited to mimicking.

Efficiency

Reducing one's personal emissions or rescuing victims of climate change directly is much more likely to strike the limits of demandingness than to promote collective action. Promotional action may even reduce costs of mitigation for example by achieving infrastructural development of public transport. In particular for the Young and the Able, Cripps believes promotional duties to be less costly than mimicking. Requiring individual agents to primarily act in isolation is not a fair option if collective action is less burdensome (Cripps, 2013, p. 157).

Independent of arguments supporting weakly collective or promotional duties, the last section of Cripps book explores the dilemma individual agents are left with when collectivities fail to fulfil their weakly collective duty (Cripps, 2013, p. 169). It underpins the importance of collective action with reference to individual psychological costs. Thereby, it also connects to the efficiency-argument.

Drawing on Thomas Nagel (1991), Cripps argues that each human being embodies three different moral standpoints that motivate her to act: (1) the "personal" standpoint, (2) the "interpersonal" standpoint and (3) the "impersonal" standpoint (Cripps, 2013, pp. 171–173). Regarding the first, human beings constitute themselves as persons with their "interests, desires, beliefs, and plans of our own" (Cripps, 2013, p. 171). Regarding the second, persons wish to establish and maintain social ties and act upon the interest in helping friends, relatives, colleagues, and so on (Cripps, 2013, pp. 171–172). Regarding the third, serious suffering of other beings, though impersonal and distant, motivates people to act (Cripps, 2013, p. 173). Cripps claims that "we cannot function properly from the impersonally moral standpoint unless we are able to respond to the serious suffering of other moral subjects" and only collective action enables individual agents to live up to more of the impersonal standpoint (Cripps, 2013, pp. 173, 175). Avoidance of collective failure is therefore in our own interest: "Collective action on climate change is something we owe to ourselves, to enable us to live fully at peace with ourselves as simultaneously moral agents and individuals with our own lives and relationships to maintain." (Cripps, 2013, p. 198).

With regard to climate change, this argument becomes more relevant by considering "marring choices" (Cripps, 2013, p. 184). In situations where agents are forced to give up at least one fundamental interest when fulfilling another, they are left with "fundamental agent-regret" (Cripps, 2013, p. 185). Cripps argues climate change is such a situation and considers a low-paid employee in the UK who wishes to see her parents who need care but live abroad. She could take a low-cost carrier flight to see them, she could not see them, or she could take an expensive train ride, which leaves her unable to financially support her own family (Cripps, 2013, p. 184). Given that she is aware of climate change, any decision will leave some regret for not having chosen the other. Cripps clarifies that this does not rule out her giving up on some interest but rather the "irreconcilability of the three perspectives is what renders it plausible to limit demandingness of individual moral [...] duties" (Cripps, 2013, pp. 185, 186).

On a general level, I believe, that making decisions is part of human nature and whatever one decides it will leave the agent with some kind of regret. Correspondingly, Cripps discusses whether "we are all marred by tragic choices" (Cripps, 2013, pp. 190–191). With regard to climate change, the matter should not be about exploring in which ways individuals are trapped in tragic choices. The central message of Cripps last section should be the following: Giving up on our current lifestyle will create individual costs – including individual psychological costs. And individual (psychological) costs may indeed be lower if everyone acted simultaneously and individuals could assume that individual actions are secured by collective action. The question with regard to climate change is rather how collective action can advance the existing structural and technological circumstances in order to prevent tragic choices to occur at all.

3.2 Objections against Individual Duties

Given that climate change is a collective action problem, objections may be raised against individual moral climate duties in general. In particular and in distinction to mimicking duties, objections may also be expressed against promotional duties. Many of the reasons discussed in the following sections have already been mentioned in the previous section. And yet, much of the following remains to be substantiated throughout this book. I will describe and discuss the key counterarguments including their variations in the following sections to provide a transparent view on possible weaknesses right at the beginning and to keep the relevant ones in view throughout this book. There are three key objections: (1) the no-competence-objection (chapter 3.2.1), (2) the fairness-objection (chapter 3.2.2) and (3) the demandingness-objection (chapter 3.2.3).[54]

3.2.1 *No-Competence-Objection: It is None of My Business*

The first counterargument may be referred to as the no-competence-objection. Assuming that climate change mitigation is successfully brought about only by collectivities, why do individuals face a duty at all? The answer is: because they are members of the relevant collectivity facing the moral duty. The Young, the Able and the Polluters, as weak collectivities, face a moral duty to act. Given that they are not yet prepared to act collectively, the duty falls back to individual members to initiate relevant action.

Cripps takes note of this argument with regard to the Polluters: "If what an individual did made no difference, why should she have any moral duties in connection to it, at all?" (Cripps, 2013, p. 71). Cripps' reply is the following: The individual "may not be personally responsible, but she is one of those collectively so." (Cripps, 2013, p. 71). Next, Cripps considers two slightly different versions. First, if individual actions are not perceptible, why do individuals count as members of those weakly defined collectivities? The answer is: because every agent is similarly unnecessary to cause the collective result. It would be inconsistent and also unfair to all remaining individuals to attribute a derivative duty to all but one individual (Cripps, 2013, p. 71). The second version considers that this kind of unfairness does not occur if all others are not obliged to act either. The argument goes: Why should it be possible to identify a potential collectivity at all if no individual action is perceptible? The response closes the circle and brings us back to the justification of potential collectivities. With

54 Certainly, there are many more objections against individual (promotional) duties. One of them is for example the problem of distance in morality. Cripps addresses the objection that the principle of beneficence is applicable only to those close in time and space (Cripps, 2013, pp. 62–63). My brief comment is the following: Though individuals may have difficulties broadening their minds to the distant in practice, in a globalised, interconnected world these effects rather constitute psychological mechanisms than serious moral objections. For a discussion on the problem of distant morality see e.g. Arneson (2004), Kamm (2004) or R. B. Miller (2004). I limit my objections to the three most serious objections.

regard to the Polluters, climate-induced harm could be avoided collectively, and each individual is a member of some potentially perceptible subset. To capture all subsets in one statement, reference to the potential collectivity as a whole is essential.[55]

3.2.2 *Fairness-Objection: Why Should I Do More and Step in for Others?*

The above explanations already touched upon the second counterargument: the fairness-objection. Fairness-objections are raised with comparative reference to others. In the case of climate change, they are usually linked to other peoples' inaction, which is also referred to as partial compliance. The fairness-objection subsumes two different versions: One follows Murphy (1993, 2000), the other follows Keith Horton (2004).

The first version of the fairness-objection, which may be referred to as the fair-share-objection, draws on the theory of Murphy and has been addressed earlier. Murphy's theory actually considers positive individual moral duties under non-ideal conditions. The challenge of non-compliance, however, equally applies to negative moral duties for collective harm. His approach may be summarised in the following way: Assuming full compliance (every agent effectively does her part to the collective scheme), agents are assigned equal shares. According to Murphy, every agent is obliged to fulfil exactly her fair share. And even under partial compliance, no agent is obliged to do more than her fair share. Murphy argues that, an agent may otherwise correctly claim: "Why should I have to do more just because others won't do what they ought to do? Why should I have to take up the slack caused by their wrongdoing? Surely I should only have to do my own fair share?" (Murphy, 2000, p. 76). Though Murphy is not concerned with cases of collective harm, his argument may apply to promotional duties in the following way: the claim is that anything but mimicking is unfair to the individual agent even if it leaves the collective duty unfulfilled.

Such an argument assumes that mimicking is less burdensome than promotion of collective action. Promotional duties will be further substantiated in chapter 6. Nevertheless, building on Cripps' examples for promotional action, it is even at this early stage safe to say that promotional action is all about raising awareness and creating political pressure. Linking to that, my first comment is that there is reasonable doubt about mimicking being in fact less burdensome.

My second comment is bound to effectiveness. As highlighted earlier, mimicking is only suitable when it is secured by collective action, which makes

55 A detailed defence of this key argument will be subject to chapter 5.2.

collective action not only necessary but more effective. If promotional actions are less burdensome and more effective, there is hardly any fairness-related reason, in Murphy's sense, to limit duties to mimicking. Further, one may take up Cripps' rejection of Murphy's use of fairness. Fairness with regard to the duty-bearer requires evaluation with regard to the normative grounds of weakly collective duties: Severe climate-induced harm to other agents. If other agents' serious suffering is the normative ground to establish duties, it is hard to justify why individual agents should not be expected to do more only because it is more than what would be required under full compliance (Cripps, 2013, p. 159).

My third comment is connected to my second and focuses on the understanding of duties with regard to partial compliance. Duties correspond with a normatively relevant moral ground. In Cripps theory, peoples' fundamental interests are the moral ground that puts agents under the duty to act in accordance. These duties exist regardless of what others do or fail to do. In the first instance, partial compliance does not affect agents' duty to act respectfully of other agents' interests. If partial compliance is a case for fairness at all, then it is a case of unfairness on part of the non-compliers. Those who do not comply violate their duty. This, however, does not leave the compliers with less of a duty or only of a fair share as Murphy assumes. On the contrary, those who do not comply are left with an increased moral burden for failing to meet their duty. Consider Cripps' example of the beach rescue (see chapter 3.1.2): Suppose four tourists find themselves in a position that requires them to organize collectively to save the child from drowning. Suppose two of them refuse to help, walk away and let the other two take the burden for them. The non-compliers violate their moral duty. Assuming that the complying tourists can still save the child without sacrificing a morally significant harm themselves (to take up on Cripps' definition of the moderate principle of beneficence), the non-compliers' behaviour does not affect the compliers' duty to help the child. In fact, the non-compliers' failure to fulfil their duty increases their own burden but does not decrease the burden of others.

Yet, Murphy's argument may apply to promotional duties in a second way. Individual members of the relevant weak collectivities may ask: Why should I promote collective action at all if others do not take the burden to promote collective action themselves? And why should I perform more promotional action than I would have to perform if others did their fair share? One reply, again with reference to Murphy's use of fairness, may be to start from the point of view of the victims of climate change. As long as people severely suffer with regard to their most basic needs, it is hard to justify why others need not to promote collective action even under partial compliance. Another reply, with regard to the general understanding of duties, may be that non-compliers' failure to fulfil their own moral duty does not affect others' duty to fulfil theirs. A third reply, which I owe to Hohl (2017), builds on the you-go-first-problem. She refers to unilateral emission reductions that are perceived as an unfair competitive disadvantage compared to nations that do not commit themselves to fight climate change (Hohl, 2017, p. 132).[56] An example for this position is China's former climate policy (Hohl, 2017, p. 132, footnote 91): Chinese negotiators made the you-go-first claim with regard to industrialised countries in the follow-up to the Climate Conference in Copenhagen in 2009 (Revkin, 2010). As Hohl correctly claims, the you-go-first-problem equally applies to individuals. She argues that a perspective of relative burdens is inadequate for any moral theory. The less others fulfil their promotional duty, the smaller an individual's relative moral burdens are. The result would be a downward spiral of individual moral duties (Hohl, 2017, p. 132).

In distinction to Murphy, Horton (2004) defends an alternative to Murphy's approach (this is the second version of the fairness-objection).[57] Inspired by Horton, critics may argue that doing more than their fair share imposes more of a cost on them, which increases inequalities. Horton highlights that fairness-objections develop their strength mainly through a combination of fair-shares and additional costs that exceed fair shares. Horton calls them "supra-fair costs" (Horton, 2004, p. 167) and claims:

> [T]he strength of that objection to principles requiring agents to do more than their fair share of some task is proportional to the size of the supra-fair cost that acting on such principle would impose. The greater the supra-fair cost is, the stronger the objection is. (Horton, 2004, p. 167).

In the light of Cripps' argument above, one reply to critics is to prove a wrong use of fairness. In the face of serious climate-induced suffering, it is hard to justify why bearing more of a cost than an equal share is at least *prima facie* unfair. Nevertheless, there is something serious about the costs imposed on individuals. The relevant question is not whether costs are imposed or whether costs exceed an equal share, but the relevant question is whether costs cross a certain threshold that imposes too much of a burden. This leads to the third objection.

56 Hohl highlights this point with regard to relative distribution of costs. Fairness-objections with regard to individual costs are analysed below.

57 Just like Murphy and Hohl, Horton discusses fair shares with regard to mimicking. In the view of promotional duties, I modify Horton's claim accordingly.

3.2.3 *Demandingness-Objection: It is too Much of a Burden*

The third counterargument is the demandingness-objection. On a rather superficial level, critics may argue that individual moral duties are generally too demanding.[58] The question is: Too demanding with respect to what? In contrast to promotional duties, mimicking duties may indeed be too demanding, for example, with respect to very limited chances of success. Passing on the burden to individuals if the solution is only to be successfully brought about by collective action is at least frustrating if not overdemanding. In another respect, put forward by Hohl (2017), moral duties are too demanding if individuals are constantly forced to think of them throughout the day (Hohl, 2017, p. 160).[59] Yet another way is to claim that omission of certain actions restricts individuals' ability to lead their lives, which leaves them unable to participate in social life. In these respects, duties may be too demanding but none of them applies to promotional duties: Promotional duties have a chance of success and promotional duties are neither obtrusive throughout the day nor do they prevent persons from leading their lives nor do they prevent social interaction. The only relevant demandingness-objections, as I see it, must be raised with respect to individual costs.

In the case of weakly collective duties, individual costs occur in two ways that correspond with the nature of derivative individual duties. In the absence of collective level action, the primary individual duty is of a promotional nature. Therefore, costs occur first of all with regard to promotional action in order to initiate collective mitigation. In the second way, however, once collective action has been initiated and is actually gaining momentum, individual costs arise from doing one's part of collective level action as a member of either one of the three relevant collectivities (Cripps, 2013, pp. 71, 79). Given what promotional duties at this early stage of the discussion cover (Cripps names writing e-mails, signing petitions, organising marches and demonstrations, etc.), it is hard to see where they impose too much of an individual cost. Especially with regard to mitigation, the relevant challenge is rather whether

58 In the climate change context this objection has mainly been raised with regard to direct omission of carbon-intensive actions (see e.g. Lichtenberg, 2010).

59 Thinking about a duty throughout the day may not seem to be a too serious objection at first. Yet, consider an individual agent who attempts to mimic. Doing her groceries, she might be torn between different purchases. Which one is the most environmentally friendly one? Going out for dinner, she might be overstrained with what to choose given that she has no control over or information on the products' life cycle. Suggesting the application of "rules of thumb", Hohl does not assess this to be too demanding (Hohl, 2017, p. 160, own translation). I consider it to be a serious obstacle to mimicking, but not to promotional duties.

participating in collective action as a member imposes too much of a cost on each individual.

Cripps notes that, in practice, correlative individual duties on these two levels occur especially for the Polluters (Cripps, 2013, pp. 71, 79). Their lifestyle causes climate change collectively and it is the Polluters' duty not to increase harm any further. The Young and the Able must ensure that mitigation is successfully initiated. Duties to the Young, the Able and the Polluters certainly interrelate with each other. Cripps does not say a lot about how they relate in practice. Potential questions could be if individuals who are members to more than one collectivity (such as 'the young polluter') have more of a duty to act and if so, to what extent? However, in order to cope with the demandingness-objection discussing these questions is the second step before the first. The relevant questions in the first place are still if individual duties generally impose too much of a cost on members of any of the three collectivities and how to cope with individual limits of demandingness at all? This first step is a complex one and I will limit myself accordingly. Chapter 4.4 will elaborate on demandingness in more detail and I will present my own proposal on reasonable costs with regard to both, promotional action and doing one's part of collective climate action. At this point, it probably suffices to indicate how Cripps limits costs in order to cope with over-demandingness. Her limitations have at least partly been mentioned in chapter 3.1. The following makes Cripps' limitations explicit.

Cooperation of the Young is to its members own very fundamental advantage, which leads Cripps to apply the *weak* principle of beneficence. Their duty to secure their own fundamental interest is limited to performing actions "without sacrificing another such fundamental interest" (Cripps, 2013, p. 48). Cooperation of the Able must meet two conditions: the "cost of action condition" and the "cost of coordination condition" (Cripps, 2013, p. 51). Collective action must neither itself impose too high costs on each individual nor must coordination to act collectively be too costly to each individual. Since both are met, the *moderate* principle of beneficence applies to duties of the Able. Their duty to act is limited to "less than morally significant costs to each" (Cripps, 2013, p. 61). With regard to limits of demandingness of the Polluters, Cripps is not as clear. On the one hand, the Polluters' duty is based on the no-harm principle as a "moral intuition so strong and so widely shared that it is hard even for the most avid libertarians to deny" (Cripps, 2013, p. 78). The no-harm principle requires bearing comparable costs. On the other hand, she writes: "I leave it open whether the limits for members of the Polluters should be taken from the individual no-harm principle, or be set in line with beneficence cases [such as the Young and the Able]." (Cripps, 2013, p. 155). That is indeed a big difference.

Given that the case of the Polluters is probably the most critical one, leaving limits of demandingness open is quite a weak point.

With a rather general regard to mitigation, Cripps asks the rhetorical question: "How much, after all, could most of us give up by way of economic income and wealth, or of emissions, without actually sacrificing anything fundamental: anything that really counts as a precondition for an individual flourishing life?" (Cripps, 2013, p. 55). She assumes that if costs were truly distributed across the collectivity, mitigation would amount for three or four per cent of resources and might increase in proportion to individual income and wealth (Cripps, 2013, p. 55).[60] Cripps strongly rejects the view that mitigation action has a too demanding impact on "pursuit of a particular plan or life, or conception of the good" (Cripps, 2013, p. 55).[61] Cripps argues that people may in many cases be able to achieve the same fulfilment by performing less-polluting actions. She believes for example that people who enjoy speed driving by car could replace their satisfaction with speed driving by bike or they could simply switch to challenging themselves by driving in the most fuel-saving way (Cripps, 2013, p. 55). Further, excluding certain actions does not mean destroying a whole life or life plan. Also, costs of mitigation for individuals are distinct from costs of mitigation for members of the respective collectivity (Cripps, 2013, p. 56). Admitting that she cannot exclude the case that certain actions infringe upon another fundamental interest, Cripps writes: "[I]f anyone could show that certain actions would genuinely require sacrificing a fundamental interest, or central capability (not just an expensive taste), then there is a case for adjusting whatever distribution of individual duties is determined as part of the collective scheme." (Cripps, 2013, p. 56). Only if there was no collective scheme that could be realised without violating some members' interest, would mitigation be too demanding (Cripps, 2013, p. 56).

I suspect that many readers will think of economic costs. That is certainly one way of thinking about individual costs. With regard to individual agents, economists usually focus on maximization. Accordingly, any value that negatively effects an individual's maximum utility or benefit is a cost. If an individual decides to speed drive by car, then speed driving by car is expected to realize the maximum outcome. Any action that deviates from speed driving with a car

60 Here, Horton's fairness-objection with regard to additional costs that exceed fair shares may take effect. As I have argued with reference to Cripps, this objection fails with regard to its use of fairness.

61 Others, for example Shue (1993), share this view.

creates substantial costs. That is true for any individual action. Cripps' claim that speed drivers may simply switch to speed driving a bike instead seems to be very different from the economic understanding of individual cost. There is certainly every reason to doubt that individuals in fact act as rational utility maximisers. However, economic thinking points to an important matter: Costs are only one side of the coin. Talking about individual costs should necessarily involve talking about individual benefits too. Considering the severe resistance among society with regard to environmental or climate protection measures from time to time, the economic understanding seems to display individuals' perception of individual loss in benefit rather than Cripps' understanding of reasonable costs.

Following Cripps, this book focuses on moral duties and on whether they are *morally* too demanding. Yet, economic costs may impose too much of a moral burden if they infringe with morally relevant values such as fundamental interests. However, moral demandingness is not limited to considerations regarding maximizing utility. Though economic considerations are important, obligations that exclusively rest upon them should not guide the basic idea of reasonable costs. In the case of mitigation, objections based on the self-centred economic model of wealth, prosperity and well-being described above distracts from the ground of morality in the case of climate change: other peoples' basic harm due to climate-induced impacts.

The tenor of Cripps' approach, as I understand it, and in that way, I agree with it, is that current mitigative action is not even close to having reached its full potential and that many objections regarding individual costs illegitimately bar the way to progress. I agree with Cripps that costs should be bearable on the collective level that is to be distributed across members. Also, the focus of the demandingness-discussion should not be exclusively put on individual reasonability of costs but rather on how collectivities may initiate changes that make it more reasonable for individuals to act. Instead of asking if costs are too demanding for individuals, we should ask: In which ways can individuals act collectively to decrease individual costs to a minimum that ensures individual action within the collective scheme?

Even though I agree with Cripps in these regards, it is a very thin line between demanding too little and demanding too much of individual agents. In this regard, it seems to me that Cripps is not able to fully cope with the significance of the demandingness-objection, in particular with regard to the Polluters. Facing the intensity of economic cost-objections, the use of expressions such as 'sacrificing some fundamental interests', 'morally significant costs' or 'comparable harm' in cost-analysis probably leaves too much scope for interpretation – at least with an interest-based foundation of the sort that Cripps uses.

One way to reason is that the use of chosen moral principles is in fact not that uncontroversial and accepted. Otherwise, critics might argue Cripps would not have such a hard time defending very demanding duties to mitigate. Cripps may reply with reference to the impossibility of drawing an exact line and quantifying an unreasonable level of costs in theory. Much of the discussion of costs is indeed subject to the respective discourse. The point is that basic normative assumptions are crucially important for the strength of tools and instruments used in a discussion. Given that the discussion questions the Polluters' lifestyle in particular, it is a big difference between having to lead a heated discussion using instruments that are built upon a rather weak foundation and using reliable tools that allow a stringent and ambitious moral discourse. In fact, I believe, the challenge with Cripps' theory is not so much a matter of moral principles chosen, the matter is rather the subjective choosing of certain interests in the first place that lacks a moral justification. Of course, this is a serious objection, which is not only to be substantiated but also a criticism that requires the provision of an alternative normative foundation that is more straight forward to cope or at least better able to cope with the limits of demandingness.

Having roughly outlined possible objections, neither the no-competence objection nor fairness-objections prove individual moral duties in general or promotional duties in particular wrong. The only relevant objection is the demandingness-objection with respect to individual costs. The key question is about the limits of demandingness but not about the legitimacy of demanding something of individuals. Demandingness is a central issue rather for being a member of the collective scheme brought about by promotional action than for promotional duties directly. In that respect, philosophers should focus on ways in which collective action may be able to decrease individual costs to a minimum to ensure individual action within the collective scheme. I have claimed that Cripps' normative foundation in application leaves too much space for interpretation. This is, however, not the only limitation that I see. The following chapter specifies the areas that I wish to analyse.

3.3 Cripps' Approach: Modification, Justification, Advancement

I wish to modify, justify and advance Cripps' theory with regard to three major areas: (1) the normative ground for selecting interests, (2) weakly collective duties and (3) the nature of promotional duties. Each area will be described in the following section. Advancements that I suggest, will be substantiated accordingly in chapters 4, 5 and 6.

3.3.1 *Modifying the Normative Foundation*

Cripps' key assumption with regard to the normative foundation is an account of fundamental human interests inspired by the capability approach. She writes:

> I am taking it as read that there are certain prerequisites to living a fully flourishing human life: that in order to do so an individual must have the capability, or meaningful opportunity, to enjoy continued life (at least to a normal human length), bodily health, bodily integrity, affiliation (relationships with others), and practical reason. This last includes being able to develop and pursue a plan for her own life, with such education as this requires. [...] These human functionings are so centrally important that it is a serious harm to an individual to be deprived of her opportunity to exercise any one of them. (Cripps, 2013, p. 7)

In the previous chapter, I claimed that Cripps is not able to fully cope with the strength of the demandingness-objection regarding individual costs by building on the described account of fundamental interests. Further, I highlighted that cost-expressions remain vague not because the chosen moral principles (no-harm principle and principle of beneficence) were controversial or unaccepted but because the normative basis of choosing certain fundamental interests is insufficiently justified. This can be seen best by considering the fundamental interest in bodily health and pursuing a plan of life as part of practical reason. There is little disagreement that climate-induced impacts violate the fundamental interest in bodily health and that climate change requires large-scale mitigation. With regard to the Polluters for example mitigation may require the collective adoption of a green lifestyle. Members of the Polluters may argue that it is their fundamental interest to pursue a plan of their own life. Because all five fundamental interests stand equally next to each other, Cripps has quite a hard time defending why mitigation does not violate pursuing a plan of life. Recalling from chapter 3.2.3, Cripps defends a difference between prohibiting certain actions and destroying a whole life planning. In my view, the difficulties to cope with the seriousness of the demandingness-objection are only symptoms or indications for a more fundamental problem. This problem is linked to the normative basis. The claim I defend in the following is that an account of fundamental human interests as defended by Cripps lacks a transparent justification of entitlements and a substantial normative criterion that enables such an account to solve cases of conflicting rights by providing a hierarchical order of entitlements.

Cripps links her account of fundamental interests to the capability approach defended by Nussbaum and Sen.[62] Even though a comprehensive description of the capability approach goes beyond the scope of this book it is useful to elaborate the connections to Cripps' theory in a bit more detail. With reference to Sen, Cripps agrees with his central motivation to provide an alternative evaluation of well-being that does not exclusively rely on the assessment of utility or availability of resources.[63] With reference to Nussbaum, Cripps uses the terms capabilities and functionings and introduces the idea of compiling a list of capabilities. Cripps picks five out of ten capabilities and argues: "I do not, and do not need to, commit myself to her list as a whole. Rather I am taking it as read only those elements which it is hardest to deny are central to human life." (Cripps, 2013, p. 8). With reference to the capability approach as a whole, Cripps defends the distinction between actually being able to live up to a central functioning and having the capability or opportunity to do so. Given the variety of concepts for living a fully flourishing life, the opportunity to exercise capabilities is centrally important. My argument unfolds with regard to Cripps' reference to Nussbaum but is not limited to her.

Nussbaum provides a normative approach that is inspired by Aristotelian thinking. One of the central components is that Nussbaum compiles a general list of "central human capabilities" that are essential to a valuable human life (see e.g. Nussbaum, 2000, p. 5). Assuming that humanity serves as a sufficient normative criterion, she takes an internal perspective and chooses elements that are essential to a human life.[64] Nussbaum calls this "internal essentialism" (see e.g. Nussbaum, 1992, p. 207).[65] The list is always open to discussion and revision (Nussbaum, 2006, p. 76). The current list of essential elements of a human

---

62 Cripps highlights that her account may alternatively be seen as list of basic needs of a self-controlled human being as defended by Raz (1986) or Steward (1996) (Cripps, 2013, pp. 8, 64). As such an account also fails to provide a hierarchy, I will only focus on the capability approach to make my argument.

63 As such the capability approach has received lots of attention across disciplines and serves as a ground for the Human Development Index (HDI) used by the United Nations Development Program (UNDP).

64 In later contributions, Nussbaum moves away from the criterion of humanity and refers to a type of Rawlsian 'overlapping consensus' to increase justification (Nussbaum, 2000, p. 76) and a kind of dignity-criterion (Nussbaum, 2006, pp. 74, 162). Cripps highlights that she agrees with Nussbaum's early contribution (Cripps, 2013, p. 212, footnote 28). Therefore, I limit the description to the criterion of humanity.

65 The *internal essentialism* goes as follows: Considering the feature of life for example Nussbaum reasons that all humans face death but generally want to live. A being that is either immortal or wants to die as soon as possible without valuing life is so different from a *human* being that she could not properly be called human. Life is therefore an essential feature of our common humanity (see e.g. Nussbaum, 1992, p. 207).

life entails ten capabilities: *life, bodily health* and *integrity, practical reasoning* and *affiliation* (these are the five features that Cripps identifies too) as well as *senses, imagination and thought, emotions, play, control over one's environment* and *other species* as the capability to relate to other natural, non-human beings (Nussbaum, 2006, pp. 76–78). Nussbaum's intention is to develop a "source of political principles for a liberal pluralistic society" (Nussbaum, 2006, p. 70). As such the list of capabilities is supposed to provide citizens with sufficient justification and profound reasoning to demand politics to include them as constitutional principles that people are entitled to. Each capability represents a minimal threshold-level beneath which a valuable human life is not feasible (Nussbaum, 2000, p. 6). Nussbaum claims that each feature covers different but equally essential aspects of a human life, which leads her to reject any hierarchically ranking or relative weighing (Nussbaum, 2000, p. 12).

Irrespective of the kind of features (be it fundamental interests, functionings or capabilities) a list entails, I have to admit that providing a list is somehow appealing. It is concrete, tangible, catchy and therefore open to many unrelated to philosophy. Given that climate policy is dominated by political and thus non-philosophical science, it gains particular attractiveness. Although the normative approach that I will suggest as a reasonable alternative in the following at least partly agrees with Cripps' and Nussbaum's lists content-wise, my criticism targets the lack of a moral justification leading to a non-intuition-based selection of entitlements and the provision of a hierarchy.[66]

Relying on "some clear-cut intuitions" with regard to the criterion of humanity, Cripps in contrast to Nussbaum, does not even provide a profound moral justification (Cripps, 2013, p. 19). As stressed above, she simply limits her list to only those features "which it is hardest to deny [to be] central to a human life" (Cripps, 2013, p. 8). With regard to Nussbaum, Rutger Claassen and Marcus Düwell (2013) have questioned whether the criterion of humanity provides sufficient normative ground for setting up a list of capabilities as entitlements and imposing corresponding duties (Claassen & Düwell, 2013, p. 495). Given that Nussbaum includes *play* or *emotions* as essential features but excludes the opposite such as cruelty, aggression or anger, the list of essential features deserving moral recognition seems to be a subset of morally recognisable capabilities (Claassen & Düwell, 2013, p. 498). Then, why do certain human features require moral recognition? And why should corresponding duties be seen as a

66 In recent contributions, Nussbaum is less and less shy to admit that her approach involves testing against intuition (Claassen & Düwell, 2013, p. 498, footnote 5). She writes: "We can agree that the capability approach does indeed rely on intuition [...]." (Nussbaum, 2006, p. 83).

reasonable ground for political obligation? Claasen and Düwell conclude the following: "'Humanity' is insufficiently discriminative to serve as the guiding principle for selecting a capabilities list that deserves political protection." (Claassen & Düwell, 2013, p. 498). Cripps may object that she is not interested in a list that covers humanity as a whole but only in those very fundamental, basic features. Both, Cripps and Nussbaum, refer to humanity, claim to cover a minimal threshold and both highlight that none of the items is weightier than another. Nussbaum defends ten features while Cripps picks five. If Cripps does not intend to provide a transparent moral justification, she will have to provide a justification for picking five out of ten. Probably, she implicitly assumes that human capabilities, functions or interests do have a hierarchy but remains silent on this matter.

In the case of conflicting entitlements, lacking a hierarchy gives rise to serious vulnerabilities and the need of relativization. In particular regard to costs of mitigation, that comes at the price of defending less ambitious or at least less straight-forward duties to collectively adopt a greener lifestyle. Facing the severity of climate-induced impacts, trading the appeal of a concrete list off against less ambitious or less concrete mitigative action is a bad bargain. I am not saying that this is Cripps' intention. Rather, I point to the fact that the lack of a transparent moral justification that enables the establishment of a hierarchy of entitlements involves substantial challenges. These challenges may be overcome by moving to a different moral theory.

The first area of modification that I suggest is the normative foundation. I suggest taking the rights-based moral theory defended by Gewirth as a basis to establish moral duties. The rights-based moral theory suggested may be seen as a close relative to the interest-based approach defended by Cripps: Gewirth also moves away from the purely utilitarian, resource-based assessment of well-being. He too intends to provide a minimal moral threshold. And though Gewirth does not provide a concrete list of specific entitlements, he provides categories of necessary entitlements. Rights may be seen as entitlements to certain interests. Though I will not make the claim that persons have rights because they have certain interests, I regard rights as instrumentally protecting certain interests. That way, even Cripps may consider Gewirth's theory as a compatible alternative.

By developing the idea of weakly collective duties and derivative promotional duties, Cripps has made a valuable and ambitious contribution to climate ethics. The second and third issues that I wish to advance are not so much areas of modification but rather of required justification, clarifying and critical discussion and necessary advancement.

3.3.2 *Justifying and Advancing Weakly Collective Duties*

Weak collectivities as well as duties assigned to them, are fundamentally new concepts. New theories require critical discussion, advancement and, if necessary, defence. That is the second broad task that I wish to dedicate myself to. Besides taking up critical statements, I mainly focus on advancing the nature of the weakly collective duties assigned to the Polluters and the Young.

In view of standard accounts of collective responsibility, the Polluters do not qualify as a collectivity in the strong sense. Nevertheless, Cripps holds them responsible resulting "from the predictable aggregation of their avoidable individual acts." (Cripps, 2013, p. 74). The questions that I wish to discuss with regard to the Polluters are the following: Are predictability and avoidablity truly fulfilled with regard to climate change? And, if actions are predictable and avoidable, why are members of the Polluters not directly obliged to refrain from increasing their contribution to climate change any further? The status quo that I have provided, with regard to the discussion on individual moral duties in chapter 1.2, summarises almost fifteen years of discussion. Even though I believe it is time to move away from proving the primacy of individual duties to change one's own behaviour in isolation and even though I consider Cripps' account to be the most elaborate, those scholars who defend direct individual duties will legitimately claim that I overlooked their efforts over the past years. Indeed, considering promotional duties without any further connection to the discussion on mimicking duties would be a great weakness of my book. The most deliberate and thorough attempt to defend individual lifestyle changes, that I have come across, has been presented by Hohl (2017). Even though she acknowledges that individuals perform their polluting actions in relation to others, she believes that duties do not require reference to the collective level. Hohl seems to be able to explain the dominant moral narrative that every individual agent ought to do her part to fight climate change. Yet, is she convincing? That remains to be discussed.

In March 2019, right before the completion of this book, school strikes for more ambitious climate policies gained attention across the developed parts of the planet. Students and pupils seemed to become aware that they are members of the Young who will suffer severe harm from climate change throughout their lifetime. Irrespective of the striking fact that young people globally kept calm for such a long time, it is very surprising that only a part of the Young (those attending school) seem to recognise the need to defend their common self-interest. I claim that young people between leaving school and turning forty have every reason to worry about climate change too. In fact, they do not only have reason to worry, they have a weakly collective duty to push mitigation efforts in all directions. I wish to advance this duty with regard to a particular aspect: the weakly collective duties of young parents with regard to their children and future descendants.

In the years following her contribution on weakly collective and individual promotional duties, Cripps has dealt with parental duties too. My defence of weakly collective duties assigned to young parents will take up her argumentation. By critically discussing her hypothesises, I will use Cripps' later contributions to defend and substantiate the strength of her idea regarding weakly collective duties of the Young. It may seem odd to some readers to use more recent contributions to defend less recent ideas. The great openness regarding other kinds of moral grounds (such as communitarianism and global justice) that Cripps has shown lately is not incorrect per se and may even be used to further approve the need for collective action to a broader audience. Yet, as I do not consider this a positive or at least not a very helpful development, I use Cripps' latest arguments to strengthen the concept of weakly collective duties.

3.3.3 *Advancing Promotional Duties*

The third area, as one of necessary advancement, is about promotional duties. Given that none of the three relevant weak collectivities (the Young, the Able and the Polluters) is in fact able to act collectively yet, promotional duties to bring about the required collective action is a heavy load to carry. Cripps largely remains inexplicit about two questions: How does promotional action look like? And how is the weakly collective duty best fulfilled?

With regard to the first question, Cripps suggests that individuals may fulfil their promotional duties in one of the following ways that I have introduced earlier: vote and lobby for those political candidates who are eager to promote international and national achievements to mitigation and adaptation; if there is no such candidate, be such a candidate yourself; write e-mails to political leaders directly; write open letters; donate to campaigns; sign petitions; organize demonstrations, marches and international summit (Cripps, 2013, p. 143).

With regard to the second question, Cripps does not intend to commit herself to a way of collective action that best fulfils the weakly collective duty. She states that a state-solution, a sub-state solution, a solution achieved by a stronger international institution or a global-level agreement between individuals are possible options. Yet, in order to specify promotional duties, it would be very helpful to have a rough idea of the nature of collective action that promotional action aims to initiate.

Though Cripps' approach is a philosophical theory not a consultancy for policy making, putting such a load on individual agents without providing further guidance, examples or recommendations for action is likely to result in failure to act on various levels. Yet, it is certainly beyond the scope of this book too, which is likewise one from climate ethics, to fully cover recommendations for action. Though I do not wish to exclude that collective action may be brought about on the level of states, sub-states, international institutions or agreements, I mainly think of collective action of the Young, the Able and the Polluters as action of groups and groups as connected individual agents who collectively stand up, raise their voice and push the existing social and political limits of global action against climate change. Nevertheless, I do not share the kind of political scepticism widely represented through climate ethics.[67] First of all, there is already widespread agreement regarding the urgency of political action.[68] In this regard, the Nationally Determined Contributions with regard to the Paris Agreement are a good sign. I am convinced that the lack of effective global climate action is rather a matter of means and acceptance than a matter of political will. As the introduction of this book already indicated, instead of bypassing the political agenda, collective action of the Young, the Able and the Polluters should be regarded as additional, supplementary and supporting climate action. However, that does not mean that individuals had to wait until governments are prepared to implement mitigation strategies.

My advancements will therefore focus on the question how promotional action looks like and it will include two major aspects. First, Cripps aims for raising political awareness and increasing pressure to act politically. I will refer to this as action through the political channel. Prior or at least simultaneous to promotional action through the political channel, as I will suggest, promotional action through the social channel is required. Members of the relevant weak collectivities ought to raise awareness among members of society and more precisely within their own collectivity. Weakly collective duties that require promotional action is a completely unknown line of thought among society and requires promotion itself. In the introduction, I claimed this book to be a plea for changing the moral narrative. Changing the narrative will not happen all by itself and neither will it happen overnight. It needs to find its place in peoples' mind – one step at a time but as fast as possible.

The second advancement takes up upon the first and finally outlines concrete recommendations for action. Facing what Cripps suggests as promotional action (e.g. writing e-mails to political leaders, organise marches and demonstrations, and so on), the key challenge that is overlooked by Cripps is

67 With reference to Maltais (2013) and Jamieson (2011), Cripps points to diverse scepticism with regard to climate policy (Cripps, 2013, p. 46). Jamieson, for example, doubts "that our political institutions are up to it" (Jamieson, 2011, p. 36).

68 Gardiner emphasizes this point too (Gardiner, 2011, p. 433).

the following: What is collective action of the Young, the Able and the Polluters, actually supposed to achieve content-wise? Given that I, as a regular individual agent, intend to fulfil my promotional duty, what am I supposed to claim when writing e-mails to political leaders? What is the purpose of the demonstration I wish to organise? What kind of campaign am I supposed to donate for? Though adaptation and compensation are important, I will limit my analysis to tasks that are central to mitigation of the three weak collectivities.

For the Young, I see two major tasks: the first is to use the extended reach of social media – here, the idea of the social channel is particularly important. The idea of celebrities and public figures having a duty to use their voice is nothing new to climate ethics (see e.g. B. Johnson, 2011; Sandler, 2010). In view of the extensive number of people being very influential, the Young are obliged to use their communication possibilities to raise awareness among members of society. The second task is education (in several ways that are to be substantiated). The Young cannot afford having a new generation growing up and adopting the same lifestyle as their parents and grandparents. So, their own task is to guarantee children to be environmentally responsible agents.

In climate ethics, the Able have traditionally been the financially able nations globally. Considering individuals as collectivities, there is no reason to limit ability to financial resources. I consider any person who is actually able to do something to promote the collective solution to belong to the Able. In addition to the financially able, the group of individuals who have a duty to deal with climate change (regardless of their age and their own lifestyle) are scientists. Scientists around the world ought to pool and focus their abilities to answer what collective mitigative action actually looks like. What is the overarching goal of mitigation? For sure, it is decarbonisation. Unfortunately, there is great disagreement how decarbonisation is achieved best. Figuring the way to a zero-carbon economy and society is definitely no task for the Young or the Polluters nor are political leaders in the position to provide a silver bullet for climate change mitigation. This is a major if not the most important task for fighting climate change and it is the task for the Able.

The major task for the Polluters is the collective decarbonisation of their own lifestyle. The last section, on the Polluter's promotional duty, is an attempt to get an idea of what can actually be required of members of the Polluters with regard to the key sources of their carbon footprint. Given that members of the Polluters basically promote against their personal lifestyle, the central claim is that the Polluters ought to call for a structural setting that aims for mainstreaming a sustainable, climate friendly lifestyle across their own collectivity. Focussing on the three key sources of consumption-related goods and services, I attempt to analyse promotional duties with regard to three domains: housing, mobility, and food. In view of the complexity of the issue and the structural dependence on fossil energy of the Polluters' lifestyle, I will show that a major promotional task of the Polluters is to target their own collectivity via the social channel. Yet, that is not their only promotional task. This book closes by taking a brief look at the question what it means to live up to what one promotes for.

CHAPTER 4

Normative Foundation

Climate change is a major threat to human well-being and there is great agreement, I assume, that climate change is wrong. Yet, there is a range of conflicting views among philosophers regarding the moral justification. Regardless of the foundation philosophers rely on, all moral theories have very different implications in practice. Relying on some "clear-cut intuitions" related to the capability approach, Cripps defends an account of fundamental human interests (Cripps, 2013, p. 19). I have criticised Cripps for the lack of transparency with regard to the moral justification and the lack of a moral principle that is able to govern cases of conflicting entitlements by providing a hierarchy. As a consequence, I have claimed, she fails to be fully convincing with regard to the demandingness objection.

Gewirth's *Reason and Morality* is a very elaborate, well-argued theory of moral rights. Based on rational considerations, he provides a substantial justification of a supreme moral principle claiming that all agents have equal moral rights to freedom and well-being. In application, the supreme moral principle provides an adequate framework to explicate and evaluate morally relevant actions and institutions in a consistent and conflict-solving way. The ethics of climate change are complex and assigning individual climate duties is even more so. In view of this complexity, I consider it important to provide a normative foundation that is able to display and manage different cases of conflict. With such a normative foundation it is possible to lead a heated discussion about individual moral duties and to evaluate proposed solutions in practice.

Yet, rights-based and interests-based approaches are not completely distinct theories. They may be regarded as close normative relatives and I have expressed my confidence elsewhere that Cripps could also agree with the rights-based argument I defend in this book. After all, as Claassen and Düwell (2013) indicate, Nussbaum and Gewirth must have had some discussions in Chicago too, even though its results have never been published (Claassen & Düwell, 2013, p. 494, note 1). Gewirth distinguishes different theories that might provide grounds for peoples' having rights (Gewirth, 1982, pp. 43–45). Among them, there are interests. Accordingly, people have certain rights because they have certain interests in common.[69] If that was true, my proposal

69 In the climate change context, Caney (2006) defends such an account building on the interest-theory defended by Raz (1986).

 | DOI:10.30965/9783957437426_005

for a rights-based foundation would be a needless detour arriving at the very heart of rights, namely interests. However, that is not the case. Regarding interests, Gewirth clarifies that it remains open "how a normative conclusion about rights can be derived from factual premises about empirically ascertainable characteristics such as having interests" and "why moral rights are generated by characteristics that all humans have in common rather than by more restrictive, inegalitarian characteristics that pertain only to some persons, or to persons in varying degrees, such as expert knowledge [...]" (Gewirth, 1982, p. 44). Certainly, rights instrumentally protect human interests. Yet, having certain interests is a necessary but not a sufficient condition for having rights. The moral justification of rights must come from a different foundation. In fact, as Gewirth defends, it does: the moral foundation for having rights is human action itself.

The first part of this chapter introduces Gewirth's theory (chapter 4.1). Central to this first section is the reconstruction of Gewirth's foundational argument leading to a clear and non-intuition-based selection of entitlements and the supreme moral principle. Next, I analyse the implications of the supreme moral principle in practice (chapter 4.2). A distinction will be made between the principle's applications to individual actions and its applications to institutions. Also, the second part explicates principles for solving cases of conflicting rights. Among them a principle that allows a specific order of precedence, or hierarchy, of rights. The third part of this chapter applies Gewirth's rights-based theory to climate change (chapter 4.3). I demonstrate that Cripps' concept of weak collectivities, weakly collective duties and derivative individual promotional duties are compatible with Gewirth's rights-based approach. Finally, in the fourth part of this chapter, I elaborate how the comprehensive rights-based framework is better able to guide discussions on reasonable costs and to evaluate specific proposed collective solutions to climate change (chapter 4.4).

4.1 Alan Gewirth's Rights-Based Moral Theory

In *Reason and Morality* (1978), Gewirth elaborates that any purposive or prospective purposive agent is rationally committed to accept a sequence of necessary judgements leading to a supreme moral principle with strong normative force: The Principle of Generic Consistency (PGC). The PGC states that every agent has claim rights to freedom and well-being. Logically derived from the concept of action, Gewirth's claim is that no prospective purposive agent can rationally refuse to accept the PGC or violate its implications in practice without contradicting that she is a prospective, purposive agent herself. In the context of climate change – and in distinction to Cripps – I attempt to show that Gewirth's foundational argument provides the following: First, a non-intuition-based selection of entitlements and second, a justified hierarchy between them. By providing a hierarchy, I will demonstrate that Gewirth's theory is able to solve cases of conflicting rights.

4.1.1 *Preliminary Remarks*

In order to understand Gewirth's argument fully, I consider it helpful to examine several key concepts and considerations, the method, as well as a first brief overview of the key steps first, before examining the sequence of necessary judgments leading to the PGC.

Action

Action is the starting point for Gewirth's theory. The great importance of action for moral judgments had already been widely acknowledged by ancient philosophers (Gewirth, 1978, p. 26). However, Gewirth claims that its central characteristic has been overlooked ever since: Acting itself has a "normative structure", which logically implicates the PGC as the supreme moral principle (Gewirth, 1978, p. 26). He defines action as behaviour done (1) voluntarily (or freely) and (2) purposively (or intentionally). Gewirth explains:

> By an action's being *voluntary or free* I mean that its performance is under the agent's control in that he unforcedly chooses to act as he does, knowing the relevant proximate circumstances of his action. By an action's being *purposive or intentional* I mean that the agent acts for some end or purpose that constitutes his reason for acting; this purpose may consist in the action itself or in something to be achieved by the action. (Gewirth, 1978, p. 27, emphasis added).

Given that human behaviour may be diverse on purposes, incentives and appearances, Gewirth does not claim to describe the nature of action. Rather, he elaborates the invariant characteristics of action that are "the common object of all moral precepts as well as of many other practical precepts that set requirements for action" (Gewirth, 1978, p. 26). Steigleder (1992) explains that in order to attribute actions to agents and to assign responsibility for performing these actions an agent needs to be aware not only *that* she is acting in a certain way but also *why* she is acting that way (Steigleder, 1992, p. 121). In this strict sense, action has the two "generic features", as Gewirth calls them, described above: voluntariness and purposiveness (Gewirth, 1978, p. 27).

Two more clarifications: First, Gewirth does not claim action to be voluntary and purposive at any given moment. Given that persons sometimes act on routine or habits, they might not always be aware of why they act or even that they act. Agency refers to the disposition of action, on persons being prospective agents at any given moment but not their being actual agents (Gewirth, 1978, pp. 111–112). Yet, and this is the second clarification, action is unavoidable for prospective agents. Readers may object that people sometimes refuse or avoid acting for example ultimately by committing suicide. The point is, avoidance to act is in itself a voluntary and purposive action: Avoidance to act does not constitute an agent's decision whether she acts or not but rather in which way she acts (Steigleder, 1999, p. 29). Even committing suicide necessarily requires action at least if the agent voluntarily chooses to commit suicide.

Rationality

From the rational analysis of the normative structure of action constituted by the two generic features voluntariness and purposiveness, it follows that a prospective agent must make certain judgments that are "logically implicit in all actions" (Gewirth, 1978, p. 26). Gewirth describes them as "modes of thinking" running in the background of actions (Gewirth, 1978, p. 42). The agent may not be aware of them, let alone that she speaks them out loudly. Yet, they can be expressed linguistically as implicit judgements (Gewirth, 1978, pp. 42–43). The prospective agent must make these judgements as a "rational agent" (Gewirth, 1978, p. 46). Rational in the "minimal deductive" sense "involving consistency or the avoidance of self-contradiction in ascertaining or accepting what is logically involved in one's acting for purposes and in the associated concepts" but also in the sense that the agent has a "minimal inductive rationality" in the sense that the agent is able to incorporate "sense perception" (Gewirth, 1978, pp. 22, 46). The argument Gewirth tries to develop is that from the concept of action it follows *logically* that prospective agents, insofar as they are rational agents, must accept a sequence of judgements, the PGC and its implications in practice.

The Dialectically Necessary Method

The method that Gewirth employs to develop his argument is called "the dialectically necessary method" (Gewirth, 1978, p. 42). The method "starts from statements or judgments that are necessarily attributable to every agent because they derive from the generic features that constitute the necessary structure of action" and implies a certain sequence of judgments (Gewirth, 1978, pp. 43–44). Since the sequence develops from the internal standpoint of the prospective agent, the method is dialectical. The dialectical method makes judgments *about the agent* making the judgment. It states for example 'A believes that X is Y'. In contrast, the assertoric method makes judgments about an object and simply states 'X is Y'. Some dialectical judgements are contingent such as 'A believes or does not believe that X is Y'. And some are necessary, which means that the agent cannot deny them without denying the necessary requirements for making the judgement (Steigleder, 1999, p. 31). Gewirth explains:

> It will be by the application of this nonarbitrary, rationally grounded method to the generic features of action that we shall see that every agent, on pain of self-contradiction, must accept a certain supreme principle of morality. In virtue of this, the [supreme moral] principle will emerge as categorically obligatory and as having a strictly rational justification. (Gewirth, 1978, p. 47)

Three Key Stages Leading to the Supreme Moral Principle

The supreme moral principle, which requires every agent to act in accord with the rights of all other agents, is derived in three stages. These three stages derive from (1) the evaluative structure of action, (2) the normative structure of action and (3) the universalization of claims made regarding the agent's rights. This last stage finally leads to the justification of the PGC. Gewirth summarizes the three stages as follows:

> First, every agent implicitly makes *evaluative* judgements about the goodness of his purpose and hence about the necessary goodness of the freedom and well-being that are necessary conditions of his acting to achieve his purpose. Second, because of this necessary goodness, every agent implicitly makes a *deontic* judgment in which he claims that he has rights to freedom and well-being. Third, every agent must claim these rights for the sufficient reason that he is a prospective agent who has purposes he wants to fulfil, so that he logically must accept the generalization [or *universalization*] that all prospective purposive agents have rights to freedom and well-being. (Gewirth, 1978, p. 48, emphasis added).

From Gewirth's definition of action, the first judgment, which any agent must rationally commit to, is that she does her action voluntarily for a purpose that she aims to achieve by her action (Steigleder, 1992, p. 130). The task of the first stage is to show that this first judgment logically implies a certain positive judgment about the agent's chosen purpose but also a positive judgment about the conditions she necessarily needs to maintain her ability to act and to achieve the purpose that she chooses to act upon (Steigleder, 1992, pp. 130–149). These necessary conditions for action are freedom and well-being as "necessary goods" (Gewirth, 1978, pp. 52–65, 127). Freedom and well-being can and will be specified separately in the next section. The second stage proves that if it is a dialectical necessity for the agent, from her internal standpoint, to hold that freedom and well-being are necessary goods for action, then the agent must, again from the same internal standpoint, hold that she has rights to freedom and well-being. According to this second stage, the agent contradicts her own being a prospective purposive agent (first judgement) if she does not claim rights to freedom and well-being (Steigleder, 1992, pp. 159–167). If the agent, from her internal standpoint, must hold that she has claim rights to freedom and well-being on the sufficient ground that she is a prospective purposive agent, she must also – in the third stage and based on the logical principle of universalizability – hold that all other agents, insofar as they are prospective purposive agents, have rights to freedom and well-being. Thus, *every* prospective purposive agent, from her own internal standpoint, must, as a dialectical necessity, accept that *every* other prospective agent has rights to freedom and well-being. Because rights correspond with strict duties, the judgment that every prospective purposive agent has rights to freedom and well-being logically implies that every agent has a strict duty to act in accord with the rights of all other agents as well as of herself (Steigleder, 1992, pp. 197–201). This is the PGC. Any agent who rationally denies the PGC or its practical implications, as a matter of consistency, denies that she is a prospective purposive agent.

Freedom and Well-Being as the Necessary Goods for Action

As mentioned above, every prospective purposive agent is rationally committed to accept the necessary conditions as necessary goods, which she requires to act for her purpose and to successfully achieve this purpose (stage I). That probably requires further explanation, and, in this context, I found Deryck Beyleveld's clarification helpful. In order to achieve a purpose, the agent needs particular "capacities and resources" (Beyleveld, 1991, p. 23). Some of them are contingent in the sense that they are needed for one or a few purposes. Now, with regard to Gewirth's argument, freedom and well-being are the *necessary* goods of the agent's agency – those goods that are required to achieve *any* purpose (Beyleveld, 1991, p. 23). In analogy to the generic features of action, Gewirth calls freedom and well-being "generic goods" (Gewirth, 1978, p. 52). Using speech marks, Steigleder stresses that 'well-being' is a *terminus technicus* (Steigleder, 1997, p. 253). I have emphasised elsewhere that Gewirth, like Cripps, Nussbaum and Sen, intends to move away from the purely utilitarian, resource-based assessment of well-being (see chapter 3.3.1). Well-being does not simply refer to the aggregation or maximization of goods as the possession of particular resources. Even though "agents often identify their well-being with their possessing certain particular goods", well-being as a necessary good for action is "to be understood as a continuum that comprises having general capacities and successfully exercising them" (Gewirth, 1978, p. 60). Thus, well-being here is a term combining all necessary conditions for prospective purposive agents to act for and to achieve any purpose. In order to simplify reading, I forgo special indication in the following. Both, freedom and well-being as necessary goods for action can be specified further.

In view of a broad range of possible actions, well-being as a necessary good from the internal standpoint of the agent, encompasses three kinds of goods: "basic goods, non-subtractive goods, and additive goods" (Gewirth, 1978, p. 54). Basic goods are the necessary conditions the agent requires to be able to act at all, such as "life", "food, clothing, and shelter" (Gewirth, 1978, pp. 54, 212). Non-subtractive goods are the necessary conditions the agent requires to maintain the "level of purpose-fulfillment" she already achieved through purposive action (Gewirth, 1978, p. 55). Non-subtractive goods including, for example, not "being lied to" or "stolen from" (Gewirth, 1978, p. 233). Additive goods are the necessary conditions the agent requires to achieve additional purposes and to achieve "an increase in his level of purpose-fulfillment" such as "education" or "income" (Gewirth, 1978, pp. 56, 244). Consider a malnourished person who suffers from severe hunger. She will hardly be able to engage in voluntary and purposive action at all – her actions will probably be driven by obtaining food. Or, consider a person who is constantly told untrue or misleading information. She will not only be unable to successfully achieve her desired goal that motivates her to act but she will also have a hard time maintaining the resources she already successfully achieved.

Next to well-being, voluntary and purposive action requires freedom. While well-being covers the more "substantive aspect" of action (the necessary conditions to achieve and maintain objects or goods at which her action aims), freedom encompasses the "procedural aspect" of action (Gewirth, 1978, p. 251). As such, freedom is the necessary condition for agents to act in a self-controlled way and in absence of external interference by any other individual, institution or thing (Gewirth, 1978, p. 52). Gewirth distinguishes three broad kinds of interference with freedom: "violence, coercion, and deception" (Gewirth, 1978, p. 252). Consider, for example, a person who is kidnaped and arrested by other persons. She will almost be unable to engage into any purposive action.

In the process of this book, the structure incorporated in freedom and well-being as necessary goods for action, which is yet to be elaborated in greater detail, will provide the ground for the resolution of conflicting rights. With respect to that, the principle of degrees of necessity for action will play an important role.

Rights

The last concept I consider important before elaborating Gewirth's foundational argument in detail is the concept of rights. The rights to freedom and well-being that every agent must logically accept to have are "strong" or "claim" rights (Gewirth, 1978, pp. 65–67).[70] Claim rights are correlative to strict duties on the part of their addressees (Gewirth, 1982, p. 2). Rights and duties in everyday language are used primarily or even exclusively in a legal or moral sense, which may cause serious misunderstandings (Steigleder, 1992, p. 166). Claim rights, in contrast to the legal or moral meaning, are "logically implied correlatives" (Gewirth, 1982, p. 2). In order for persons to have claim rights to freedom and well-being, freedom and well-being must neither be required by law (on the contrary, persons have them even if they are not protected by law) nor are they initially moral rights that normatively count for every person. In contrast to such a "supra-individual claim to validity" as Steigleder describes it, claim rights are logically implied demands, which every prospective purposive agent has to all other prospective purposive agents (Steigleder, 1992, p. 166, own translation). The claim in claim rights is not necessarily something that a person insists to have, the claim is an entitlement a person has as her due (Steigleder, 1997, p. 254).

4.1.2 *The Foundational Argument: Deriving the Supreme Moral Principle*

The previous chapter examined Gewirth's concept of action, explained the importance of rationality as well as the dialectically necessary method, it briefly introduced the three key stages of the argument as well as the importance of freedom and well-being as necessary goods for action and roughly explained the meaning of claim rights. This chapter elaborates the sequence of dialectically necessary judgments any prospective agent is logically committed to make in more detail. Regarding the reconstruction, I largely follow Steigleder (1992) and his unpublished translation of the sequence into English.[71]

According to the generic features of action, an agent voluntarily acts for a purpose that she considers the reason of her acting. Thus, the first necessary judgment that any agent must logically make, from her own internal standpoint, is the following:

70 Gewirth contrasts claim rights with "liberties" that are "merely permissible or all right" to have and that do not correspond with strict duties imposed on other agents (Gewirth, 1978, pp. 67, 68).

71 Following Gewirth, the method is introduced as first person judgments of the agent.

(1) "I do action A for purpose P."

A represents any action that a purposive agent may voluntarily choose to perform, and P stands for the purpose or end that she attributes to this action. If the agent denied this judgment, she would deny her being an agent. Starting from this first necessary judgment, the normative structure of action implies a sequence of dialectical necessary judgments. One more clarification is required: Taking up on my explanation on rationality above, the sequence is not simply a semantic implication from one judgement to the next. The key judgements follow from the agent's rational reflection of herself as an agent – from her internal viewpoint (Steigleder, 1997, p. 252).

Stage I: The Evaluative Structure of Action

In the first stage, Gewirth proves that action has an evaluative structure (Steigleder, 1999, p. 35). The first necessary judgment implies that the agent, in a second judgment, necessarily attributes a positive value to her own purpose of acting. Gewirth describes this positive evaluation as a kind of "desire, at least in the intentional sense of wanting" to achieve a certain goal that constitutes the purpose of her action A (Gewirth, 1978, pp. 50, 49–52).

(2) "I want P."

Accordingly, her wanting to achieve her purpose implies that the agent considers the purpose of her action as good. Yet, that does not mean that the purpose is morally acceptable. Rather, the judgement that the purpose is good expresses that the purpose is worth the agent's action as part of the positive evaluation (Steigleder, 1992, pp. 131–133). Further, Steigleder clarifies that the agent does not state the purpose to be good because she is acting towards it but that she is acting towards it because she believes it to be good (Steigleder, 1992, p. 132). Moreover, the agent's positive evaluation extends to considering her purpose as an object that she positively evaluates. Here, it is worth stressing that the purpose may not generally be considered a good but that the agent, from her internal perspective, considers the purpose a good for herself (Steigleder, 1992, p. 134).

(3) "P is good."

(4) "P is a good."

Because the agent positively values the purpose of her action as a good, she wants her action to be successful. Performing her action successfully must therefore be considered to be a good too. It follows that if the agent positively values the purpose and her acting as goods, she, as rational agent, must also consider her ability to act at all as a good.

(5) "I want A to be successful."

(6) "My acting successfully is a good."
(7) "My ability to act is a good."

Because the agent's ability to act is a means of achieving something that she positively values, she also positively values those goods, which she necessarily requires for being able to act at all. Thus, the evaluative structure implies also that any agent is logically committed to consider the necessary conditions for agency, and for successful agency in particular, as necessary goods for action. The necessary goods are fundamentally the same for any agent and are summarised by freedom and well-being (Gewirth, 1978, pp. 52–65, 127). Recall that well-being encompasses the substantive aspect of action and, as such, well-being includes all necessary conditions to act at all and to act successfully at all – except for freedom as a necessary good for action. Thus, the last two dialectically necessary judgments derived from the evaluative structure of action are:

(8) "My freedom and the other necessary conditions to act at all and to act successfully at all are necessary goods."
(9) "My freedom and well-being are necessary goods."

Stage II: The Normative Structure of Action

In the second stage, Gewirth shows that action has a deontic structure leading the agent to claim rights to the necessary conditions of agency. The agent's wanting the necessary goods is no 'wanting' in the hypothetical sense of a wish or hope. The agent's intentional wanting the necessary goods is "unconditional and, related to that, continuous and unrestricted" (Steigleder, 1992, p. 161, own translation). Thus, the necessary judgement (9) logically implies the following judgment:

(10) "I need my freedom and well-being necessarily and I want to have them unconditionally and without exception."

Neither freedom nor well-being are generally guaranteed. Both exist and are maintained only by non-interference from the outside. Steigleder further emphasises that freedom and well-being are potentially at risk of interference at all times (Steigleder, 1997, p. 250). This constant risk of interference may be formulated in a premise from which it follows that the agent does not want her freedom and well-being to be interfered at all.

Premise P_1: "My freedom and well-being can be interfered with from the outside."

(11) "I, unconditionally and without expectation, do not want my freedom and well-being to be interfered with from the outside."

Because of this unconditional wanting, it follows:

(12) "Unconditionally and without any restrictions, it ought not to be the case that my freedom and well-being will be interfered with."

Possible sources of interference with the agent's freedom and well-being may be diverse. From the agent's viewpoint, there is a difference between interference by a higher force (such as for natural disasters) or interference by other agents. In the latter case, interference is their responsibility.[72] Instead of 'it ought not to be the case', the judgment may be reformulated as a judgement against all other agents (Steigleder, 1992, pp. 162–163). Because the agent depends on other agents to maintain the necessary goods of action, she must logically commit to raising a normative claim to freedom and well-being towards all other agents. The central point of the second stage is that if the agent holds, from her internal standpoint, that freedom and well-being are necessary conditions for her ability to act (successfully) at all (judgement (9)), then she must, as a dialectical necessity, make the judgment that nobody ought to interfere with her freedom and well-being. Logically, the agent is committed to claim rights to freedom and well-being and to claim others to have a strict duty not to interfere with her freedom and well-being (Gewirth, 1978, pp. 63–68).

Premise P_2: "My freedom and well-being can be interfered with by other agents."

(13) "Nobody ought to interfere with my freedom and well-being."

(14) "Everyone has a strict duty not to interfere with my freedom and well-being."

(15) "I have a right to my freedom and well-being."

Connecting to my explanations above, claim rights in the deontic phase are neither moral nor legal but remain for now within the perspective of the agent's own self-interest, which is why Gewirth calls them "prudential rights" (Gewirth, 1978, pp. 69, 71–73).

Stage III: The Universalization and the Supreme Moral Principle

In the third stage, Gewirth shows that *every* agent has claim rights to the necessary conditions of agency and that *every* agent has a duty to act in accord with the rights of all other agents. This last part of the sequence of judgments justifies the necessity to move from claim rights, though they are dialectical and not moral, to a supreme moral principle. To understand this transition, the Logical Principle of Universalizability is of particular importance: "[I]f some predicate P[*] belongs to some subject S because S has the property Q (where

72 Recall from above that their voluntariness and purposiveness as prospective purposive agents allows the assignment of responsibility (Steigleder, 1992, p. 121).

the 'because' is that of sufficient reason or condition), then P[*] must also belong to all other subjects $S_1, S_2, \ldots, S_n$ that have Q." (Gewirth, 1978, p. 105).[73] The agent (S) claims rights to freedom and well-being (Q) as the necessary conditions to act (successfully) at all, *because* she is a prospective purposive agent (P*).

(16) "I have a right to my freedom and well-being, because I am a prospective agent."

In committing that being a prospective purposive agent is a sufficient condition for having rights to freedom and well-being, the agent must also commit that every prospective agent ($S_1, S_2, \ldots, S_n$) has rights to freedom and well-being (Q) in virtue of being a prospective purposive agent (P*). Gewirth writes:

> For on the one hand in holding, as he logically must, that he has the rights of freedom and well-being because he is a prospective agent, he accepts that being a prospective purposive agent is a sufficient condition of having these rights; but if he denies the generalization, then he holds that being a prospective agent is not a sufficient condition of having these rights. (Gewirth, 1978, p. 112)

Since the agent is logically committed to accept the generalised statement that every prospective agent has claim rights to freedom and well-being against all other agents, she must also accept, on pain of self-contradiction, that every agent – including herself – has a strict duty not to interfere with the freedom and well-being of any prospective purposive agent (Gewirth, 1978, p. 133).

(17) "Every prospective purposive agent has a right to his freedom and well-being."

(18) "Every agent has a strict duty not to interfere with the freedom and well-being of any prospective purposive agent."

(19) "I have a strict duty not to interfere with the freedom and well-being of any prospective purposive agent."

The transition from judgment (16) to (17) and (18) requires further explanation: Just like any previous judgement, (16) "I have a right to my freedom and well-being, because I am a prospective agent", is a necessary judgement made from the agent's internal standpoint. Claim rights in (16) are prudential. In contrast, judgments (17) and (18) are *moral* judgments because they normatively count for every agent. Likewise, claim rights and corresponding duties are moral. Accepting mutual rights and duties gives way to morality (Gewirth, 1978, p. 129). The transition from prudential to moral is "in the first instance,

73 Gewirth indicates the predicate with P. In order to avoid confusion with the purpose P, I indicate the predicate with P*.

not motivational but logical" (Gewirth, 1978, p. 146). Gewirth explains that judgement (17)

> as said by the original agent is a moral judgment precisely because its criterion is not his own interest or prudential purpose but rather the interest or prudential purpose of persons other than himself as a speaker. He thus has to accept a judgment that he must favorably consider the prudential purposes or interests of persons other than himself. The agent is logically compelled to make this transition from a prudential to a moral judgement, because if he did not he would be in the position of denying what he had previously had to affirm, namely, that being a prospective agent is a sufficient justifying condition for having rights to freedom and well-being. (Gewirth, 1978, p. 147)

Morality limits an agent's range of actions where they "affect persons other than their agents" (Gewirth, 1978, p. 129). Gewirth refers to such "interpersonal actions" as "transactions" (Gewirth, 1978, p. 129). The term 'transactions' highlights the "asymmetric distribution of roles" (Steigleder, 1992, p. 199, own translation) between the agent or "actor" and her "recipients" (Gewirth, 1978, p. 129). While the agent "actively" engages in the transaction, her recipients remain "passively, by undergoing or being affected by the agent's action towards him" (Gewirth, 1978, p. 129). The point here is that the agent's active position determines whether her recipients will be affected with regard to their freedom and well-being. "[W]hile transactions may be individuated in different ways, what is decisive so far as concerns morality is their impact on [the recipients'] freedom and well-being, and more generally on their conditions of agency" (Gewirth, 1978, pp. 131–132). Because the agent logically must accept the rights of her recipients, the final necessary judgment is:

(20) "Every agent has a strict duty always to always act in accord with the generic rights of her recipient as well as of herself."

This is the dialectical form of the PGC. Steigleder emphasises that having a right does not imply that others act in accord with it but that the agent is normatively entitled against all other agents (Steigleder, 1992, p. 157). The PGC is the necessary principle in two ways:

> It is formally or logically necessary in that for any agent to deny or violate it is to contradict himself [...]. The principle is also materially necessary, or categorical, in that, unlike other principles, the obligations of the PGC cannot be escaped by any agent by shifting his inclinations, interests, or ideals, or by appealing to institutional rules whose contents are determined by convention. (Gewirth, 1978, p. 135).

Thus, the PGC's name results from the fact that it "combines the formal consideration of *consistency* with the material consideration of rights to the *generic* features or goods of action" (Gewirth, 1978, p. 135, emphasis added). Because every agent ought to do what she is logically committed to accept as a matter of consistency, the PGC may be reformulated in assertoric form:[74] "Act in accord with the generic rights of your recipients as well as of yourself." (Gewirth, 1978, p. 135). Because no agent can rationally deny the PGC without rationally contradicting her being a prospective purposive agent, the PGC is "unchallenged as a criterion of moral rightness, and conformity with its requirements is categorically obligatory" (Gewirth, 1978, pp. 48, 135).

4.1.3 *Two Additional Remarks*

Before I elaborate the implications of the PGC in practice with regard to its direct and indirect applications, I wish to emphasise two more considerations.

Negative and Positive Duties

Up to this point, I followed Gewirth in concentrating on negative duties of non-interference first. However, the agent does not only raise claims against others not to interfere with her freedom and well-being, but she is also aware that she might occasionally depend on other agents' assistance to maintain her freedom and well-being (Steigleder, 1992, p. 159). If an agent's right to freedom and well-being cannot otherwise be maintained through her own power, others ought to act positively to assist her to restore or maintain the preconditions of her prospective agency. The claim to assistance, however, must consider the specific circumstances an agent might end up with by helping. The right to assistance can therefore not be claimed against others in the same way as the right to non-interference (Steigleder, 1992, p. 235). Considering that every agent must logically commit herself to certain duties, an agent negatively ought to refrain from interfering with others' freedom and well-being. Positively, she has a duty to assist others to enjoy freedom and well-being as necessary goods of action if (a) others cannot do so from their own effort and (b) the agent can assist at no comparable cost to herself (Gewirth, 1978, pp. 134–135). Gewirth clarifies: "By comparable cost is meant that he [the agent] is not required to risk his own life or other basic goods, and similarly with the other components of the necessary goods of action." (Gewirth, 1978, p. 218). The positive duty to assistance "is not a matter of charity to be begged for but rather a matter of right" (Gewirth, 1978, p. 210). Thus, the PGC justifies not only negative but also

74 It might be objected that the transition from dialectical judgements to assertoric judgement (and thus to the PGC in its assertoric form) is invalid. Gewirth (1978, pp. 152–161) and Steigleder (1992, pp. 210–227) elaborately show that this objection fails to be fully convincing. Therefore, I assume the validity of the assertoric judgement.

positive duties. Duties with regard to well-being and freedom will be specified in further details in the next section.

Duties to Oneself

For the sake of completeness, I wish to mention that the PGC's implication regarding duties to oneself may be critically discussed. Gewirth provides a number of objections (Gewirth, 1978, pp. 333–334). Indeed, the PGC's implication is at first sight "other-directed" (Gewirth, 1978, p. 333). Primarily, my freedom and well-being impose a duty on others not to interfere with my rights without my agreement and to assist me if I cannot protect my freedom and well-being myself and others can do so at no comparable cost.

Nevertheless, according to Gewirth, duties to oneself may unfold in two ways: First, if there are duties to oneself, there must be rights to oneself too. According to the PGC, the agent must act in a way that respects her own rights to freedom and well-being as a recipient of her own action. A person who "degrades himself", "demeans himself" or acts "unfair to [himself]", performs actions that are imposed on herself so that a person can be both agent and recipient (Gewirth, 1978, p. 336). Concerning this matter, Gewirth draws a distinction between being an "immediate agent" and "long-term recipient" (Gewirth, 1978, p. 336). Gewirth's example is the following: If a person gets drunk, she anticipates the headache the next day and may tell herself "I'll regret this later" (Gewirth, 1978, p. 336). At this thought, the agent considers herself as a recipient of her of getting drunk and critically addresses herself as the agent acting upon her (Gewirth, 1978, p. 336). The prescription of the PGC to respect one's own right to freedom and well-being in acting, requires the agent to consider her own freedom and well-being as a recipient of her own action in the longer run. In order to act with respect to her own rights in this long-range perspective, the person has a duty to herself to limit consumption of alcohol to a low level.

The second way of duties to oneself, which Gewirth elaborates, develops with regard to prudential virtues. I will return to prudential virtues in the following chapter. It is fairly easy to see that developing certain character traits, mindsets or attitudes is essential or at least very beneficial with regard to fulfilling the requirements of the PGC. Gewirth writes:

> For if the agent leads a personally intemperate, fearful, self-brutalizing, dissolute, and unintelligent life, if he is slothful, ignorant, improvident, and lacking in self-respect, then he is less likely to respect other persons and to be able to fulfill adequately his strict duties to them. Hence, he has a duty to avoid such a life and to inculcate in himself the opposite qualities that are conducive to his fulfilling the PGC's requirements: such prudential qualities as being self-respecting,

> self-aware, temperate, courageous, provident, and well-informed. (Gewirth, 1978, pp. 334–335)

Given that there is a duty to others, there is at least a duty *regarding* oneself to cultivate certain appropriate mindsets or virtues. But is it a duty *to* oneself? In defence for a duty to oneself, Gewirth formally defines: "If A has a duty to do X, and if his doing Y is necessary or highly conductive to his doing X, then A has a duty to do Y if doing Y is in his power and if it does not involve violating any of his other duties." (Gewirth, 1978, p. 334). Based on that, Gewirth argues the following: By recognizing herself as a purposive agent, the agent must acknowledge that if developing certain virtues (doing Y) is conductive to her respecting the PGC (doing X) in virtue of others' being purposive agents, she must also rationally acknowledge that developing virtues (doing Y) is a duty to herself in virtue of her own being a purposive agent (Gewirth, 1978, p. 335). That is quite in contrast to Gewirth's earlier remarks on virtues, which I will introduce in the following chapter. There, he argues that there is no direct duty to develop certain virtues but only an indirect duty with regard to others. However, if an agent had a duty to herself, then there was a direct duty to develop certain mindsets. The clarification of this question, I am afraid, goes beyond the scope of this book and I will leave the discussion regarding duties to oneself open. Given that the requirements of the PGC primarily prescribe duties with regard to certain kinds of actions, I limit myself to the statement that the development of certain virtues is necessary or at least very useful to fulfil the duty to others and that the duty to others at least assigns a duty regarding oneself.

4.2 Practical Applications of the Foundational Argument

The foundational argument defends that every agent has equal rights to freedom and well-being (insofar as they are necessary goods for action) and that every agent has the same strict duty to act in respect of her recipients' rights to freedom and well-being. Gewirth writes:

> The PGC is an egalitarian universalist moral principle since it requires an equal distribution of the most general rights of action. It says to every agent that just as, in acting, he necessarily manifests or embodies the generic features of action – voluntariness and purposiveness – and necessarily claims the generic goods as his rights, so he ought to accept that his recipients, too, should manifest or embody these same generic features and have these same generic goods as their rights. The agent must hence be impartial as between himself and other persons when the latter's freedom and well-being are at stake, so that he ought to respect their freedom and well-being as well as his own. (Gewirth, 1978, p. 140)

In practice, however, it could be the case that one agent's rights to freedom and well-being conflicts with another agent's rights to freedom and well-being. Consider a hiker for example who is trapped in a snowstorm.[75] With her last ounce of strength she happens to find an untenanted alpine hut, obtains access and burns the owner's furniture in order to survive the freezing cold night. The hiker's right to life and bodily integrity conflicts with the tenant's right to property. The challenge is: Whose right has precedence? And, given that all rights are equal, on which ground is precedence of one right over another legitimate?

Gewirth and Steigleder convincingly argue that precedence must and should not be an arbitrary decision. Steigleder regards it as a serious misunderstanding to believe that the PGC is a "magic formula" that provides an *a priori* guidance for moral action (Steigleder, 1992, p. 243, own translation). Given that Gewirth's foundational argument justifies the PGC as a supreme moral principle, the central task of applied ethics is to clarify "*in which way, which* and *whose* interests or abilities are to be considered" (Steigleder, 1999, p. 19, own translation). Gewirth writes that

> the principle [the PGC] must also undergo an a posteriori inquiry to ascertain whether it is adequate for explicating and evaluating all morally relevant actions and institutions in a consistent and conflict-resolving way. Such an inquiry is especially important because, as a complex principle combining rights of freedom and of well-being on the part of both agents and recipients, the PGC faces the problem of the inherent consistency of these combinations. And because the principle is maximally general in scope, whereas morality is ultimately a matter of how persons ought to act in particular situations, the particular moral significance of the principle emerges only in what it requires of persons in particular cases. (Gewirth, 1978, p. 199).

The PGC's practical implications are the task of this chapter. Gewirth distinguishes between the "direct applications" and the "indirect applications" of the PGC (Gewirth, 1978, p. 200). The direct applications concern the PGC's requirements regarding "interpersonal actions" or "transactions" (Gewirth, 1978, p. 129). The second type of applications elaborates the PGC's requirements imposed on social rules and institutions. Because the PGC's requirements are, in the first place, imposed on organised institutions and these institutions (insofar as they are legitimated or prescribed by the PGC) set rules for transactions only in the second place, this part of the analysis is referred to as the indirect applications (Gewirth, 1978, p. 200). The indirect applications may modify the direct applications (Gewirth, 1978, p. 206). The first part of this

75 The example is taken from Feinberg (1980, p. 230).

chapter elaborates the direct applications before examining the indirect applications in more detail.

4.2.1 *Direct Applications*

With regard to transactions (the object of the direct applications), the PGC states that every agent has a strict duty to always act in accord with the rights to freedom and well-being of her recipient as well as of herself. Freedom and well-being as the necessary goods for action have roughly been introduced above: Well-being encompasses three kinds of goods: "basic goods, non-subtractive goods, and additive goods" (Gewirth, 1978, p. 54). Freedom is the necessary condition for agents to act self-controlled and in absence of external interference ("violence, coercion, and deception") by any other person, institution or thing (Gewirth, 1978, pp. 52, 252).

Up to this point, I have left one further aspect rather implicit. Regarding freedom and well-being as necessary goods for action, Gewirth differentiates two interpretations: First, the "generic-dispositional" interpretation and second, the "particular-occurrent" interpretation (Gewirth, 1978, p. 61). In the particular-occurrent interpretation, well-being and freedom are the necessary conditions for achieving a *particular* purpose of a *particular* action. In the generic-dispositional interpretation, in contrast, well-being is the necessary condition for achieving *any* purpose of *any* action the agent might want to achieve. I have touched upon this difference with regard to Beyleveld in the preliminary remarks: he explains that while some goods are contingent, some others are necessary (Beyleveld, 1991, p. 23). While the former refers to certain particular purposes, the latter refers dispositionally to any purpose a prospective purposive agent may want to achieve. The difference is probably easiest to see from cases of interference: As mentioned earlier, not "being lied to" is a good of well-being (Gewirth, 1978, p. 233). Now suppose it is agent A's birthday. To celebrate, her friends organise a surprise party. Under false pretences, agent A is navigated to the party. In this particular situation, her friends interfere with agent A's occurrent right to not being lied to. A different situation is if A is systematically being told the untruth in a wider social context. In the latter case, A is *dispositionally* unable to maintain goods she already achieved, let alone that she will probably be able to attain additional purposes. Well-being and freedom are "not only necessary *goods* for action but also *necessary* goods" (Steigleder, 1992, p. 144, own translation). Throughout the foundational argument, freedom and well-being must receive a generic-dispositional interpretation (Gewirth, 1978, p. 63).

In accordance to freedom and well-being as the two necessary goods for action, the PGC encompasses two further principles, one with regard to well-being, the "principle of Common Good", and one with regard to freedom, the "principle of Equal Freedom" (Gewirth, 1978, pp. 211, 250). According to the former, "every transaction must be for the good of the recipient as well as of the agent" and according to the latter, "both the agent and his recipient [must] participate freely in the transaction in which they are involved" (Gewirth, 1978, pp. 211, 250). The following first adds some more details to the nature of well-being as a necessary good before, based on that, elaborating rights and duties with regard to well-being.

Rights and Duties with Regard to Well-Being

Recall that well-being encompasses three kinds of goods: (1) basic goods, (2) non-subtractive goods, and (3) additive goods. Basic goods are the "general necessary preconditions of action" and include "certain physical and psychological dispositions ranging from life and physical integrity (including such of their means as food, clothing, and shelter) to mental equilibrium and a feeling of confidence as to the general possibility of attaining one's goals." (Gewirth, 1978, p. 54). In order to successfully maintain purposes and to successfully achieve additional purposes, the agent requires further necessary conditions: Non-subtractive and additive goods. Non-subtractive goods encompass the agent's "retaining and not losing whatever he already has that he regards as good" in order not to have her "level of purpose-fulfillment lowered" (Gewirth, 1978, p. 54). Necessary non-subtractive goods involve for example not "being lied to, cheated, stolen from, defamed, insulted, suffering broken promises, and having one's privacy violated" and not having to suffer from "dangerous, degrading, or excessively debilitating conditions of physical labor or housing or other strategic situations of life when resources are available for improvement"(Gewirth, 1978, p. 233). Additive goods are the resources an agent requires "to achieve the purpose for which he acts and hence to gain something that seems to him to be good" and to increase her level of purpose-fulfilment (Gewirth, 1978, p. 56). Necessary additive goods include "freedom, knowledge, education, and income" as well as "self-esteem" (Gewirth, 1978, pp. 244, 245).

The three elements of well-being as a necessary good for action provide a hierarchy. Gewirth explains the hierarchy in the following way:

> Although every agent regards his whole well-being as a necessary good because of its relation to his purpose-fulfillments, its components fall into a hierarchy determined by the degree of their indispensability for purposive action. The basic capabilities of action, whereby the agent has *basic goods, are the most necessary of all*, since without these he would be able to act either not at all or only in certain very restricted ways. *Among these* basic goods there is also a hierarchy,

> *headed by life* then including various other physical and mental goods, some more indispensable than others for action and purpose-fulfillment. Of the other two kinds of capabilities, the *nonsubtractive rank higher than the additive* because to be able to retain the goods one has is usually a necessary condition of being able to increase one's stock of goods. Thus the hierarchy of well-being corresponds somewhat to the 'hierarchy of needs' developed in psychological theories. (Gewirth, 1978, pp. 62–63, emphasis added).

Guided by the criterion "degrees of necessity for action", the hierarchy within the three kinds of well-being provides the key frameworks for solving conflicting rights with regard to transactions (Gewirth, 1978, p. 342). In cases of conflicting rights, the PGC prescribes that rights to goods that rank higher with regard to the necessity for action override rights to goods that rank lower in the hierarchy (Gewirth, 1978, pp. 342–344). Corresponding to the basic goods of well-being, agents have "basic rights" (Gewirth, 1978, p. 212). Corresponding to the non-basic goods of well-being, agents have rights to non-subtraction and addition. Thus, basic rights have precedence over rights to non-subtraction and rights to non-subtraction have precedence over rights to addition. Returning to the example of the hiker trapped in a snowstorm, the response to conflicting rights is: the hiker legitimately obtains access, burns the owner's furniture in order to survive the freezing cold night because her right to life and her right to bodily integrity is more needed for action than the owner's right to property.

Well-being may be interfered with by others in various, way more complex ways than the way initiated by the hiker. Gewirth differentiates between "basic harm" (Gewirth, 1978, p. 212) as opposed to "specific harm" (Gewirth, 1978, p. 230). Basic harm encompasses interferences with agents' basic rights – the rights to the basic goods of well-being. Gewirth explains:

> Basic harm [...] includes killing, maiming, and other sorts of physical injury, such as depriving of food, clothing, and shelter. They also include brainwashing, terrorizing, and other forms of extreme psychological pressure [...]. The advances of modern technology have provided many further, though more long-range, means of basic harm, including polluting the environment and contaminating food with carcinogens and other lethal materials. (Gewirth, 1978, p. 212)

The PGC generally prohibits such basic harm, which puts agents under the duty to at least act in a way that does not interfere with other agents' basic rights. Yet, an exception to the general prohibition of basic harm is the right to self-defence and, under specific conditions, the duty to help others defend themselves. Suppose agent A is violently attacking agent B and imposes basic harm on her. According to the PGC, B has the right to defend herself even if that imposes basic harm on A. Even though both impose basic harm, B's basic harm imposed on A by actions of self-defence do not have the same moral status as A's basic harm imposed on B by actions of attack. Because B, as the recipient of A's action, has her basic rights violated by A, the attacker, in the first place, actions of self-defence actually only re-establish the equality of rights, which was interrupted by A before. Gewirth argues that these actions of defence "restore an equilibrium of mutual nonharm" (Gewirth, 1978, p. 213). Thus, even though A and B might both impose basic harm on each other, the basic harm that B, the recipient, inflicts on A in return is legitimate. Likewise, if A is violently attacking B and agent C witnesses the attack, C has a positive duty to help B even though that involves imposing basic harm on A, if B cannot help herself and C is able to help B at no comparable cost (so without being imposed basic harm herself). Note, however, that actions of defence and self-defence must be in proportion to the basic harm imposed on the recipient B. Actions of defence should always aim to harm the attacker as little as possible but if the only means to save B's life is to kill A, defence may even including killing (Gewirth, 1978, p. 213).

Specific harm, as opposed to basic harm, is the interference with or loss of goods other than basic goods (Gewirth, 1978, p. 230). Given that recipients may regard a broad range of non-basic goods as goods, is any interference with non-basic rights a violation of the PGC in practice? Indeed, if the PGC were to prohibit all specific harm, any interference would be wrong leading to an unlimited number of claims. Then, the agent would be tied up with regard to her own generic rights by her recipients' non-basic rights (Gewirth, 1978, p. 231). The difference between the generic-dispositional and the particular-occurrent interpretation of well-being provides the solution. Following the generic-dispositional interpretation of well-being, specific harm in the relevant sense concerns *necessary* non-subtractive goods and *necessary* additive goods: Those goods that are the necessary conditions for *any* voluntary purposive action. Necessary non-subtractive goods are capabilities that enable an agent to act for the purpose of maintaining her level of purpose-fulfilment. With regard to specific harm to capabilities encompassed by necessary non-subtractive goods, Gewirth writes that "it is when actions universally tend to attack or diminish these capabilities that they inflict nonarbitrary harms in the sphere of non-subtractive goods, over and above the basic harms that come from having one's basic goods attacked." (Gewirth, 1978, p. 233).

The challenge of legitimate interference is a bit more complex with regard to specific harm in the sphere of additive goods such as education, property, wealth or income opportunities (Gewirth, 1978, pp. 244–245). However, agents' conditions and abilities vary considerably with regard to social conditions, talent, health, genetic disposition, etc. (Gewirth, 1978, pp. 246–247;

Steigleder, 1992, p. 254). How are agents supposed to act in accord with their recipients' additive right to education for example? The point is, agents' particular-occurrent additive well-being varies indeed substantially. However, agents' generic-dispositional additive well-being is invariable (Gewirth, 1978, pp. 55–56, 59). With regard to additive well-being, and concerning wealth and income in particular, the PGC does not necessarily require a transfer of resources. Rather, the focus is on providing equal access to resources in order to enable self-development. Viewed dispositionally, additive rights are rights to equal opportunity rather than rights to equal property of particular resources (Gewirth, 1978, pp. 246–247). With regard to education for example, agents have equal rights to access to education, but agents do not have equal rights to a specific level of education. However, taking education as an example indicates already that the primary duty with regard to additive well-being is not in the hands of individual agents in isolation but rather, as will be shown in the indirect applications, in the hands of institutions. Yet, individuals have a moral duty to support respective institutions.

For reasons of simplicity, I have largely limited the considerations concerning well-being to negative rights and duties. However, I believe the following should be stressed, the generic right to well-being does not only encompass the negative right not to be interfered with well-being by other agents. It also includes the positive right to assistance if the agent cannot do so by her own effort and others can provide assistance at no comparable cost. Considering the case of a violent attack, I have touched upon corresponding positive duties with regard to basic rights. Yet, positive duties extend to non-subtraction and addition. The hierarchy provided by the elements of well-being apply to positive rights and duties too. If agent A is the first on the site of an accident with a number of injured persons, her duty is (next to calling additional assistance) to help those persons who are injured most – those persons who have the least capacities for purposive action.

A detailed description of basic, non-subtractive and additive well-being is subject to careful considerations and analysis of the ethical discourse. Compared to basic well-being, the description of non-subtractive and additive well-being gets more complex and relevant circumstances are way more diverse (Gewirth, 1978, p. 245). In fact, 'non-subtractive' and 'additive' as referring to goods are "second-order, relational, aggregative expressions" (Gewirth, 1978, p. 55). Non-subtraction and addition refer to the relation between two different states: Non-subtraction refers to the agent's ability to preserve an amount of goods from status A to status B. The status quo remains unchanged and since the agent positively evaluates maintaining the status quo in comparison to losing goods, she has a non-subtractive good here. Addition refers to the agent's ability to increase her amount of goods from status A to status B. Since improving the status quo is better than maintaining the status quo, the agent has an additive good (Gewirth, 1978, p. 55). The transition from non-subtractive to additive goods and sometimes even from basic to non-subtractive to additive goods is (yet within boundaries) continuous.[76] Having access to drinking water in order to maintain life is a basic good. However, for a person who is about to die of thirst, maintaining life is a non-subtractive good while actually getting to drink some water is an additive good. For most cases, however, non-subtractive goods will exceed the scope of basic goods and additive goods will exceed the scope of non-subtractive goods (Gewirth, 1978, pp. 55–56). Thus, the specific contents of non-subtractive and additive goods are relative in two ways: First, they depend on the status quo and second, they depend on what agents consider to be goods (Gewirth, 1978, p. 55). In spite of these difficulties, the direct applications of the PGC persistently follow the hierarchy provided by the elements of well-being.

In preparation of the argumentation defended in this book, I wish to briefly elaborate two further aspects, which Gewirth emphasises with regard to additive well-being: "self-esteem" (Gewirth, 1978, p. 241) and, linked to that, prudential "virtues of character" (Gewirth, 1978, p. 242). The latter has been briefly introduced in the previous chapter with regard to duties to oneself. Starting with the former, "self-esteem" is a central good of additive well-being (Gewirth, 1978, p. 241). In order to develop additive capabilities and to achieve new purposes ('increasing one's level of purpose-fulfilment'), an agent necessarily requires a sense of valuing herself. In stage I of the foundational argument, the agent positively evaluates the purpose of her action.

> When such evaluative purposiveness is more than incidental and transient, the agent has an abiding self-esteem in that he views the worth of his goals as reflecting his own worth as a rational person whose life, freedom, and well-being are worthy of protection and development. Without such self-esteem, his ability to achieve further goals become problematic, and with it his prospect of taking satisfaction in what he accomplishes thereby. (Gewirth, 1978, pp. 241–242)

76 Gewirth further differentiates between an inclusive and an exclusive meaning of additive goods. While the exclusive meaning includes only those achievements that are not non-subtractive or basic goods, the inclusive meaning encompasses any good an agent aims to add to her level of purpose-fulfilment (Gewirth, 1978, p. 56). According to the exclusive understanding, access to drinking water is a non-subtractive or basic good, not an additive good. Since the exclusive meaning of additive goods is more substantial, Gewirth applies the exclusive understanding.

According to the equality of rights, the PGC requires regard of the self in two dimensions: With regard to one's own self and with regard to the self of others. If the agent holds herself in high esteems, she must also attribute such esteem to the self of other persons (Gewirth, 1978, p. 242). Mutually attributing esteem to one another's selves gives way to "duties of respect", which require that "each person must refrain from feeling or exhibiting contempt towards others" (Gewirth, 1978, p. 242). In acting, no other agent must be "insulted, belittled, or patronized, nor must they be discriminated against on grounds of race, religion, or nationality" (Gewirth, 1978, p. 242). Duties of respect do not correspond with particular actions. Rather, interpersonal actions ought to express mutual respect and support for each other.

Linked to the agent's own self-worth, there are "virtues of character" (Gewirth, 1978, p. 242). These virtues help to strengthen the agent's self-worth. In addition, they are mindsets "that underlie and help to motivate actions" (Gewirth, 1978, p. 242). Just as self-esteem, the virtues of character are additive goods too, which explains why Gewirth also refers to them as "additive goods of character" (Gewirth, 1978, p. 244). According to Gewirth, the relevant virtues of character "may be summarized under three of the traditional four cardinal virtues." (Gewirth, 1978, pp. 242–243). First, courage as the ability to assess fear and to act correspondingly. Courage also includes "determination in overcoming obstacles rather than finding the easy way out" (Gewirth, 1978, p. 243). Second, temperance as the disposition to practice self-control by moderating personal "appetites and inclinations" (Gewirth, 1978, p. 243). And third, prudence as a mindset that allows the assessment of different courses of action to achieve a certain end, to determine the most adequate course of action that combines personal competences and desires, and "to distinguish among both particular and more general ends themselves" (Gewirth, 1978, p. 243). In order to be able to mediate between one's own competences and desires, prudence also encompasses "self-knowledge and knowledge of one's natural and social environment" (Gewirth, 1978, p. 243). Gewirth clarifies that the relevant virtues of character are "not moral but prudential" (Gewirth, 1978, p. 243). A person could for example act courageously but for a generally immoral purpose. Only if a person acts courageously, moderately, or prudentially and these actions are in accord with the requirements of the PGC, they become moral virtues (Gewirth, 1978, p. 243).

The right to acquiring and maintaining virtues of character as additive goods requires an environment where no other agent interferes with the agent's developing such virtues and where developing such virtues is actually supported or at least favoured. The negative duty of others is primarily

> to refrain from actions that hinder the development of the prudential virtues in other persons. Such actions may take a variety of forms: promoting a climate of fear and oppressions: encouraging the spread of physically or mentally helpful practices such as excessive use of alcohol, tobacco, drugs, or pornography; contributing to misinformation, ignorance, and superstition, especially as these bear on persons' ability to act effectively in pursuit of their purposes. (Gewirth, 1978, p. 243)

By performing such actions, agents interfere with their recipients' right to "additive goods of character" and thus, with their right to additive well-being (Gewirth, 1978, p. 244). Yet, agents also have a positive duty to provide assistance by "setting an example of the practice of the virtues in his own actions and supporting a [...] milieu in which these virtues are respected and fostered" (Gewirth, 1978, p. 244).

Now, does living up to these virtues in practice impose a duty to develop these virtues in oneself? With regard to virtues as additive goods, Gewirth writes, "[i]t cannot *directly* be said [...] that the agent has a moral duty to develop these virtues in himself" (Gewirth, 1978, p. 243, emphasis added). Yet, *indirectly* and with regard to others, "the agent has a moral duty to develop [the virtues] in himself and to practice the prudential virtues on his own" (Gewirth, 1978, p. 244). As I have stressed in the previous section, that is in contrast to Gewirth's analysis of duties that an agent has *to* herself (see chapter 4.1.3). If an agent had a duty to herself, as Gewirth seems to attempt to show (Gewirth, 1978, pp. 334–335), then there would be a direct duty to develop certain virtues. With regard to duties to oneself, I have limited myself to the statement that there are at least duties *regarding* oneself, which is in accordance with Gewirth's remarks on duties concerning virtues as additive goods described here. Agents have duties to other agents. And since developing certain mindsets or virtues is beneficial to one's ability to fulfil these duties, agents ought to do their best in order to cultivate virtues. Gewirth summarises:

> Although the PGC directly requires certain kinds of actions rather then [*sic*] virtues, it also requires that the agent act in such ways that he develops the moral and prudential virtues in himself, because of the assistance such development provides for his fulfilment of the various moral duties of action prescribed by the PGC. (Gewirth, 1978, p. 244).

Rights and Duties Regarding Freedom

Freedom and well-being are closely related with regard to two aspects: First, freedom as the generic feature of agency, is itself a basic good of well-being because it is a necessary condition to be able to act at all (Steigleder, 1992, p. 255).

Second, the right to well-being, as a necessary good for action, is the right to enjoy the goods of well-being in a way that is free from interference (Gewirth, 1978, p. 250). Thus, interference with well-being is usually also an interference with freedom. Apart from this two-fold relation, freedom has an independent status with regard to the "procedural aspect" of action (Gewirth, 1978, p. 251). Gewirth explains:

> To interfere with someone's freedom is to interfere with his control of his behavior, including his participation in transactions; such interference hence affects the procedural aspect of behavior. To interfere with someone's well-being, on the other hand, is to interfere with the objects or goods at which his behavior is aimed; it hence affects the substantive or purposive aspects of his behavior. This contrast holds even where well-being is interpreted, as here, so that it [well-being] consists in having the general abilities and conditions required for maintaining and obtaining goods. For the emphasis still falls on the relation of the general abilities to all the objects or purposes that for any prospective agent are goods required for or attainable by action. (Gewirth, 1978, pp. 251–252)

The right to freedom, prescribed by the PGC, implies that every agent has the right to act without interference by others – as long as she does not interfere with her recipients' rights to the necessary goods of action.

> In general, the PGC limits the freedom of agents only where they coerce or harm other persons, that is, where the agents violate the rights of their recipients to freedom and well-being. An agent should be free to perform any action, to engage in any transaction, if and only if his recipients are left free, through their voluntary consent, to participate or not participate in that transaction and if and only if he does not inflict basic or specific harm on them. (Gewirth, 1978, pp. 270–271)

Recall that the term 'transaction' highlights the "asymmetric distribution of roles" (Steigleder, 1992, p. 199, own translation) between the agent and her recipients (Gewirth, 1978, p. 129). While the agent "actively" engages in the transaction, her recipients remain "passively, by undergoing or being affected by the agent's action towards him" (Gewirth, 1978, p. 129). As regards the right to freedom of the recipients, it requires their "unforced, informed consent", so the recipients' agreement to interference must be of their own free will and with knowledge of all relevant circumstances (Gewirth, 1978, p. 252). Linked to that, Gewirth distinguishes three kinds of interference with the recipients' right to freedom: "violence, coercion, and deception" (Gewirth, 1978, p. 252). Violence encompasses "physical and psychological compulsion" (Gewirth, 1978, p. 252). With regard to coercion, the recipient expresses her agreement, yet, she is given "compulsory, undesirable alternatives" and is forced to choose between a very bad and an even worse option (Gewirth, 1978, p. 252). In the case of deception, the recipient expresses her agreement, yet only as a result of "falsehood or misinterpretations" caused or provided by the agent on intention (Gewirth, 1978, p. 252). Examples of interference with the right to freedom are "robbery with assault and battery, a holdup by a gunman, and fraud; or [...] rape, forced prostitution (or 'white slavery'), and seduction" (Gewirth, 1978, p. 252).

Just as well-being, freedom has two interpretations: In the particular-occurrent interpretation, interference with the right to freedom is an interference with the agent's ability to perform a particular action, at a specific date. In the generic-dispositional interpretation, interference with the right to freedom results in the recipients' general inability to perform any action, or as Gewirth puts it: an interference with her "long-range control" (Gewirth, 1978, p. 253). So, just as with well-being, freedom has degrees of necessity for action too. In cases of conflict, the right to dispositional freedom ranks higher than the right to occurrent freedom.

The recipients' expressed agreement to interference is "the necessary and sufficient condition of a transaction's being morally justified" (Gewirth, 1978, p. 256). However, expressing one's agreement in person is sometimes neither possible nor necessary. Gewirth distinguishes "verbal" and "nonverbal" consent (Gewirth, 1978, p. 256). In cases of silent consent, the agent may assume that the recipient would have consented to the relevant interference if she had had knowledge of the situation. To provide an example, Gewirth considers a person A crossing a busy road (Gewirth, 1978, p. 260). The person overlooks a car and is about to be overrun. A pedestrian B may interfere with A's right to freedom by violently pulling her back from the street causing her to stumble and fall. Assuming that crossing the street is not a voluntarily attempted suicide, pulling someone back who is about to be overrun by a car is a silently agreed interference with her right to freedom. But not only that. The PGC may even positively *require* A, the pedestrian, to pull B back. In certain circumstances, it is a very thin line between fulfilling positive and not violating negative moral duties (Gewirth, 1978, p. 260).

Applications of the PGC in cases where the right to freedom conflicts with the basic right to well-being are particularly complex. In fact, one of these complex cases is committing suicide. Others are enslavement or maiming with the recipients' consent. Unfortunately, a satisfying assessment goes beyond the scope of this work. I limit myself to a few remarks.[77] Often, committing suicide is the result of desperation, hopelessness or more generally, the result of psychic syndromes such as depression. Thus, in many cases, there is considerable doubt about the choice being voluntary. Interference with the agent's

77 For a more elaborate discussion see Gewirth (1978, pp. 264–267) and Steigleder (1992, pp. 258–259).

right to freedom might then be morally justified in order to secure her right to basic well-being. The same is true for agents who consent to be maimed or enslaved. If they are, for example, so desperate about their conditions that harm is the desired option, then there is considerable doubt that the choice is voluntary. However, if there is no doubt about their voluntariness, the agent consenting to be harmed or to harm herself ought not to be interfered with by others (Gewirth, 1978, p. 265). Yet, in the case of suicide, the PGC sets further limits to the right to freedom of self-harmer: In harming herself, she ought not to interfere with others' freedom and well-being (Gewirth, 1978, pp. 264–265). It is for this reason that an attempted suicide of a parent who has two dependent children must be assessed differently than the attempted suicide of a single person. In committing suicide, the parent certainly violates her own basic well-being, but she also interferes with her children's well-being. In the case of persons consenting to be harmed, the PGC, as will be elaborated in the indirect applications, requires a moral general critique of social institutions. If circumstances permit that harm is the most desired option, the PGC requires to initiate "institutional arrangements that help to assure for all persons the right to freedom and well-being and hence minimize the circumstances that lead some persons to try to surrender one or both of these rights of action by inflicting harm on themselves" (Gewirth, 1978, p. 267). The PGC's further implications regarding social rules and institutions is subject to the indirect applications below.

Before moving on, one important final remark has to be made: Again, for reasons of simplicity, I have limited the considerations concerning freedom to negative rights and duties. However, I believe this should be emphasized here too, the generic right to freedom does not only encompass the negative right to non-interference. The right to freedom also includes the positive right to assistance if the agent cannot do so by her own effort and others can provide assistance at no comparable cost. Thus, agents have a strict duty not to interfere with their recipients' freedom, but they also have a duty to help others to restore their freedom if they cannot do so by their own effort, but agents can do so at no comparable cost. If, for example, agent A get kidnaped and agent B witnesses her getting kidnaped, B has a duty to help A to escape from her kidnappers if A cannot do so by her own effort and B can do so without getting kidnapped herself.

4.2.2 *Indirect Applications*

Transactions as interpersonal actions cover quite a range of cases. Nevertheless, actions may have effects that are beyond control, which are hard to be specified by interpersonal actions only. Actions are performed in the framework of institutional arrangements and as such, actions may heavily interfere with their recipients' rights to freedom and well-being. Gewirth considers a judge who sentences a defendant to prison (Gewirth, 1978, p. 302). The judge clearly interferes with the defendant's right to freedom. Yet, readers' moral intuition probably already tells that the judge is legitimately interfering. This chapter examines this case and, more generally, it elaborates the requirements of the PGC imposed on social rules and institutions. Because the requirements of the PGC are imposed on transactions only through social rules and institutions, this part of the analysis is referred to as the indirect applications (Gewirth, 1978, p. 200). The indirect applications are probably strongest for their justification of the minimal as well as the welfare state. Yet, the analysis of the indirect applications in this chapter does not focus on the justification of state institutions. This chapter aims to provide the ground for locating and justifying Cripps' idea of weak collectivities from a rights-based perspective in the next chapter. The remarks are therefore only a rough overview of the profound indirect application that Gewirth provides in *Reason and Morality* (Gewirth, 1978, pp. 272–365).

Gewirth defines institutions as follows:

> In general, an institution is a relatively stable, standardized arrangement for pursuing or participating in some purposive function or activity that is socially approved on the ground (whether justified or unjustified) of its value for a society. Because it is standardized, an institution is constituted by rules that define what persons are required to do if they are to participate in the respective activities or functions; and these requirements are obligations persons have qua such participants. (Gewirth, 1978, p. 274).

Further, he draws a distinction between "functional" and "organisational" institutions:

> A functional institution consists in the purposive standardized activity itself with its rules and requirements, while the organizational institution includes also structured groupings of persons associated for pursuing and especially for regulating the standardized activity and enforcing the rules. (Gewirth, 1978, p. 275).

Some functional institutions like education or religion have corresponding organisational institutions. For instance, "education has schools and colleges, [and] religion has churches" (Gewirth, 1978, p. 275). Some other institutions are exclusively functional: "[T]elling someone the truth" or "keeping a promise" for example have no corresponding organisational institution. The indirect application of the PGC applies to organisational institutions (Gewirth, 1978, p. 277).

Institutional criticism, as a central task of normative ethics, should assess the legitimacy of existing institutions. Yet, as Steigleder (1992) highlights, institutional criticism is neither the only nor the primary task of normative ethics. In view of cases where relevant institutions exist but fail to ensure freedom and well-being, normative criticism should target their transformation. And with regard to cases where institutions do not exist, normative criticism should claim their establishment and development. The primary task of normative ethics is therefore the moral justification of institutions. Based on that, the secondary task is institutional criticism (Steigleder, 1992, pp. 260–261).

Gewirth differentiates two modes of moral justification: the "procedural" and the "instrumental" justification (Gewirth, 1978, p. 282). In view of freedom as a necessary good for action, the former justifies institutions "through procedures of voluntary agreement" (Gewirth, 1978, p. 282). In view of well-being as a necessary good for action, the latter justifies institutions if they are "instrumental toward the well-being that underlies the purposiveness of human action" (Gewirth, 1978, p. 282). Procedural justification may be "optional" or "necessary" (Gewirth, 1978, p. 283). Instrumental justification may be "static" or "dynamic" (Gewirth, 1978, p. 292). Accordingly, four different cases may be distinguished: (1) the "optional-procedural justification" (Gewirth, 1978, pp. 282, 282–290), (2) the static-instrumental justification (Gewirth, 1978, pp. 292, 292–304), (3) the necessary-procedural justification (Gewirth, 1978, pp. 304, 304–311), and (4) the dynamic-instrumental justification (Gewirth, 1978, pp. 312, 312–327).

(1) Optional-Procedural Justification – the Case of Voluntary Associations

In view of the range of possible kinds of social cooperation, Gewirth differentiates between "public" and "private associations" (Gewirth, 1978, p. 289). The former are "compulsory", the latter are "voluntary" (Gewirth, 1978, p. 289). The optional procedural justification relates to the voluntary associations. Examples of voluntary associations range from informal sporting competitions such as a rugby match over employer-employee-relationships and scientific communities to marriage (Steigleder, 1992, p. 263). Based on the assumption that the institution's purpose and content of rules does not infringe with nonmembers' rights to freedom and well-being, voluntary associations are justified if all members voluntarily express their agreement to be a member and to accept the agreed rules (Gewirth, 1978, p. 284). The formation itself is optional and its implications are morally justified if the formation is morally legitimated through the procedure of voluntary agreement. Consider a rugby match as an example: Athletes voluntarily agree to form two competing teams and to follow the rules they voluntarily agree upon. Suppose, one of these rules justifies bodily harm on purpose. In this case, infringement even on athletes' basic well-being is morally justified. Gewirth notes that voluntariness with regard to membership or acceptance of rules may be problematic to assure for example when it comes to monopoly status of associations (Gewirth, 1978, p. 289).

(2) Static-Instrumental Justification – the Case of the Minimal State

Voluntary consent as in optionally-procedurally considerations, is a sufficient condition to justify social rules in the case of voluntary associations. But what if social rules fail to provide adequate protection of agents' well-being? Suppose that the majority of people living in a certain area decide to form associations in order to protect their well-being. They voluntarily agree on social rules, which control their members' actions with regard to their recipients' well-being. In this ideal case, no member will be harmed. Suppose further that some strong individuals, who are able of self-protection, refuse to participate. Established social rules no longer apply to those who refuse to participate. The direct application of the PGC requires the deniers to respect others' right to well-being. Yet, because established social rules do not apply to them, non-interference with their recipients' well-being cannot be guaranteed. With reference to practical evidence, Gewirth argues that coercive regulations of the criminal law are necessarily required (Gewirth, 1978, p. 291).

Rules and institutions are instrumental to guarantee peoples' right to well-being. Starting from the situation where no agent has been harmed and agents equally enjoy their right to well-being, this first phase of the instrumental justification is static because it simply protects the existing equality of rights (Gewirth, 1978, pp. 291–292). Rules of the criminal law are statically-instrumentally justified if they are established and announced by legislative authorities, if they are carried out by executive authorities and if legislative and executive authorities are made up for this purpose. The static-instrumental justification morally authorizes the minimal state with its criminal laws (Gewirth, 1978, pp. 292–293, 304).

Gewirth rather extensively elaborates the structure and regulations of criminal law including equal treatment of all people before the law, the proportionality of penalties (e.g. a murderer is not to be killed but sentenced to spent time in prison) and the protection of innocent defendant (Gewirth, 1978, pp. 294–300). Even if a detailed reconstruction goes beyond the scope of this book, it provides sufficient information to analyse why the judge's interference with the defendant's right to freedom is morally justified. Even if both have an equal right to freedom, Gewirth explains, the judge possesses a "more specific quality" legitimised by the PGC: "the quality of being a judge" (Gewirth, 1978, p. 302). And that is in contrast to the defendant. The PGC authorises the additional competence assigned to the judge because the rules of the criminal law as well as the authorities made up for the purpose of exercising the law are instrumentally justified by the PGC (Gewirth, 1978, p. 302).

(3) Necessary-Procedural Justification – the Method of Consent

In the static-instrumental justification of the minimal state, each person rationally agrees to criminal law and its rules as an instrumental necessity of their freedom and well-being. However, the existence and acceptance of the minimal state is not left to persons' optional consent. In contrast to voluntary associations, its existence is a rational necessity (Gewirth, 1978, p. 304). In spite of the criminal law with its rules, the coexistence of many people requires further rules and institutions, which govern the cooperation between people and ensure their well-being. In fact, as will be shown in (4), the PGC further requires the development of the welfare-state with its laws and officials.

All institutions and rules are coercive in the sense that persons are necessarily subject to their authority. Also, through exercise of discretion, persons who announce the relevant institutions of the welfare state and its rules hold "great power" (Gewirth, 1978, p. 305). Institutions having authority are morally justified if they are freely elected by those persons over whom authority is exercised. Its rules are morally justified if they meet all relevant criteria of the PGC and if the procedures that govern who is in the positions to determine the rules meet the procedural requirements (Gewirth, 1978, p. 306).

> For determining these officials and laws, the PGC requires that the voluntary consents of the persons subject to them must be invoked as fully as possible; in this way as much freedom can be preserved as is compatible with the moral necessity of having states and laws. (Gewirth, 1978, p. 306)

Being under authority is not necessarily compatible with voluntary consent, which could possibly lead to chaos, rebellion or "anarchy" (Gewirth, 1978, p. 306). Therefore, the PGC prescribes the existence of a constitution. Gewirth explains:

> This constitution provides for a decision procedure that depends on certain freedoms for all members of society. The procedure requires that the use and availability of these freedoms must be decisive in determining the specific laws and officials of the state. Thus, while the constitution is mandatory and hence is not itself subject to optional consent by individuals, it provides that such consent must be used as fully as possible within the political process itself for determining the laws and officials. (Gewirth, 1978, p. 306)

The main purpose of the constitution is "the equal distribution of civil liberties" (Gewirth, 1978, p. 307). The PGC requires that every person is able to perform any action she wishes to perform as long as she does not interfere with others' freedom and well-being. Likewise, the state should refrain from interferences with a person's right to freedom unless the state aims to prevent coercion and harm to other persons (Gewirth, 1978, p. 308). The specific actions, which the state must not interfere with, regard each person's "civil liberties" and include "speaking, publishing, and assembling and associating with others" (Gewirth, 1978, p. 308). Except for criminals, the PGC prescribes that civil liberties apply equally to every prospective purposive agent. More specifically, "each person has an equal right to participate freely and actively in the political process and to be protected by the state in that participation" (Gewirth, 1978, p. 308). The "method of consent" comprises the existence and exercise of civil liberties throughout the political process (Gewirth, 1978, p. 308).

Gewirth notes that "the connection between the democratic procedures of the method of consent and the requirements of distributive justice is not logical but contingent" (Gewirth, 1978, p. 322). A majority of citizens may consent to support morally wrong policies even though they may lift restrictions to the power of state authorities. The slave laws in the nineteenth's century or censorship laws in China are examples. Gewirth recommends that in such cases, citizens should try to peacefully resolve these injustices by further procedures of a method of consent. Only if they are ineffective, may civil disobedience be justified. However, civil disobedience should always be "non-violent" and no "kind of power play" but rather a "public, principled appeal" that focuses on specifying how the relevant, criticised laws or policies violate the requirements of the PGC (Gewirth, 1978, p. 322).

(4) Dynamic-Instrumental Justification – the Case of the Supportive State

Starting from the situation where no agent has been harmed and agents equally enjoy their right to well-being, the first part of the instrumental justification was static because it simply protects the existing, occurrent equality of rights (Gewirth, 1978, pp. 291–292). The criminal law and its rules are justified because they are instrumental towards well-being. However, individuals' dispositional well-being may differ considerably:

> Many people lack adequate food, housing, necessary medical care, and other basic goods [...]; they may be subject to degrading conditions of dependence, danger, and diseases in their situation of life and work; they may also not have the means to increase their capabilities of purpose-fulfilling action, such as adequate education and income. (Gewirth, 1978, p. 312)

These conditions violate a key requirement of the PGC: equality of well-being in a dispositional sense. Targeting these dispositional inequalities, the second part of the instrumental justification is dynamic and requires the development of the welfare-state (Gewirth, 1978, p. 312). Assuming that individuals cannot protect their right to well-being by their own effort, the welfare-state ought to restore peoples' dispositional inequalities. Thus, the task of the welfare-state, and I have highlighted that with regard to additive well-being already, is not simply to "distribute or redistribute wealth" (Gewirth, 1978, p. 313). Rather, the central characteristic of the welfare-state is twofold: it aims to reward individual effort while not being blind to the disadvantaged (Gewirth, 1978, p. 313).

In order to limit interventions of the state in the long run, institutions should focus on developing those conditions that enable people to help themselves without creating or reinforcing dependencies (Gewirth, 1978, p. 313). Generally, redistribution may be initiated by voluntary associations or individuals independent of the state. However, Gewirth notes that the practical limits of individual or voluntary associations to actually create dispositional equality are probably rather narrow. Public redistribution (e.g. by taxes) is therefore an instrumental necessity required by the PGC (Gewirth, 1978, pp. 314–315). Given that public redistribution is "limited by the resources actually available in a given society", other states (if they have sufficient resources) might be required to assist (Gewirth, 1978, p. 314).

One final remark: I have already stressed the difference between utilitarianism and Gewirth's theory in the preliminary remarks with regard to the use of well-being. Though the difference between the direct and the indirect application corresponds with the difference between act-utilitarianism (applied directly to individual actions) and rule-utilitarianism (applied to rules and indirectly to individual actions), the indirect application points to a further difference: While utilitarians simply seek to maximize utility or goods regardless of distributional considerations, the PGC emphasises the equality between agents' rights and stresses distributional considerations (Gewirth, 1978, pp. 200–201). Thus, rights-based moral theories, as I have emphasised earlier with reference to Steigleder (2016a), indeed account for consequences but consequences of a special kind: Those that focus on the effects to each person individually (Steigleder, 2016a, p. 255).

4.2.3 *Solving Cases of Conflicting Rights and Duties*

The indirect and direct applications demonstrate that the PGC provides a very powerful, yet complex, tool. Gewirth stresses that the complexity of "the system of the PGC" is necessary as "it reflects the complex structure of morality itself." (Gewirth, 1978, p. 339). But is the PGC fully able to solve situations of rights and duties being in conflict? Gewirth has quite extensively investigated on this question (Gewirth, 1978, pp. 338–354). Implicitly, the applications of the PGC have touched upon all his considerations. The following is therefore a rather brief summary that intends to explicate the three implicitly assumed criteria that have already been guiding through the application. Among them is the criterion of the "degrees of necessity for action" that has been mentioned several times throughout this chapter (Gewirth, 1978, p. 345). The three criteria constitute three different starting points that must be considered in different situations (Gewirth, 1978, p. 342). None of them excludes the others and they must be applied in the indicated order if more than one criterion appeals to the respective situation.

Criterion (1): "Prevention and Removal of inconsistency"

The PGC requires equality of rights and implies common recognition of freedom and well-being. "If one person or group violates or is about to violate the generic rights of another and thereby incurs transactional inconsistency, action to prevent or remove the inconsistency may be justified." (Gewirth, 1978, p. 342). Thus, though the PGC generally prohibits interference with freedom and well-being, coercive or harmful action may be justified to restore equality of rights between agents (Gewirth, 1978, p. 342). The criterion of "[p]revention and [r]emoval of [i]nconsistencies" applies to transactional cases for example with regard to self-defence or defence of others that were considered above (Gewirth, 1978, p. 342). But the criterion also applies to institutional cases of legal interventions encompassed by criminal law and the welfare state (Gewirth, 1978, p. 343). If possible, legal intervention should be used in order to remove transactional inconsistencies. In addition, prevention and removal should follow the principle of proportionality: coercion or harm that aims to correct inconsistencies may never exceed the severity of coercion or harm imposed in the first place (Gewirth, 1978, p. 342). In the case of bodily harm for example the victim should, if possible, report to the authorities instead of taking the law into her own hands. The offender on the other hand should not be inflicted with major bodily harm herself by the authorities but rather should she be condemned to pay a fine, monetary compensation for the victim's harm or, in cases of grievous bodily harm, should be sent to prison.

Criterion (2): "Degrees of Necessity for Action"

In view of the direct applications elaborated above, the criterion "[d]egrees of [n]ecessity for [a]ction" probably requires the least amount of explanation (Gewirth, 1978, p. 343). While criterion (1) applies to goods of equal degree, the criterion (2), degrees of necessity for action, applies to situations where goods of unequal degree conflict with one another (Gewirth, 1978, p. 343). Gewirth emphasises:

> Since every agent has the duty to respect his recipients' rights to the goods that are necessary conditions of action, one duty takes precedence over another if the good that is the object of the former duty is more necessary for the possibility for action, and if the right to that good cannot be protected without violating the latter duty. (Gewirth, 1978, p. 343).

As mentioned earlier, this applies to transactional cases as in the example of the hiker trapped in a snowstorm. But the criterion of degrees of necessity for action also applies to the institutional context of the criminal law. Even if rules determined by voluntary associations are procedurally justified, rules of criminal law take precedence over voluntarily agreed rules in the framework of voluntary associations (Gewirth, 1978, p. 343). The rugby match for example is not a legal vacuum. Even if participants were to agree to the rule that there are no rules, criminal law would apply.

Criterion (3): "Institutional Requirements"

The third criterion deals with the question in which way and to what extent the indirect application of the PGC may override and modify the direct application. Procedurally justified rules as may coerce agents only as long as they voluntarily consent to be part of the association and to accept the rules (Gewirth, 1978, p. 344). Complex societies rationally require rules that may coerce or harm people only if the required rules are necessary to prevent unjust coercion or harm to others. Taxes for example are indeed coercive. However, taxes are justified if they do not exceed an amount required for the protection of others' dispositional well-being and if they are justified through the method of consent.

The order given by the three criteria is based on the "successively less direct ways of fulfilling the PGC's central requirement that agents must act in accord with the generic rights of their recipients." (Gewirth, 1978, p. 345). Criteria (2) and (3) show that conflicting duties may not only arise if duties have already been violated. They may also arise with regard to the prevention of interference or with regard to the fulfilment of "institutional requirements" (Gewirth, 1978, p. 345). Gewirth further clarifies:

> The three criteria for resolving conflicts of duties have many further ramifications in both the individual and the social spheres. But amid the various conflicts, it is important to keep in mind that the PGC's standard or central requirement is the equality of generic rights and hence mutual respect for freedom and well-being among all prospective purposive agents. Departures from this mutual respect

> are justified only where they are required either to prevent or rectify antecedent departures, or to avoid greater departures, or to comply with the social rules that themselves reflect such respect in the ways indicated above. When such justifications are not to found, the moral rules that directly follow from the PGC as its direct applications set actual or conclusive duties. (Gewirth, 1978, p. 345).

4.3 Weakly Collective and Individual Promotional Duties from a Rights-Based Approach?

The previous two chapters showed that many different considerations "enter into and derive from the PGC" (Gewirth, 1978, p. 338). Gewirth summarizes the PGC and its implications as follows:

> Thus the principle [the PGC] provides for rights both to freedom and to well-being, and for each of these as pertaining both to each agent and to all his recipients; within the right to well-being it provides for three different kinds of goods that may be viewed either occurrently or dispositionally; it also provides for two kinds of applications, direct and indirect, bearing respectively on individual transactions and on group interactions; within the indirect applications it provides both for procedural and for instrumental justifications, with the former in turn divided into optional and necessary kinds and the latter into static and dynamic varieties, the dynamic again being divided into aggregative and distributive-differential provisions. (Gewirth, 1978, p. 338)

This chapter elaborates Cripps' idea of weakly collective duties and individual promotional duties from a rights-based perspective. In order to set the stage, I consider it helpful to recall Cripps' normative claims in contrast to Gewirth's theory. There are four key differences, which may be summarized as follows:

Interests vs. Rights

While Cripps defends an interest-based approach, Gewirth takes rights as a starting point for moral evaluation. Though rights instrumentally protect human interests, interests are only a sufficient but not a necessary condition for moral entitlements (see introduction to chapter 4). In the first place, persons have rights because they are their due not only because they are in their interest.

Intuition vs. Rationality

Following the capability approach with regard to the moral justification of entitlements, Cripps relies on common intuition (see chapter 3.3.1). In contrast, following the principles of logic, Gewirth provides a rational justification of entitlements. The evaluative structure of action, as the necessary starting point of moral considerations, rationally implies that agents have certain rights. I have criticised Cripps for the lack of transparency with regard to the moral justification of certain entitlements. The foundational argument, though complex, provides such transparent justification.

List of Entitlements vs. Framework of Entitlements

Cripps provides a concrete list of five fundamental interests that are required "for a fully flourishing human life": life, bodily health, bodily integrity, affiliation and practical reason (Cripps, 2013, p. 7). As already mentioned elsewhere, a concrete and tangible list is appealing in particular when it comes to subjects such as climate policy that are dominated by scientists and experts unfamiliar with philosophy (see chapter 3.3.1). A list is widely accessible to a broad audience. Yet, I have claimed that Gewirth's theory is better equipped. The rights to freedom and well-being, as necessary goods for action, are entitlements to these goods as a whole. In addition, the rights to freedom and well-being subsume a substantial framework that encompasses a range of entitlements. Thereby, the framework allows the consideration of a variety of personal circumstances. For heated discussions, such as with regard to climate change, where entitlements are in conflict, it is a great advantage to be able to take into account entitlements other than only the basic or fundamental entitlements.

Balance of Entitlements vs. Hierarchy of Entitlements

Cripps' five fundamental human interests stand unchallenged next to each other as a balanced list where no entitlement takes precedence over another. In contrast, the three components of well-being as well as the two different interpretations (occurrent and dispositional) provide a hierarchy within freedom and well-being as necessary goods as a whole. That way, Gewirth's theory is able to provide a ground for resolving conflicts. The lack of a hierarchy of entitlements was another criticism to Cripps' theory. The PGC and, linked to the PGC, freedom and well-being as necessary goods for action provide a hierarchical order of rights.

Despite all differences regarding the normative foundation, I claim that Gewirth's rights-based moral theory is compatible with Cripps' theory regarding weakly collective duties and derivative individual promotional duties. As a first step, I directly jump to the moral principles defining the weakly collective duty of the Young, the Able and the Polluters and I wish to modify them in accordance to the PGC. Recall that Cripps applies the no-harm principle and the principle of beneficence in two versions as "accepted principles" (Cripps, 2013, pp. 18–19). The no-harm principle formulates a negative duty to non-interference. The beneficence principle stipulates a duty to assistance with regard to others. The PGC, as I have stressed several times, encompasses positive and negative duties too. Yet, the PGC does not rely on common acceptance. Instead, the PGC is a rationally derived necessity and necessarily involves positive and negative duties. The fact that both, Cripps and Gewirth, account for duties of non-interference and duties of assistance is a good starting point.

Based on the no-harm and beneficence principle, Cripps applies the weak principle of beneficence to the Young, the moderate principle of beneficence to the Able and the collectivized no-harm principle to the Polluters. Recall further that the limits of demandingness remain rather blurry (chapter 3.2.3). An overview of Cripps' moral principles and the limitations of costs might be helpful:

Table 2 Moral principles and limitation of costs according to Cripps (2013)

Weak collectivity	Justifying principle	Limitation of costs
The Young	weak principle of beneficence	"without sacrificing another such fundamental interest" (Cripps 2013, p. 48)
The Able	moderate principle of beneficence – cost of action condition – cost of coordination condition	"less than morally significant costs" (Cripps 2013, p. 61)
The Polluters	collectivized no-harm principle	– according to individual no-harm principle, "without suffering comparable harm" (Cripps 2013, p. 11) or – "in line with beneficence cases" (Cripps 2013, p. 155)

The different versions of the beneficence principle in connection to cost-expressions such as 'at less than significant cost' or 'at minimal cost' demonstrate that Cripps tries to capture cases of conflict only after setting a normative foundation. Based on Gewirth, I suggest reformulating Cripps' principle of moralised collective self-interest (the Young), the collectivized moderate principle of beneficence (the Able) and the collectivized no-harm principle as follows:

> **Modified weak principle of beneficence**
> Members of the Young have a moral duty to cooperate to secure their own generic rights to freedom and well-being as long as this is possible without members of the Young having to sacrifice some other equally important generic right.
>
> **Modified moderate principle of beneficence**
> Members of the Able have a duty to cooperate to prevent others from interference with their generic rights to freedom and well-being if others cannot protect themselves and members of the Able can do so at no comparable cost.
>
> **Modified collectivized no-harm principle**
> Members of the Polluters have a moral duty to organize themselves as necessary to prevent others from interference with their rights to freedom and well-being resulting from the predictable aggregation of their avoidable individual acts if members of the Polluters can do so without having an equal right violated.

Readers may object that it remains open whether the very idea of weak collectivities fits into Gewirth's theory. I am convinced that Gewirth's rights-based theory is able to accommodate weak collectivities. But more than that, I claim that his rights-based theory or, more specifically, the applications of the PGC provide a basis for specifying the nature of collective action and promotional action in further detail. I wish to start by broadly locating Gewirth's theory in the context of duties and climate duties in particular.

The philosophical discussion regarding duties is shaped by two positions: Individual duties and collective duties. In general, the latter is the duty to any collectivity taken as a whole. It has commonly been applied to organized institutional collectivities such as states or corporations. The applications of the PGC follow these two standard ways of thinking with regard to individual agents on the one hand and state or organizational institutions on the other. Until Cripps' contribution in 2013, the discussion on climate duties was (and it still is) characterised by these two positions. On the one hand, there are institutional accounts. On the other hand, there are individual accounts. Cripps has criticised Sinnott-Armstrong, who has famously defended an institutional account, for exclusively focussing on the state as a collectivity. As I have stressed earlier, she claims that there is something between the individual-level and the state-level: individuals as collectivities (Cripps, 2013, p. 141). Given that no set of individuals intentionally causes climate change, strong accounts of collective responsibility do not provide a solution either. This leads Cripps to defend weak, non-intentional collectivities: Sets of individuals form a weak collectivity "if and only if those individuals are mutually dependent for the achievement or satisfaction of some common or shared purpose, goal of fundamental interest […]" (Cripps, 2013, pp. 29–30). Weak collectivities have the weakly collective duty "to cooperate" (the Young and Able) and "to organized themselves as necessary" (the Polluters) (Cripps, 2013, pp. 48, 61, 74) in order to initiate collective action. New challenges, and climate change is one of them, require either the modification of existing accounts or the development of new theories. Cripps' approach is a new theory. In the late 1970s when Gewirth published *Reason and Morality*, climate change was nowhere near the public agenda. Simply, there was no need for an approach that captured collectivities other than organised institutions. Yet, the very idea that individuals and collectivities do not operate in separate spheres is part of Gewirth's theory already. He even rejects an extreme individual position and an extreme institutional position when he writes: "Thus, although the extreme individualist position is mistaken that the generic moral rules directly constitute the whole of moral rightness, the generic rights of individuals supply the ground for the authority of institutions to set morally justified rules and obligations." (Gewirth, 1978, p. 281).

Recall Gewirth's definition of an institution: "In general, an institution is a relatively stable, standardized arrangement for pursuing or participating in some purposive function or activity that is socially approved on the ground (whether justified or unjustified) of its value for a society. […]" (Gewirth, 1978, p. 274). Based on that, Gewirth differentiates between voluntary associations as institutional arrangements whose formation is optional and state institutions whose existence is not left to people's optional consent. Their existence is a rational necessity in order to protect peoples' rights. Now, the duty to protect peoples' rights to freedom and well-being is a duty that each and every individual has. Yet, in order to fulfil this duty individuals are mutually dependent with each other. Climate change therefore requires the development of certain associations. Their formation is not optional but a rational necessity. In contrast to voluntary associations, I suggest referring to those associations that are required to effectively protect peoples' rights from climate-induced harm, as necessary associations.[78] In general, I regard as necessary associations any collective body that could bring about required regulations of collective action if it was organised or structured. Necessary associations may be regarded as the rights-based equivalent to weak collectivities. Yet, there is no association that would fit the scheme of an organised, stable institution. The individual duty

78 Even though weak collectivities may be considered as necessary associations, I continue to refer to the Young, the Able and the Polluters as weak collectivities in the following.

to initiate collective action is a derivative duty to members of the necessary associations.

Though with regard to organisational institutions in a fully cooperative sense of the state, Gewirth does not remain silent about how to initiate required changes either. He hints to citizens having to resolve injustices by procedures of a method of consent. If these are ineffective, non-violent civil disobedience may be justified. Disobedience should raise public awareness and focus on specifying policy with regard to the inequality of rights (Gewirth, 1978, p. 322). How can this be applied to climate change? I think the main point is that if existing institutions fail to protect individuals' rights, other individuals are not left off the hook. They are required to first, peacefully initiate social change among themselves and second, to initiate political change with regard to institutions that implement rules and regulations that apply to all.

Cripps is very open minded about the forms of collective action, the forms of promotional action and about the levels of social organisation at which action could occur. Recall that collective action encompasses anything from a global agreement between states, the establishment of a new global institution to "some global-level agreement between individuals or sub-state collectivities, perhaps coordinated by NGOs or some other voluntary body" (Cripps, 2013, p. 143). Collective action with regard to mitigation seems to describe any systematic path to zero-carbon emissions. Given that promotional action ought to initiate collective action, it is almost impossible to make concrete suggestions with such a variety of possible forms and levels in view. Indeed, according to Cripps, promotional action ranges from campaigning and voting for committed politicians, writing open letters to organising demonstrations and calling "for adherents to the kind of norm-changing 'global citizen movement' [as] suggested by Dale Jamieson" (Cripps, 2013, p. 143). Most promotional actions that Cripps names aim at changing or implementing laws and rules. Others "bring about mitigation through new 'green' behavioural norms" (Cripps, 2013, p. 165). It seems that Cripps suggests either a top-down change that is largely coercive or a bottom-up change that largely bypasses institutional settings. I have stressed earlier that in order to act effectively against climate change, the aim should be to use synergy effects between society and political institutions. Based on Gewirth's theory, my claim is that the question is not either top-down or bottom-up, but both, top-down and bottom-up. Climate action of individual agents is about pushing the effects of collaboration between individuals and institutions.

However, the kind of bottom-up change that I have in mind does not incorporate mitigation through the changing of behavioural norms as defended by Jamieson (2011). Assuming that political institutions are not prepared to tackle climate change as a challenge of "multiple, overlapping and hierarchically embedded collective action problems", Jamieson draws on the "respect for nature" as a virtue that could stimulate a norm-changing movement within society (Jamieson, 2011, pp. 34, 36). He writes:

> The only way to break through on this problem [climate change and related political failure], which is the world's largest and most complex collective action problem, is through actions of a morally motivated global citizens' movement that acts as a highly committed political interest group. Such a movement would stigmatize coal, meat eating, trophy houses, overheating and overcooling, large living spaces, and private automobiles. It would celebrate living lightly with dignity and elegance, relying on nature's own energy, rediscovering food and pleasures of eating, the joys of living with nature and other people, and the satisfaction of effective political activism. I think (and hope) that we may be witnessing the birth of such a movement. (Jamieson, 2011, p. 36)

I agree with Jamieson that societal action is required, and I share his hope for seeing individual agents together and truly succeeding to overcome their own inaction. However, I disagree that such societal action should be a norm-changing movement that exclusively focuses on individual behavioural change and that largely bypasses political institutions. I have expressed earlier that, in my view, the lack of effective global climate action is rather a matter of means and acceptance than a matter of political will. Even though chapter 6 will substantiate this claim, I wish to make a brief remark. Technical means to zero-carbon emissions are in many cases too expensive or not advanced enough. Economic means to zero-carbon emissions such as emission-trade or tax-systems are either still ineffective or they are not feasible. The Polluters, and mitigation is all about decarbonizing their lifestyle after all, are basically caught in a structure that heavily relies on fossil fuels. If political inaction was a matter of political will and the Polluters were organised to act together, then individuals should bypass climate governance, trying to put a social agreement with each other on the individual level into place in order to change behavioural norms. However, first, the Polluters are not at all prepared for collective-level action and, second, if changing behavioural norms does not mean turning development back to pre-industrial times, changing behavioural norms will not be enough to successfully mitigate climate change. Given that Cripps has expressed doubts about virtue-based mitigation in modern societies too (Cripps, 2013, p. 134), it is surprising that she includes such a virtuous-driven social change as promotional action. Taking action among individuals does not prescribe virtuous-driven social change. Taking collective action among individuals means to me that individuals together stand up, raise their voice and push the existing economic, technical, social and political limits of global climate action. In that regard, individuals would do well by using existing institutions and not bypassing them.

However, pushing the limits of collective action first of all requires social acceptance. The thing is, whatever collective action incorporates, collective climate action will – right now – include regulations. And regulations are coercive. In order for regulations to be sustainable, it requires at least some support among those individuals who are under the authority of those regulations. Granting the necessary backing for certain regulations is different from primarily living up to virtues that are in line with implemented regulations. Consider the Polluters, for example, it is one thing not to actively fight against regulations that aim for decarbonising the Polluters' lifestyle, and it is quite another to primarily make individual behavioural changes in order to personally reduce carbon-emissions. Suppose, it turns out that reducing private car use is an effective means for collective action. Suppose further that taxes on petrol are an effective means to reduce private car use. Individuals do not need to be enthusiastic about taxes on petrol, but they need to grant some support. Calling for adherents, as Cripps puts it, requires first of all raising social awareness for individuals' common weakly collective duty to push collective action. That is the kind of promotional action that individuals can do together in the first place. Once they are affiliated with each other, individuals can push the existing limits together. The point is that existing institutions are part of the solution, but they are not *the* solution and individuals do not have to wait until existing institutions are prepared to implement mitigation strategies.

4.4 Providing Guidance on Reasonable Costs of Individual Action

Chapter 3.2.3 claimed that Cripps' approach is not fully able to respond to rendering the limits of demandingness for individual agents. It also suggested that the only relevant demandingness-objection must be raised with respect to individual costs. Drawing on Gewirth's rights-based moral theory, the following shows that the ethical discourse is more comprehensive and therefore better equipped to provide guidance on costs that members of the relevant collectivities can reasonably be expected to carry. I will focus on the most controversial collectivity: The Polluters. This chapter asks: Do duties of the Polluters impose too much of an individual cost?

4.4.1 *A Framework for Discussion*

Before moving on, it might be helpful to recall what kind of duties the Polluters actually have (see chapter 3.1 and 3.2.3). Given the Polluters' weakly collective duty for action, Cripps argues that members of the Polluters have individual corresponding duties on two levels:

(1) acting promotionally in the process of initiating collective action and
(2) acting as part of the collective scheme that secures the Polluters' common goal (Cripps, 2013, pp. 71, 79).

Given that I have limited the analysis of this book to mitigation-duties, I think of these two levels in the following way: (1) Theoretically derived from the weakly collective duty, promotional duties are obligations in the absence of collective mitigation. Because collective mitigation does not exist yet, the duty falls back to the individual level in order not to leave the weakly collective duty unfulfilled. On a general level, promotional activity is anything that helps to initiate morally required collective action to limit warming. (2) In the presence of a collective mitigation scheme, in contrast, doing one's part is an individual obligation together with others on the collective level. The collective duty connects to individual agents as members of the collectivity.

Chapter 3.2.3 highlighted a couple of further issues that connect to this section: First of all, mitigation is a collective task. Therefore, costs must be primarily bearable on the collective level (Cripps, 2013, p. 56). That being said, corresponding individual duties on both levels highlighted above impose costs on individual agents as members of the collectivity. Second, Cripps does not explain whether the limits of demandingness for members of the Polluters are to be determined from the individual no-harm principle or from beneficence cases (Cripps, 2013, p. 155). I highlighted this to be quite a big difference, which, given the complexity and controversial nature of the Polluters, should not be left open. Third, Cripps makes a case for adjusting individual burdens across the Polluters. She stresses that only if no collective scheme could possibly be realised without violating some members' interest, would mitigation be too demanding (Cripps, 2013, p. 56). Following Cripps but adopting the PGC, we may specify that mitigation is too demanding only if (a) costs violate members' generic rights and (b) redistribution of costs across the collectivity is unable to cushion such infringements. Given that costs to members of the Polluters may occur with regard to (1) promotional duties and (2) doing one's part of collective mitigation, a first attempt to specify the relevant research questions to explore the legitimacy of the Polluters' duties is:

(1) Does *acting promotionally* cause costs that violate members' generic rights? And if so, is there any adjustment across the Polluters that may reduce costs so that no member has her generic rights violated?
(2) Does *acting as part of the collective scheme* cause costs that violate members' generic rights? And if so, is there any adjustment across the Polluters that may reduce costs so that no member has his generic rights violated?

The reply to either question is anything but simple. Assessment of costs on both levels involves several fields of tension. The trouble with providing guidance on acceptable individual costs is that fields connect with each other. Because a comprehensive answer, as I believe, has to fully consider each field the following briefly lays out the four key areas:

Individual vs. Collective

The first field of tension is between the collective and the individual level. It is the constant challenge of this book and one of the key problems of this chapter. In short, mitigation is of a *collective* nature but involves *individual* costs that should be bearable on the *collective* level while not violating any *individual* generic right.

Costs vs. Benefits

Any action that is connected to costs may also be connected to benefits. At least in the sense that costs occur through the setting aside of benefits. So, costs are only one side of the coin. Benefits are the other side. The balancing of costs and benefits is the second field of tension. Readers familiar with economics will probably immediately think of Cost-Benefit-Analysis (CBA). Inspired by the economic assessment of costs and benefits, Gewirth (1990) suggests a first approximation: "[O]ne should estimate the costs of performing the action and the benefits expected to result from it, and one should not do the action unless its expected or actual benefits are greater than its costs" (Gewirth, 1990, p. 205). Transferred to corresponding individual duties that may mean: members of the Polluters should act (1) promotionally and (2) as part of the collective scheme as long as the expected or actual benefits of acting that way are greater than its costs. But that is not the whole story. Morally estimating costs and benefits from a rights-based approach must not be understood as a purely utilitarian assessment. Peoples' generic rights are untradeable values, which should be protected even against the greatest benefits. Considering the assessment of individual costs with regard to corresponding individual duties, that means that benefits resulting from promotion and doing one's part may never outweigh members' violation of generic rights. Nevertheless, a comprehensive assessment from a rights-based approach should at least consider the benefits. To complicate things, subjects carrying costs or having benefits are rarely only collective or only individual. Individual benefits, for example, may result in collective costs and collective benefits may lead to individual costs.

Economic vs. Moral

Gewirth (1990) has suggested to differentiate between the economic and the moral CBA. The economic CBA measures costs and benefits exclusively in monetary terms, which involves "economizing" moral values such as life or health (Gewirth, 1990, pp. 219, 209). It thereby problematically assumes that values are always comparable. Some values, such as life or freedom, however, may be incommensurable. Closely connected to this is the matter that the economic CBA, similar to the utilitarian assessment of well-being, focuses on the cumulative outcome and disregards distributive effects. In contrast to economising and maximising the overall outcome, the moral CBA considers economic costs with regard to its moral dimension (Gewirth, 1990, p. 209). If, for example, collective mitigation affects the economy in a way that interferes with persons' general abilities of agency, economic costs have a moral dimension that should be considered. I do not intend to discuss details between the economic and moral CBA here or elsewhere in this chapter. Rather, I focus on an assessment that basically moralises economic costs if they interfere with agents' generic rights. And yet, as my example already suggests, economic costs too occur on the individual and the collective level.

Members vs. Non-Members

Up to this point, I have followed Cripps in focusing on moral costs of individual agents as members of the Polluters. However, costs and benefits do not only occur with regard to members of the Polluters. The results of collective mitigation might affect non-members' generic rights too. That is particularly evident for economic costs: To stay at a superficial level, imagine the unlikely case that collective mitigation would turn out to be best fulfilled my abolishing coffee in Europe. Coffee sold in Europe is usually imported from countries located in Africa, Middle- or South-America. Abolishing the coffee trade will not only hit the European economy and European individual agents hard, more importantly, it is likely to affect the coffee sector in the world's poorest countries whose citizens do not belong to the Polluters.[79] Though coffee is abolished collectively, and economic costs occur on a collective level, it will affect individual agents by losing their jobs. Thereby economic costs interfere with peoples' income opportunity. That is indeed an infringement upon the generic right to well-being. Considering costs and benefits of non-members does not necessarily lead us to drop corresponding duties of the Polluters all together. But a comprehensive moral assessment must not neglect costs and benefits of non-members. To keep the discussion manageable, I will first discuss each research question with a focus on costs and benefits of the Polluters' members before broadening the perspective to non-members.

79 Following Cripps, chapter 5.2.4 will defend that there is a need for a threshold level that specifies who belongs to the Polluters and who does not.

4.4.2 *Reasonability of Promotional Action*

To start with promotional duties, the criterion for reasonability might be formulated in the following way: Promotional duties are *prima facie reasonable* if benefits outweigh the costs and if there is a collective distribution that ensures that costs do not violate members' generic rights. They are *in fact reasonable* if benefits to non-members outweigh their costs too and if there is a way to ensure that costs do not violate non-members' generic rights.

Promotional duties are a challenge and, as opponents of individual duties correctly claim, empowering agents to act collectively against climate change may cause considerable costs. These costs may infringe on members' generic rights. But do these costs violate members' generic rights?[80] The preliminary answer is 'no'. Even though Cripps does not elaborate the content of promotional duties in great detail, she gives an impression of what promotional duties entail: Raising political awareness and accelerating political action.[81] Chapter 6 will suggest raising social awareness in addition. There, I will also advance the nature of promotional duties. For now, emphasis should be put on the broad range of possible promotional actions.

With regard to promotional duties, individual agents bear costs in two ways: First, performing promotional activities themselves involves costs. In order to participate in a climate march there is, for example, the train ticket to the venue or a day's leave may be taken. Considering non-monetary costs, people who are not very comfortable with public communication might have to leave their own comfort zone in getting in touch with others. Also, a climate march may attract people with interests other than the promotion of collective climate action such as left- or right-wing extremists who take advantage of public attention. Individuals who participate for the purpose of climate promotion may then naturally feel uncomfortable. The second kind of costs occurs with regard to omitted choices. Maybe individual preferences for spending an afternoon are different from participating in a climate event. Then, marching for the purpose of climate awareness is connected to, what economists refer

80 Regarding the distinction between violation and infringement of rights, I follow Thomson (1986). She defines: "[s]uppose that someone has a right that such and such shall not be the case. I shall say that we infringe a right of his if and only if we bring about that it is the case. I shall say that we violate a right of his if and only if both we bring about that it is the case and we act wrongly in doing so." (Thomson, 1986, p. 51). For critique and discussion see e.g. Oberdiek (2008).

81 Recall that she names examples such as writing e-mails to political candidates, signing petitions, participating and organize demonstrations, marches or international summits for political action (for more examples see chapter 3.1.4).

to as, opportunity costs. It is hard to see where promotional action involves individual costs that violate generic rights. The costs of writing e-mails, signing petitions, participating or organising demonstrations or a climate march, and so on hardly count as costs that violate individuals' rights to well-being or freedom.

One might, however, think of a single parent who wishes to promote collective action by participating in a climate march. Her child is entitled to have her generic rights protected. If the single parent leaves her child unattended while participating, she fails to fulfil her parental supervisory responsibility. Failing to live up to ensuring her child's generic rights is not in accordance with the PGC. Given the range of promotional actions, the single parent might instead initiate changes online by using social media reach or creating petitions – also online. In anticipation of chapter 5.3 and chapter 6, the single parent might possibly also fulfil parts of her promotional duty by raising her child to be an environmentally conscious person. I would consider that a case where costs of promotion are to be distributed across the collectivity. Those, who have the ability to carry the cost of promotion itself and its opportunity costs might perform the more active, personal kind of promotional activity, others (such as the single parent) may then take on the rather passive, impersonal promotion of collective mitigation.

Turning to the other side of the coin, the major benefit of promotion is the initiation of collective mitigation in the first place. Considering the urgency of collective action against climate change and assuming that mitigation will ensure rights of all persons globally, overcoming the state of collective inaction is indeed a huge benefit. And the benefit applies to members and non-members equally. Against a common misconception, climate impacts will affect peoples' generic rights globally, not only the poorest people as non-members. Cripps reduces this point to the Young. But it is also in the Polluters' own interest to trigger large-scale mitigation. That given, it is somehow tempting to conclude that promotion for the purpose of securing the Polluters' own generic rights can hardly be too costly for its members. Given the possible benefits of promotion, given the wide range of promotional activities and the possibility to distribute costs across the Polluters, promotional duties to the Polluters' are reasonable. They are reasonable not only *prima facie*, but there is good reason to assume they are in fact reasonable.

Generally, I doubt that promotional duties are the true challenge. The critical problem is, as I believe, linked to (2). Is doing one's part of collective mitigation too much of a burden? Do costs that occur in that regard violate people's generic rights? And if so, is there any way to distribute costs across the Polluters in order to protect all members from having their rights violated? Just as for promotional duties, I will first focus on members of the Polluters before examining non-members.

4.4.3 *Reasonability of Doing One's Part: Assessing Air Travel for Tourist Purposes*

Assuming that the dominant narrative of climate mitigation in readers' minds is still strongly connected to mimicking, I would like to begin by stressing once more that, for reasons of effectiveness, the Polluters' collective mitigation activity is *not limited* to reducing their own aggregated carbon emissions (see chapter 3.1.4). Likewise, doing one's part of the collective scheme must not be confused with reducing one's personal emissions in isolation. Because there is no collective scheme right now, mimicking in isolation would not only fail to fulfil the weakly collective duty, but it would also be too much of an individual cost. The primary collective task of the Polluters is to establish a framework for action that reduces individual costs to an amount that does not violate individuals' generic rights (e.g. through structural or technological changes). The nature of collective action is complex. Any straight-forward answer is extremely vulnerable to objections. This is probably the reason why concrete recommendations for action are hardly found in the climate ethics discourse. Philosophers usually limit themselves to rather uncontroversial suggestions – often with reference to future research. As I have said earlier, I will not be able to specify the full content of collective action here or elsewhere in this book. In fact, chapter 6 will claim this to be a duty of scientists globally. The challenge is that doing one's part of the collective scheme inherently depends on what collective action incorporates. That given, are we unable to determine whether doing one's part of collective mitigation imposes too much of a cost? Because I am not in the position to tell what collective mitigation involves, I will indeed not be able to make a case for any kind of participating duty. However, the discussion may be narrowed to a concrete example.

Even though the Polluters' collective action remains rather blurry, the idea of mitigation, that I focus on, suggests targeting large-scale decarbonization. After all, mitigation is primarily about decarbonization. Yet, reducing the Polluters' aggregated individual emissions is not sufficient to cope with climate change altogether. Mitigation requires structural and technological changes; market solutions and mitigation should also target actors across all levels, not only individuals. Though the Polluters' way of life may not be the only reason for climate change, the Polluters' lifestyle – collectively – is part of the problem. Mitigation may not be limited to the Polluters' emissions, but it is related to it. Thus, at some point the Polluters' collective carbon footprint will have to come under scrutiny too.

Given the requirements of the PGC, the ethical discourse is, as I believe, prepared to cope with the complexity and vagueness of collective mitigation by examining the legitimacy of concrete recommendations for collective action. Does a certain policy, instrument or measure initiated by collective action impose too much of an individual cost? I will limit my analysis to the discussion of a single tangible example that I will take up upon throughout this book. It is an example that seems to encounter intense resistance among society: the limitation of air travel for tourist purposes. Just to be clear, the discussion is not concerned with whether individual agents should refrain from flying in isolation. The discussion is concerned with duties imposed on individual members by the collective limitation of air travel. The relevant question for the following is: Do individual, participative duties connected to the collective limitation of tourist air travel impose an amount of individual costs that violates members' generic rights? And if so, is there any way to distribute burdens across the collectivity in order to avoid rights violations? Analogous to promotional duties, these duties are *prima facie reasonable* if benefits outweigh the costs and if there is a collective distribution that ensures costs not to violate members' generic rights. They are *in fact reasonable* if benefits to non-members outweigh their costs too and if there is a way to ensure that costs do not violate non-members' generic rights.

In order to cover the reason for choosing air travel, I have to anticipate findings from chapter 6. There, I provide at least seven reasons that strongly recommend analysing air travel: First, long-distance travel is high in emissions and high in volume (ATAG, 2018; UNWTO, UNEP, & WMO, 2008; World Bank, 2019).[82] Second, growth rates for aviation are projected to grow rapidly until the middle of the century, which is largely inconsistent with mitigation targets (IATA, 2018; UNWTO, 2017, p. 11). Third, distances travelled increase while duration of stays decrease (Alegre & Pou, 2006; P. Peeters & Landré, 2012). Fourth, aviation makes up the biggest share of emissions from tourism (at least right now) (P. Peeters & Dubois, 2010; P. Peeters & Landré, 2012; UNWTO et al.,

82 To get a first sense for how high in emissions air travel actually is, it might be helpful to consider numbers provided by one of the carbon-offsetting calculators. *Atmosfair* for example calculates an amount of 0.5 tons (t) CO_2 per person for a return flight from Frankfurt to Mallorca, 3.6 t CO_2 per person for a return flight from Frankfurt to New York City and more than 10 t CO_2 per person for a return flight from Frankfurt to Perth. For comparison, driving 12,000 kilometers per year in a medium size car produces about 2 t CO_2 (status: March 2019). Regarding the volume, it probably suffices to specify the number of flights: In 2017, there were 41.9 million scheduled commercial flights worldwide, which is more than 114,000 flights per day (ATAG, 2018, p. 5). Note, however, that the number of flights for tourist purposes is only a smaller part.

2008). Fifth, aviation is different from other modes of transport. In contrast to car or rail transport, which might be transformed into electric mobility, there is hardly any reasonable large-scale technological alternative to fuel-reliant civil aviation. Though there are efforts towards fuel-efficiency and biokerosene, these technologies will only slow growth in emissions from aviation but will probably not be sufficient to reduce them to zero. Sixth, for some destinations there is no reasonable alternative to air travel. Cripps' idea that collective action has to initiate structural changes in order to decrease individual costs (e.g. reliable, fast train connection to decrease the cost of reducing car traffic volume) is not very helpful for overseas travel destinations. There are no possible structural changes that will help to decrease individual costs of not being able to fly internationally. This is probably why behavioural response of consumers tells a completely different story. The seventh reason for choosing air travel is that pro-environmental attitudes are hardly reflected in the travel behaviours performed by individual agents (see e.g. Davison, Littleford, & Ryley, 2014). In order to bring aviation in line with urgently needed, ambitious mitigation targets, the volume of air travel requires limitation until large-scale technological innovations allow for environmentally friendly aviation. An eighth reason might be added, which makes tourist air travel particularly interesting for the moral analysis: Given that costs should be bearable on the collective level, individual costs that turn out to be too demanding should be adjusted across the Polluters. With air travel, however, there is no possible adjustment across the collectivity. If the limitation of tourist air travel turns out to be a legitimate part of the collective solution, members of the Polluters are all equally required to carry the cost. From a moral perspective, it is implausible to assume that some tourists legitimately fly more while others are expected to fly less or not at all.

The area of discussion basically extends between two extreme positions. From the first position, one might start by considering the devastating climate impacts as what they are: threats to basic well-being such as life, bodily health and integrity, security, and so on. Assuming that these threats should be minimised by all means, one might then refer to the PGC's criterion of degrees of needfulness for action. It requires that primacy be given to the right that is more needed for action. Considering that flying for a holiday in the sun does not nearly have the same moral weight than life, health and bodily integrity, it is tempting to conclude flying to be nothing but a luxury. And such a position has a point: what, after all, could be more important to minimise threats to persons' most basic well-being? Anyone who makes a strong case for individual duties will probably find this view appealing. The problem with such a position is that it disregards impacts on the individual level. It overlooks agency to include freedom as a right to control one's own behaviour. The first position tends to be constantly ready to interfere with persons' freedom in order to minimise violations to persons' basic well-being. The right to freedom is exactly the starting point for the second position. Limiting air travel is regarded to be an interference with someone's control of her own behaviour. And thus, it is perceived to be a violation of her right to freedom. If it was a violation of people's right to freedom, limiting air travel might indeed impose too much of an individual cost. However, such a position neglects the collective effect of individuals performing their behaviour. The key challenge to the following discussion is, again, the relation between the individual and the collective level.

My aim is to consider the position against the limitation of air travel from its strongest side. Only if the following shows that the ethical discourse is equipped to convincingly prove the opposite side wrong can it make a strong case for the limitation of tourist air travel. To start from the individual level, I commit myself to take it seriously that there is something about tourist air travel that truly constitutes a generic right. Does the collective limitation of air travel result in costs that violate agents' right to freedom and well-being? Though the above suggests limiting the discussion to freedom only, I will start with well-being and then proceed with freedom.

On the individual level we may distinguish the economic and moral dimension of well-being. Economically, the limitation of air travel hardly counts as an individual monetary cost. After all, the amount spent for a flight can be spent otherwise. Purchasing power and material prosperity remains unchanged. Morally, however, agents choose to fly for a purpose. The purpose of flying for a holiday might, at best, be to broaden the mind, to evolve new perspectives, increase cultural exchange, reduce barriers, bend stereotypes, and so on. It may even help to develop a sense of humility. And sometimes the purpose might be as simple as getting a break to get back to work energized and full of power to master everyday life. According to Gewirth, persons have a right to maintain or to increase their level of purpose fulfilment. Does that constitute a generic right to air travel?

A first response, with regard to the nature of generic rights, could be that there it is a serious misunderstanding to conceptualise a right to air travel as a claim right. Remember that claims are entitlements that persons are due. As such, they logically correspond with positive and negative duties. Drawing on the distinction of everyday use of the term 'having a right' suggested by Wesley N. Hohfeld (1919), the right to air travel may probably be a privilege rather than a claim against others. However, reflecting on purposes named above, air travel is a societal achievement and purposes might be more than a privilege. A second response, which is my suggestion, is that no purpose mentioned above is inherently connected to air travel. To my understanding, most purposes may be fulfilled by travel in general not by air travel in particular. People may broaden their mind, evolve new perspectives, experience cultural exchange, reduce barriers, bend stereotypes and so on by travelling. Individuals may have a right to travel in order to increase their level of purpose fulfilment. But they do not have a right to a certain distance travelled nor to an especially comfortable or fast way of travel. Limiting air travel for tourist purposes must not be confused with prohibiting the general ability to travel or mobility altogether.

Readers may now reflect on their own purposes of air travel. They might come up with seeing friends and family who live abroad, doing voluntary work or having a lifelong dream fulfilled. First of all, my aim is to target the rapidly increasing volume of tourist air travel. That is mainly to tackle excessive travel itineraries such as those involving several weekend trips to European metropolises, summer holidays on a Mediterranean island and at least one vacation in an overseas destination per year. Except for overseas-destinations, people may travel but they will have to bear the burden of travelling a few thousand kilometres by train or bus. Then, spending a *weekend* in a city that is several hundred kilometres away or a *week* in a country that is more than a day-drive away does not seem too reasonable after all. Besides, I would not consider flying for the purpose of voluntary work to be a sort of tourism. Also, seeing friends and family every few years or having a lifelong dream fulfilled does not justify a travel itinerary that involves several touristic flights per year. The Polluters are those individuals whose emissions exceed a level such that were everyone else only contributing to that amount, there would be no harm (Cripps, 2013, pp. 59, 72–73). Maybe a flight to see friends and family every few years, calculated down to an average annual level may one day be compatible with a per capita amount of emissions that is below the threshold level for further climate change. Right now, however, the way we lead our daily lives already exceeds the threshold several times. So, flying for a holiday is not an option compatible with successful mitigation.

Yet, in an increasingly interconnected world it is indeed tough to turn back a development that was firmly rooted in peoples' minds a long time ago. Air travel has become a part of ourselves, an expression of identity. That applies to a majority of the Young in particular. It is almost self-evident that young, educated people living in the industrialised countries travel as backpackers, spend semesters abroad, do voluntary work, maintain friendships across borders, speak several languages, obtain cultural competencies, and so on. Many young Polluters probably see themselves as urbane, open-minded, liberal people who are able to feel at home almost anywhere in the world. Naturally, limiting practices that are inherently connected to the way we see ourselves, feels like a loss. A loss so serious that it is perceived as something we have a right to.

Purposes of air travel cover a wide range. In fact, I cannot exclude that there are cases where the purpose of air travel may not be fulfilled by travel in general. That being said, I suggest taking a look at generic rights in Gewirth's theory: Generic rights actually protect the general ability of agency. They do not protect subjective, personal preferences. Next to basic goods, well-being incorporates non-subtractive goods such as being told the truth, not being betrayed or stolen from, not suffering from inhuman working or living conditions (Gewirth, 1978, p. 233). Additive goods further encompass access to education, income-opportunities or self-esteem (Gewirth, 1996, p. 14). Though non-subtractive and additive goods are relational expressions that depend on agents' status quo, we get a feeling for how off target we actually are when discussing air travel as part of well-being. Maybe individuals have a generic right to air travel on the ground of freedom – that remains to be elaborated in the following. But on the individual level, based on well-being, it is reasonable to assume that the limitation of tourist air travel does not interfere with peoples' well-being.

Considering Cripps' rather intuitive normative foundation of fundamental human interests, critics may argue that the above is a result that does not require the detour involving Gewirth's rights-based theory with a high degree of complexity. This criticism is certainly correct. Flying for the purpose of a holiday in the sun is anything but a fundamental human interest. Given that Cripps chooses five fundamental interests (Cripps, 2013, p. 7), members of the Polluters, on an individual level, are neither suffering a loss of life, bodily health, bodily integrity, affiliation nor practical reason by limiting tourist air travel. The first point where Gewirth's rights based moral theory extends beyond Cripps' normative foundation is with regard to freedom. I highlighted above that a position in defence of individual duties is at risk of constant readiness to interfere with persons' right to freedom. Regarding Cripps' position, that is in some way understandable. Freedom is not an independent functioning of the capability approach. And likewise, as Cripps' theory is inspired by the capability approach, it is none of Cripps five fundamental human interests either. Achieving certain human interests intrinsically incorporates being free from interference. Being free from interference is a necessary requirement for having one's fundamental interest in for example life protected. Yet, freedom has an independent status too. According to Gewirth, interference with someone's freedom is an interference with "the procedural aspect of agency" not only with the "substantive or purposive aspects" of agency (Gewirth, 1978,

p. 251). Capturing the independent status of freedom does not only allow the elaboration of individual costs more comprehensively. Considering freedom independently is also a necessary feature for examining opponents' key argument against individual duties.

Turning to the generic right to freedom, it might be helpful to briefly recall Gewirth's explanations. Generally, individual agents are free to perform any action they wish. The PGC limits personal freedom only where performing an action interferes with other persons' freedom or well-being (Gewirth, 1978, pp. 270–271). Individual agents, passionate about air travel, might claim that they will not allow others to restrict their freedom. On an individual level, these agents have a point. By taking the plane for a holiday, agents do not directly interfere with other persons' well-being or freedom. It is only on the collective level that they contribute to climate change, which interferes with persons' basic well-being. The causal inefficiency on the individual level itself is the reason for the weakly collective duty of the Polluters. Only with one another, members of the Polluters interfere with others' generic rights, which is why the limitation of air travel is necessarily of a collective nature. Individual agents are not obliged to refrain from flying in isolation. Only as part of the collective scheme, members of the Polluters have a *prima facie* duty to participate. Here, the impossibility of separating the collective and the individual level becomes particularly evident. The limitation of air travel does in fact interfere with persons' right to freedom on an individual level. Yet, collective interference is not only legitimate but morally required on the ground of the individuals' right to basic well-being. Because there is no collective setting prepared to prevent people from having their rights protected, individuals should primarily promote for collective action. And promotion, as we have seen, does not impose too much of a cost.

Up to this point, one might conclude the following: Given that mitigation is of a collective nature and measures are applied to the collectivity of the Polluters, having selected the limitation of air travel for tourist purposes, members have a *prima facie* duty to participate in the collective limitation of air travel.

Now, that I have elaborated moral costs of members of the Polluters, it is time to broaden the perspective to economic costs. Traditionally the economic calculus is concerned with maximizing welfare (not well-being) of a particular society or of humanity as a whole (not for particular individuals).[83] To stay in the terminology of this chapter, economists usually ask whether

83 I owe this clarification to Gewirth (1990, p. 213) who stresses reference to the collective level and welfare with regard to Mishan (1976).

economic benefits outweigh economic costs on the collective level. Gewirth has criticized the economic CBA for overlooking how benefits are distributed (Gewirth, 1990, p. 213). The same is true for costs. The benefits of tourist air travel will be the costs of limiting air travel. According to the economic CBA it seems: As long as welfare caused through air travel outweighs its costs, air travel should not be limited. An adequate economic CBA of air travel exceeds the scope of this book. But a full analysis is not the interest of this chapter anyway. This section is concerned with whether eliminating the benefits of air travel imposes a cost that infringes with peoples' generic rights.

Air travel is indeed an achievement, and from an economic perspective, the aviation industry is an integral part of the world economy. Limiting air travel is a tough burden to the global economy in particular with regard to the value of speed and connectivity. This chapter discusses air travel for tourist purposes, so the significance may be narrowed down to air tourism. In order to develop a feeling for its significance, I would like to consider a practical example: Lufthansa is the largest German airline and one of the biggest carriers in Europe. Despite passenger operation, Lufthansa (just as many other airlines) has several subsidiaries. Next to smaller, formerly independent airlines that Lufthansa wholly or partly owns, Lufthansa has several affiliates such as Lufthansa Cargo, which covers Lufthansa's freighter fleet, Lufthansa Technik, which is "the world's leading independent provider of maintenance, repair and overhaul services (MRO) for civilian commercial aircraft" (Hagenbring, Brodowski, & Winter, 2018, p. 56), LSG Sky Chefs as "the global market leader in airline catering" (Hagenbring et al., 2018, p. 155) and several more units. Lufthansa has almost 130,000 employees worldwide with less than half of them directly working in passenger operation (Hagenbring et al., 2018, p. 24). Given that Lufthansa is only one airline of several hundred worldwide, we get a sense for the significance of economic costs. Limiting air travel will cause considerable economic costs on the collective level. Large-scale job-reduction as a consequence of limiting tourist air travel impacts agents on the individual level. That way, economic costs have a moral dimension. Even the assumption that some share of employees working in aviation industry will still be required for business passenger operation or airfreight and some others might find jobs in industries other than aviation does not fully cure the problem: Limiting air travel for tourist purposes will cause a significant number of staff to lose their jobs and with it their income opportunities. That is indeed an infringement of their right to additive well-being.

To make things worse, consequences for the aviation industry do not even tell the whole story. Air travel only accounts for a part of the revenue from international tourism. Tourism has become the number one industry in many countries. That is particularly true for developing countries that build their national economic welfare on international tourism. What is true for aviation-staff, is true for any employee working in tourism. According to the Air Transport Action Group (ATAG), a coalition of representatives from the aviation industry, the aviation industry provides 65.5 million jobs worldwide. Only 10.2 million jobs are created by services that directly serve passengers such as airline-services, baggage-handling, cargo, check-in, and so on. 36.7 million jobs are from aviation-induced tourism (ATAG, 2018, p. 13).[84] Because the ATAG is certainly an organisation that is rather devoted to aviation industry and calculation concepts and reference parameters are also subject to criticism (see e.g. Sewill, 2009), the statistics themselves should be treated with caution. The qualitive statement about the economic significance of aviation-industry, however, remains unaffected. Aviation provides a significant amount of employment possibilities worldwide. Consequently, limiting air travel means that a substantial number of people will pay the price of avoiding further climate change by losing their jobs. Does that invalidate the legitimacy of limiting air travel?

First of all, the moral dimension of economic costs is not an argument against the Polluters' duty. It merely underlines the importance of right-securing institutions. Modern welfare states might be able to cushion economic costs. Cripps, with reference to Shue (1993), states that mitigation would amount for three or four per cent of resources if costs were truly distributed across the collectivity (Cripps, 2013, p. 55). The estimation is rather blurry, and it remains unclear what kind of costs she actually refers to. If the amount of costs remains equally small when considering the economic costs of mitigation for members of the Polluters, that is indeed a bearable amount of costs. By limiting air travel, collective mitigation does not only impact the Polluters' own members. Collective mitigation affects non-members, the world's poorest, too. Because the world's least developed countries usually lack an established social system, the costs of limiting air travel will hit their citizens disproportionately hard. It seems that Cripps does not bear economic costs for non-members in mind. Certainly, I am not in the position to estimate a total share of resources that the Polluters would have to give up carrying costs for members and non-members. My suggestion is the following: Instead of considering economic costs of non-members as an argument against the Polluters' duties,

84 To complete the statistics: 10.8 million jobs are from aviation indirectly (services from companies that supply aviation such as aerospace technology, jet fuel, airport retail, etc.) and 7.8 million jobs are induced by aviation industry (jobs created when aviation-employees spend their wages). For full statistics and further information see ATAG (2018, pp. 13–15).

one should focus on establishing an institutional arrangement that is able to ensure non-members' generic rights too. According to standard institutional accounts, transnational and national institutions could help bear corresponding costs. From the perspective of this book, weak collectivities are in the position to take responsibility too. One option is that the Polluters themselves help carry these burdens. Given that the Polluters pay the direct costs of mitigation for themselves, another option is that the Able take the burden for expanded economic impacts of climate mitigation. Direct impacts of not being able to fly for a holiday may not be distributable across the Polluters. All members are required to participate equally. The expanded economic impacts of limiting air travel for members and non-members, however, may be subject to adjustment across the collectivity and even across the relevant collectivities.

To complete the picture, one must also consider the benefits of limiting air travel. It should have become clear by now that climate change is part of the cost of aviation. Climate change violates people's generic rights. Thus, limiting air travel will protect peoples' generic rights, which is indeed a huge benefit. Given that the world's poorest countries are the most vulnerable, climate mitigation is (though not exclusively) in their own interest. With regard to economic costs of non-members, there is rather little reason to assume that benefits of air travel truly outweigh its climate-induced costs. But somehow, it seems, that the moral assessment of economic costs ends up in a dilemma. Limiting air travel is likely to infringe on peoples' generic rights. Forgoing the limitation of air travel infringes upon peoples' generic rights too.

Cripps' normative foundation values "only those elements which it is hardest to deny are central to human life" (Cripps, 2013, p. 8). Recall that these are limited to life, bodily health, bodily integrity, affiliation and practical reason (Cripps, 2013, p. 7). Actions that violate one of these fundamental human interests are too much of an individual cost. With such a rough basis, however, she is neither able to capture the independent value of freedom (that was stressed already) nor is she able to capture non-fundamental or non-basic costs. This is the major drawback of a theory. At this point the PGC unleashes its full potential. The PGC provides guidance that goes beyond Cripps' normative basis. It provides a hierarchical ordering of moral rights, which emphasises the degree of necessity for action. On the one hand, climate change infringes with peoples' basic well-being (life, health, bodily integrity, etc.). Given that income-opportunities are additive goods, the costs of climate mitigation on the other hand might interfere with peoples' additive well-being.[85] Equipped with the

85 Note that losing income-opportunity in less developed countries where there are no supportive institutions may result in interferences with non-subtractive and even basic well-being.

PGC, precedence of one right over another is legitimate if it is more necessary for action. Life, health, bodily integrity, and so on are more necessary abilities of agency than income opportunities. The necessity for action requires more weight to be out on basic well-being. But that does not mean to let people who suffer violations of non-basic well-being carry the burden on their own. Again, their costs underline that the right-securing function of institutions in the widest sense is indispensable.

The above elaborated a single possible measure of the collective scheme only. It is an attempt to demonstrate how many different facets a comprehensive analysis has to consider. And I am afraid, the analysis is not even exhaustive. Individual climate duties are complex. That is true for duties to do one's part of collective mitigation in particular. Any further discussion on individual duties should resist jumping to conclusions. A serious discussion about the limits of demandingness should neither underestimate the costs of individual members nor should it overestimate the limits with regard to freedom in particular. There is no brief and clear answer to what can reasonably be demanded of individuals in the presence of a collective scheme – not least because the nature of collective mitigation, despite its aim to target decarbonisation, is rather blurry. By considering freedom as an independent value, by valuing kinds of well-being other than basic well-being and by considering non-members too, I have tried to show that the ethical discourse, based on Gewirth's rights-based theory, is equipped to comprehensively analyse the limits of demandingness for members of the Polluters.

The research question of this chapter was: Do duties of the Polluters impose too much of an individual cost? Such a general question requires a detailed analysis. Given that the Polluters' duties involve (1) promotion and (2) participating in the collective scheme, two corresponding research questions were developed: Does acting promotionally cause costs that violate members' generic rights? And does acting as part of the collective scheme cause costs that violate members' generic rights? With regard to the analysis, four fields of tension were defined: individual vs. moral, costs vs. benefits, economic vs. moral, members vs. non-members. Focusing only on the reasonableness of individual costs with regard to members of the Polluters results in an incomplete analysis. Costs to non-members have to be taken into account too. Yet, costs of non-members should not be used as an argument against the Polluters' duty. Rather, they are an additional element that needs to be distributed across the Polluters and even across the Able. To conclude, is it reasonable to expect members of the Polluters to carry the cost of promotional action? Yes. Neither costs of promotion itself nor opportunity costs of promotion infringe on the Polluters generic rights if costs are distributed across the collectivity. Is it reasonable to expect members of the Polluters to carry the cost of participating in collective mitigation with regard to limiting tourist air travel? Considering costs for the members of the Polluters themselves, probably yes. Individual costs of members are unlikely to violate their generic right to well-being. Though the limitation of air travel infringes on their individual generic right to freedom, collective interference is morally required in order to protect persons' generic rights to basic well-being. I have stressed that the limitation of air travel must not be confused with a general prohibition of travel or mobility. Given that many polluting states are equipped with rights-securing institutions, there is only limited reason to assume that economic costs violate members' generic rights. I suggested that economic costs of collective mitigation and with it the establishment of relevant institutions may be subject to adjustment across the collectivity.

Three final remarks: First, political and social feasibility of limiting air travel in practice are two additional challenges, which this chapter did not focus on but that I wish to briefly address. Politically, envision a government that intends to limit air travel. Socially, think of a group of people trying to convince others that the limitation of air travel is morally required. It does not require much imagination to suspect that political leaders will not be re-elected and that social groups will probably be isolated and derided. In terms of political feasibility, the first step is certainly to increase economic costs of overseas air travel to at least the costs of the externalities they produce. Thereby, individual opportunity costs of flying decrease, which makes limitation of air travel more likely. Second, I have limited the discussion to the most critical weak collectivity – the Polluters. Next to the Polluters, members of the Young and the Able will carry substantial costs too. Reasonable costs in their regard require discussion and substantiation too. Also, with air travel I have picked only one out of many different sectors that produce significant amounts of carbon. Third, and finally, I wish to stress that this chapter does not intend to present tourist air travel as something inherently bad. The problem with air travel regarding climate change is not air travel as such. The problem is the heavy reliance of aviation on fossil kerosene. As soon as there are reliant, scalable and green alternative aircraft engines, there is no further need to limit tourist air travel. But as long as there are no technical projects that are further advanced than the research phase, limiting air travel is not only morally permissible but also morally required.

CHAPTER 5

Weakly Collective Duties

Certainly, the idea to assign weakly collective duties is one of the key elements of Cripps' theory. Because no collectivity is capable of intentional action yet, the three relevant weak collectivities face weakly collective duties "to cooperate" (the Young and Able) and "to organiz[e] themselves as necessary" (the Polluters) (Cripps, 2013, pp. 48, 61, 74). This chapter elaborates the idea of weakly collective duties in more detail. The first part of this chapter defends Cripps' idea against various possible criticisms (chapter 5.1). The second part jumps directly to the most controversial collectivity and examines the weakly collective duty of the Polluters (chapter 5.2). Because my claim with regard to the weakly collective duty of the Able is basically the result from discussing the Polluters and many different scholars have written about the Able already, I will limit the case of the Able to a brief remark directly following the case of the Polluters. The third section elaborates the weakly collective duty of the Young (chapter 5.3). There I will focus on a particular group of persons: young parents.

5.1 Defending Weakly Collective Duties

Assigning duties to collectivities or groups is nothing new and, in a first step, collective duties are, quite uncontroversially, obligations to any collectivity taken as a whole. That was a key aspect of chapter 2. In a second step, however, collective duties are highly controversial because they connect to different notions of collective responsibility. In order to fully understand criticism against weakly collective duties I consider it important to take a second, more elaborate view on standard views of collective duties and different concepts of collective agency and collective responsibility.

Philosophers mainly disagree about two questions:

(1) Is it actually possible to assign duties or responsibility to collectivities at all?

(2) What conditions must be fulfilled to assign collective duties or responsibility?

With regard to the first, I have previously assumed that applying duties to collectivities is widely accepted. In view of criticism put forward against Cripps' idea of weakly collective duties the matter does not seem to be as clear to all.

 | DOI:10.30965/9783957437426_006

In order to provide a framework for locating criticism, I wish to roughly clarify the two main positions with regard to collective responsibility. The two main positions have been presented under various names. I refer to them as group-agency on the one hand and shared-agency on the other. Philosophers who defend group-agency claim that duties can be assigned only to the group as a whole and are non-distributional. In contrast, philosophers who defend shared-agency claim that duties must be directly distributed to members of a group. Though cases are often way more complex in practice, reconsider the example of a group of teenage friends pushing someone's car over a cliff to see the point (see chapter 2). Representatives of group-agency would claim that the group as a whole is responsible for destroying the car. Representatives of shared-agency would claim that each teenager is directly responsible for some share. The key question seems to be whether responsibility is always reducible to individual members or whether responsibility must in some cases be assigned to the group.

Larry May (1992) is a proponent of shared-agency. May is concerned with responsibility of individual agents for harmful cases of collective inaction as well as cases of joint undertakings within an organised, institutionalised collectivity such as members of a community or citizens of a state. One of the problems he elaborates is racism. Similarly, Cripps stresses that another possible example could be Germans during the Nazi regime (Cripps, 2013, p. 176). Regardless of whether individuals participate or not, May suggests that individuals share responsibility between them for letting harm occur within their community (May, 1992, p. 38). In contrast to holding the group responsible as a group, May proposes the division of collective responsibility either as some straight share of the collective responsibility or according to different roles within the collectivity instead of assigning full individual responsibility or no responsibility at all (May, 1992, pp. 37–42). With regard to actions that were intentional themselves but without intention to contribute to collective harm, May writes "[a] person who engages in such behavior should not be assigned the same degree of responsibility as a person who intends to contribute to a joint action to cause harm" (May, 1992, p. 39).

In contrast to May, Elizabeth Ashford (2006) is a proponent of group-agency. She is concerned with cases of human rights violations for which multiple agents are responsible and no single agent can be identified. One of her examples is global poverty. Ashford's motivation is to recognize the significance and sensitivity of harm-related situations that cannot be captured by demonstrating responsibility or duties of a single individual towards another. Because focussing on the impact of individual actions in isolation misses the share of moral responsibility altogether, the victims' claim is "against the whole group of agents" (Ashford, 2006, p. 228). Yet, her focus is on ways in which institutionalized action may cause harm.

Even though Cripps draws on several of May's ideas with regard to the Able, she believes that her account of weakly collective responsibility is located closer to Ashford's approach (Cripps, 2013, p. 219, footnote 27). Cripps (2011a) has distinguished her model of weakly collective responsibility (the Polluters) from May's concept of shared responsibility. The main point with regard to climate change is that there is no organised, institutional collectivity that causes climate change. Cripps further asks how assigning pieces of harm should work, when there are no roles played in practice. She believes that dividing shares is just as unappealing: Assuming that an individual of a set of n individuals has a duty of $1/n$ does not help at all if the individual personally (qua individual) did not intentionally cause anything. The main difference with regard to Ashford is that collective action in the case of climate change is no institutionalised action. Climate change is the result of individuals situated in relation to one another so that the individual achievement of their goals has an aggregated, collective impact.

In a first step, I agree with Cripps that her account of weakly collective duties is closer to Ashford's approach. After all, the weakly collective duty is a duty to the weak collectivity as a whole because no individual agent can prevent further climate change on her own. Yet, in a second step, the weakly collective duty falls back to individual agents, though not as a share of the total, in order not to leave the weakly collective duty unfulfilled. Though the weakly collective duty is not reducible in the first step, it derives an individual duty. I therefore think of Cripps' approach as a middle path between shared-agency and group-agency.

While many philosophers continue to prove or deny the mere existence of group agency, others focus on the second discussion about the conditions that must be fulfilled in order to assign collective duties or responsibility. The discussion aims to clarify if and if so which normative criteria of individual duties apply equally to groups or collectivities. The strongest controversy probably stems from collective intention. Gilbert's account that was introduced in chapter 2 is an indispensable part of this discussion and her view has shaped the dominant intentionalist view over decades: Sets of individuals form collectivities either because members think of themselves as being a member of the collectivity or because the collectivity jointly intends or intended to act in a certain way. As I have shown in chapter 2, in the case of climate change, the dominant intentionalist view fails.

Criticism arises with regard to both discussions. The first three objections, put forward by Hohl (2017), are mainly concerned with Cripps' defence of weakly collective responsibility (the Polluters) and the possibility of weakly collective agency.[86] The fourth objection focuses on intentionality as a requirement for collective duties. The fourth and fifth objections are concerned with the more general concept of weakly collective duties.

By putting Cripps' approach on a par with accounts of group-agency, Hohl (2017) claims that group-based approaches are generally mistaken in focussing on groups rather than individuals as the only way out of collective harms (Hohl, 2017, pp. 30–34).[87] Considering one of Cripps' earlier contributions where Cripps presents a draft for the weakly collective responsibility claim of the Polluters (2011a), Hohl expresses two main methodological doubts:[88]

> The key problem with weakly collective responsibility is that it *fails to address real moral agents* and therefore it *fails to be an effective impetus for action*. Loosely connected groups of agents are, as Schlothfeld (2009, p. 101) claims, simply *not "normatively addressable"*. [...] Weakly collective duties would have to be *immediately distributed to members* to initiate an impetus for action. Normatively, identifying the group to face duties or to be responsible has only little force. (Hohl, 2017, p. 33, own translation, emphasis added).

In the strongest understanding, there are two arguments: First, loosely connected groups of agents cannot count as real moral agents. And second, weakly collective duties or responsibility fail to be an effective impetus for action if they are not "immediately distributed to members" (Hohl, 2017, pp. 32–33). I understand Hohl's expression *immediately distributed to members* as a kind of shared responsibility that is directly assigned to individual members of the group.

My response in defence of weakly collective duties is the following: It might indeed be vague to apply duties to randomly chosen non-organized groups of agents. However, this is not the case for weak collectivities. Their members' mutual dependence to secure a fundamental interest (the Young) or to achieve a common goal or purpose (the Able, the Polluters) constitutes them, according to the non-intentionalist model, as a collectivity in the weak sense (Cripps,

86 Hohl defends an account of individual responsibility. The following chapter discusses her theory.

87 Hohl discusses Cripps account in line with theories defended by F. Jackson (1987) and Ashford (2006).

88 The argument leading to the collectivized no-harm principle in Cripps book from 2013 is, according to Cripps, "an updated version" of the claim put forward in Cripps (2011a) (Cripps, 2013, p. 219, note 25). In fairness to Hohl, I would like to stress that Cripps (2011a) only indirectly calls for rethinking collectivities as non-intentionalist collectivities. Nevertheless, if Hohl does not regard the Polluters to count as a normatively addressable moral agent, she will probably not regard the Young and the Able to qualify as such either.

2013, pp. 29–30). Reference to *weak* collectivities implies Cripps' awareness that the Young, the Able and the Polluters are not real moral agents (in the sense of organised) *yet*. Insofar, Hohl's critique is correct. That does not mean, however, that the Young, the Able and the Polluters fail to be normatively addressable. Because no collectivity is capable of intentional action yet, the weak collectivities face weakly collective duties "to cooperate" (the Young and Able) and "to organize themselves as necessary" (the Polluters) (Cripps, 2013, pp. 48, 61, 74). Since they are not real moral agents yet, the duty falls back to those individuals who form the collectivity. In the first place, weak collectivities are addressable via their respective members. Individuals as members of the relevant collectivities face a promotional duty to achieve the ability to act collectively. Further, weakly collective duties are grounded in three different normative principles: the principle of moralized collective self-interest (the Young), the moderate principle of beneficence (the Able) and the collectivized no-harm principle (the Polluters). Each normative ground is far from being randomly chosen. On the contrary, they broadly follow moral principles, which are well acknowledged in the discourse about climate duties on the state level (e.g. Ability-to-Pay-Principle, Polluter-Pays-Principle). Claiming that the Young, the Able and the Polluters are too loosely connected to be normatively addressable, requires the arrangement of mutual dependence as a sufficient ground for *weak* collectivities. In that regard, Hohl fails to provide the proof.

The impetus for action is in fact twofold: Action is primarily of a collective nature and, as a derivative, of a promotional nature. The primary addressee of the weakly collective duty is the weak collectivity, not its members individually. However, because the weakly collective duty would not be met otherwise, individuals have a duty to promote collective action of the weak collectivity. The individual duty is a derivative of the weakly collective duty, but it is not some share of the weakly collective duty, which is directly distributed to individual members as Hohl assumes. Doing one's part of the collective scheme is an option only if there is collective level action. If the duty on the collective level was immediately distributed as some share of the total, individuals had a duty to mimic what would be required under a collective scheme. As chapter 3 showed with regard to the weak bridge, mimicking will not make a difference in the first place, leaving the weakly collective duty unfulfilled.

The third objection is the following: Hohl (2017) criticises that it remains open what individuals ought to do in view of the weakly collective responsibility and she argues further that, for group-based accounts, the relationship between collective and individual duties remains unclear (Hohl, 2017, p. 33). Her key claim is that even though the moral evaluation of individual actions requires reference to the actions of others, the moral justification of individual duties does not require holding the group responsible first. Again, with reference to Cripps' earlier contribution, Hohl considers a version of the weak bridge where a person (Fred) is swimming underneath the bridge (Hohl, 2017, p. 34).[89] To cite Cripps (2011a), the version of the weak bridge goes as follows:

> [T]here are three people approaching the bridge. We all step onto it at the same time, despite warning signs. I might tell myself that my acting (or not acting) as I did made (or would have made) no difference to Fred, because two people besides me stepped onto the bridge and two would have been enough to break it. Each of the others might say the same. Does this render the harm causeless, or at least render it impossible to ascribe moral responsibility? Surely not. (Cripps, 2011a, p. 180)

Hohl agrees with Cripps that the harm is not causeless even though all three agents claim to make no difference. However, Hohl, in contrast to Cripps, claims that each agent ought to refrain from crossing the bridge personally (Hohl, 2017, p. 34). Even though the next chapter will elaborately defend Cripps' theory against Hohl's approach, I wish to briefly introduce her points: With regard to the version of the weak bridge above, the boundaries of the collectivity are clearly defined. Also, stepping onto the bridge is an "all-or-nothing action" (Cripps, 2011a, p. 181, 2013, p. 72). Climate change is no such case. Millions of people contribute to the increasing of global carbon emissions. Individual contributions are a matter of "doing too much or too little of something" (Cripps, 2011a, p. 181, 2013, p. 72). Based on that Cripps claims an additional and "ineliminable" reference to the collectivity as a whole (Cripps, 2013, p. 34).

Given that the weakly collective duty applies a derivative promotional duty to individual members, Hohl's claim concerning the lack of relationship between weakly collective duties and individual duties is not acceptable. Also, given the existence of promotional duties, it becomes rather clear what individual agents ought to do in view of weakly collective responsibility. However, and insofar I would agree with Hohl, the nature of promotional duties remains indeed largely unspecified in Cripps earlier contribution.

The fourth objection is that collective duties requires collective intentionality. My response is basically a summary of Cripps' argument presented in chapter 3.1.2. The standard concept of collective intentionality overlooks that for some cases, such as climate change, there is no intention to cause climate change and there is no intention to act collectively at all. And nevertheless, climate change is the result of individuals placed in a situation where they in

89 In the earlier contribution, Cripps refers to this example as a footbridge.

relation to one another, by achieving their own personal goals, produce climate change all together. In this very weak sense of being grouped as individuals, individual agents together face a weakly collective duty. Simply, in view of severe climate induced harm, collective action is required even though there is no intention. Cripps' non-intentional model does not claim that there are no cases of collective intention, but the claim is that intentionalist models are insufficient to explain some cases such as climate change.

There are two further – contradicting – objections to non-intentional collectivities. One claims that mutual dependence is too weak to establish a requirement for collectivityhood. The other stresses that mutual dependence is too strong as a condition. Cripps responds to both (Cripps, 2013, pp. 36–37). The following will briefly reproduce her response.

The fifth objection goes that mutual dependence is too implausibly weak to truly establish a collectivity. In order to prove that mutual dependence is not too weak, Cripps presents two examples: the runners and the climbers. In the example of the runners, neighbours arrange to go for a run together in order for all to get fitter. Even though they would get fitter together, they would not plausibly constitute a collectivity. Cripps explains that the confusion here is between mutual advantage and mutual dependence. Running together is to their mutual advantage but the neighbours are not mutually dependent to either go for a run or get fitter. In contrast, two climbers are mutually dependent to climb a mountain: one climbs the mountain while another secures the rope and the other one's safety and vice versa. Mutual dependence of a set of individuals is a sufficient condition if and only if (a) individuals acknowledge a shared goal in the sense of wanting to achieve something together (like the climbers wanting to climb the mountain) or, without acknowledgement, if (b) the achievement of each individual's goal requires cooperation or if (c) individuals are mutually dependent through a fundamental interest each individual has (Cripps, 2013, p. 37).

Regarding the sixth objection it is unnecessary to set mutual dependence as a requirement for collectivityhood. In political philosophy, the state is probably the most central collectivity. With reference to an earlier contribution, Cripps argues that "the state is a collectivity through actual mutual dependence for the securing of certain fundamental interests for all" (Cripps, 2013, p. 36). In the earlier contribution, Cripps has argued that, in contrast to intentionalist approaches, the non-intentionalist model is able to include separatists who neither consider themselves to be a part of the country nor support the national government (in fact they rebel against the government) (Cripps, 2011b, pp. 9–10). One could probably think of separatists from Catalonia or Basque country. Cripps explains:

> The non-intentionalist [model] can accommodate such individuals by reference to a common interest. Those within a state are mutually dependent through its institutional framework. No matter how little they care about one another, the central capabilities each needs to live a flourishing life can, under the circumstances, only be secured together. (Cripps, 2011b, p. 10)

The criticism that Cripps intends to address in this regard could go as follows: the state as a collectivity is a set of institutions, which regulates the life of its citizens. As long as the majority of citizens endorse state institutions, the state remains a collectivity that is able to protect fundamental interests even though not all citizens might support the state for example by paying taxes. Insofar, reference to mutual dependence is unnecessary.

Cripps rejects their view as too narrowly focussing on omission. If some citizens omit to pay taxes or refuse to support the state, the state might still be able to secure fundamental interests of all citizens if the majority supports the state (Cripps, 2013, p. 36). However, successful protection of fundamental interests requires every citizen to also forgo certain actions. Cripps considers a case where those who do not support the state plan to poison the drinking water to demonstrate their disagreement and opposition (Cripps, 2013, p. 37). The state will no longer be able to secure everyone's fundamental interests. According to Cripps, reference to mutual dependence is a necessary requirement to grasp the state as a typical collectivity that secures everyone's fundamental interests. Thus, it is not enough to refer to a majority of people supporting the state (Cripps, 2013, p. 37).

I believe that the discussion regarding weakly collective duties as a new kind of obligation is only at its beginning. The philosophical discussion will probably have to prove that they can be further sustained against potential objections that may arise in the coming years. Yet, I hope to have shown that the idea of weakly collective duties sustains against a range of possible criticisms and that weakly collective duties make sense with regard to climate change. The following directly jumps to the most discussed collectivity: The Polluters.

5.2 The Case of the Polluters

Recalling from chapter 3, Cripps establishes the collectivized no-harm principle as the primary normative ground for moral duties of the Polluters:

> A set of human beings have a moral duty to organize themselves as necessary to prevent serious harm (deprivation of fundamental interests) to another human being or human beings resulting from the predictable aggregation of their avoidable individual acts. (Cripps, 2013, p. 74)

Chapter 4.3 suggested to reformulate the principle with reference to Gewirth's PGC:

> Members of the Polluters have a moral duty to organize themselves as necessary to prevent others from interference with their rights to freedom and well-being resulting from the predictable aggregation of their avoidable individual acts if members of the Polluters can do so without having an equal right violated.

The Polluters' moral duty to act derives from their weakly collective responsibility for climate change. The first section of this chapter will elaborate Cripps' concept of weakly collective responsibility. It will defend that predictability and avoidance as necessary requirements of weakly collective responsibility are fulfilled (chapter 5.2.1). Given that climate change is the result of individual actions in aggregate, readers might ask why climate change is not a matter of individual contribution. If it was a matter of individual contribution, then refraining from individual polluting actions might be a moral duty. Hohl (2017), whose theory was part of the previous chapter already, has defended such a contribution theory for collective harms. Her approach suggests that individuals ought to refrain from contribution directly. The second section of this chapter will defend Cripps' argument for weakly collective responsibility against Hohl's theory (chapter 5.2.2). Having made a strong statement for the collective solution and promotional duties, section three takes up the opposite view and elaborates the strongest reading of theories in defence of mimicking: if everyone did their part for mitigation and emitted to a sustainable extent only, climate change would not exist. This chapter will analyse the scope for mimicking within Cripps' theory. Further, it examines the question whether promotion of the Polluters' collective action must incorporate mimicking too – at least as a limited case (chapter 5.2.3). Following up on this, section four comments on 'emitting too much' and the need of a threshold-level for emissions. A threshold level allows an examination of who actually counts as a member of the Polluters (chapter 5.2.4).

5.2.1 *The Concept of Weakly Collective Responsibility: Predictability and Avoidability*

The Polluters' moral duty to act collectively stems from their weakly collective responsibility for serious climate-induced harm. As chapter 2 showed, that is in contrast to traditional collective theories, which attribute responsibility based on intention. Regarding climate change, there is no collective intention (Cripps, 2013, p. 68). The lack of intention is a challenge. It is a challenge that requires the reference to *collective* duties instead of *individual* duties but none that requires the sacrifice of the concept of responsibility altogether (Cripps,

2013, p. 76). The collectivized no-harm principle states that the Polluters' moral duty is a result of "the predictable aggregation of their avoidable individual acts" (Cripps, 2013, p. 74). Thus, predictability and avoidance are the necessary conditions for holding the Polluters responsible and assigning moral duties. This chapter will discuss Cripps' claim for the weakly collective responsibility of the Polluters and will defend that the predictability and avoidability conditions are fulfilled.

Cripps sets out the so-called "three-part sufficient condition for weakly collective responsibility" of the Polluters (Cripps, 2013, p. 69):

(1) Individual agents "acted in ways which, in aggregate, caused harm" and were either aware or could reasonably be expected to have foreseen that they "would, in aggregate, cause harm" even though "each [individual agent] only intentionally performed his own act" (Cripps, 2013, p. 69).

(2) Individual agents were either aware or could reasonably be expected to have foreseen that there "were enough others similarly placed (and so similarly motivated to act)" for combined action to cause harm (Cripps, 2013, p. 69).

(3) Individual contributory actions "were avoidable at less than comparable cost to the individuals" (Cripps, 2013, p. 69).

The first two conditions specify 'predictability'. The third condition defines 'avoidability'. Cripps claims that the first and second condition allow for excusable ignorance of the situation (1) and of the motivation of others (2) while the third excludes settings in which avoiding to perform the contributing action comes at "unreasonable cost" (Cripps, 2013, p. 69) of giving up the capability for continued life, health, bodily integrity, affiliation, or practical reason (Cripps, 2013, p. 70).

To capture the condition's significance, recall Cripps' thought experiment for the Polluters, the swimming teenagers, from chapter 3: A number of teenagers, independent of each other, pass a lake. Again independently, they jump into the lake, start swimming and splashing and cause major turbulences. Due to these turbulences, a child that is swimming in the middle of the lake – also independent of others – is put at risk of drowning. Each teenager intended only her personal actions of swimming and splashing and yet, collectively, their individual actions in aggregate cause the child to drown (Cripps, 2013, p. 61). In order to save the child, cooperation of all is required. Now consider cases of excusable ignorance and "unreasonable cost" (Cripps, 2013, p. 69).

Predictability vs. Excusable Ignorance of the Situation and of the Motivation of Others

To condition (1): The collective result was reasonably foreseeable by each or at least each of them could have reasonably been expected to have foreseen that their splashing in aggregate would harm the child.[90] To condition (2): Any of the teenagers could reasonably be expected to have noticed that there are enough others splashing and swimming in the lake. The point of (1) and (2) is that the harm is not pure coincidence, but the collective result was predictable.

To make a case for excusable ignorance, Cripps suggests imagining the child was not easy to see when the teenagers jumped into the lake (Cripps, 2013, p. 69). Maybe because she was diving through the water. In this case, the child's harm is tragic but accidental. The same is true if turbulences occurred under the water surface and actions were not foreseeable to cause harm at all. Or maybe it is a cold day in autumn and one of the teenagers has good reason to assume that the others will not jump into the freezing cold lake. In cases of excusable ignorance, where the collective harm is not predictable, there is no weakly collective duty.

Avoidablility vs. "Unreasonable Cost"

To condition (3): Refraining from splashing or limiting the number of teenagers swimming does not require the acceptance of too much of an individual cost. None of the teenagers' fundamental interests are seriously violated. To make a case for too demanding costs, suppose there was a sudden forest fire around the lake and the only available option for the teenagers to save themselves from the danger was to jump into the water, which causes the turbulences and harms the child. Here too, there is no weakly collective responsibility. The avoidability-condition is not fulfilled.

The swimming teenagers are a thought experiment illustrating the Polluters' situation. In view of the extensive number of thought experiments in Cripps' approach, Fragnière (2015) notes that "thought experiments can be useful to isolate certain dimensions of a complex problem and help us think about it, [however] they can also be deceptive by concealing the true nature and complexity of an issue" (Fragnière, 2015, p. 699). Thus, predictability and avoidablity require much more discussion with regard to climate change. The Polluters, just as the swimming teenagers, are mutually dependent to prevent climate-induced harm but are the predictability and avoidance conditions truly fulfilled?

90 Critics may correctly claim teenagers to be a bad example. Minors, in particular, may at times hardly count as rational agents: they might be testing limits or acting to desire and mood, which makes it hard to expect them to anticipate their actions' consequences. The teenagers could indeed be replaced by adults, which would make the point of weakly collective responsibility even stronger. In order not to complicate things, I chose to stick with Cripps' example.

Predictability in the Case of Climate Change

To condition (1), Cripps writes: "Although alarming numbers continue to deny it, all but the worst-off emitters can reasonably be expected to be aware of the scientific consensus on anthropogenic climate change." (Cripps, 2013, p. 76). The IPCC's reports demonstrate that the scientific consensus is indeed clear.[91] But can regular individuals as members of the Polluters be expected to be aware? Ten years ago, the American market research and survey company Gallup, Inc. initiated the first major cross-national study, which asked more than 200,000 individuals in 128 countries (according to the company's own information representing over 90% of the world's population) about their perception of climate change (Pugliese & Ray, 2009). They found that 62% of people worldwide knew about climate change. In industrialized countries shares were 90% or more. In developing countries, especially in African countries less than 25% knew about it but noticed significant changes in local weather patterns (Pugliese & Ray, 2009). Since this first large-scale study, shares have been extensively surveyed – mainly on national or local level. Findings largely support the thesis that people in industrialised countries are aware of anthropogenic climate change and perceive it as a serious risk (Brenan & Saad, 2018; Kim & Wolinsky-Nahmias, 2014; Knight, 2016; Leiserowitz et al., 2018; Pidgeon, 2012; Reser, Bradley, Glendon, Ellul, & Callaghan, 2011; Whitmarsh, 2011). Given the high-level of awareness, many surveys actually contradict Cripps' assumption of *continuing* numbers of people who deny climate change. Though on a high-share level, they found evidence that climate scepticism in (advanced) industrialised nations is actually *growing* (Brenan & Saad, 2018; Leiserowitz, Maibach, Roser-Renouf, & Smith, 2011; McCright & Dunlap, 2011; Ratter, Philipp, & von Storch, 2012; Scruggs & Benegal, 2012; Tranter & Booth, 2015).

In this context, the impact of political leaders' positions on individuals' perception is certainly not to be neglected. Conducting their annual environmental survey for the US, Gallup most recently surveyed perceptions of US citizens in particular and found that 66% still believe that climate change is happening, 64% are convinced that climate change is human induced, and 60% think climate effects already have an impact (Brenan & Saad, 2018).[92] Generally, opinions of US citizens are politically polarized under president Trump:

> President Donald Trump, who has called global warming a "hoax", may have contributed to this widening divide [the polarization of opinions] by reversing a number of governmental actions to address the issue. These include the

91 See also Cook et al. (2016) who found that there is a 97% scientific consensus.

92 Others even found a share of 70% of US citizens who believe climate change is happening (Leiserowitz et al., 2018).

> announcement that the U.S. will withdraw from the 2015 Paris climate accord, the removal of climate change from the list of top U.S. national security threats and elimination of the terms "global warming" and "climate change" from U.S. websites and lexicons. (Brenan & Saad, 2018).

Yet, Gallup found a net-decline in all shares of individual perception (such as awareness of the climate science consensus, perception of present impacts of climate change and worry about climate change). On the one hand, political leaders impact peoples' perception and if leaders do not take climate change seriously, individuals have reason not to take it seriously either. On the other hand, however, individual agents are citizens. At least in democratic countries, citizens are far from being politically immature. Here, citizenship comes with responsibility and empowerment to shape one's own opinion and to distant oneself from political positions if required. Media coverage, accessibility of data and online information are features that may help shape individual perceptions. Citizenship comes with the duty to choose media that is widely known for accurate reporting and to gather detailed information beyond national borders. But again, in times when political leaders declare high-standard press and media to be enemies of the state, one should probably not demand too much of individual agents. Is the predictability-condition (1) not as clearly fulfilled, after all?

Predictability as awareness, as I see it, can reasonably be assumed. Drawing on the survey mentioned above, on an absolute level, the share of people who are aware of climate change are still high in countries whose citizens count as members of the Polluters even though shares of climate scepticism among individuals are alarmingly growing. Chapter 6.1 will highlight that next to promotional action through the political channel (raising awareness and pressure on political leaders), promotion through the social channel (raising awareness among society) is required. The social channel links to ensuring that scepticism does not increase any further and that the awareness-condition (1) may legitimately be assumed in the long-run.

To condition (2), the predictability of others acting alike, Cripps writes: "We are also aware of how others live, in the developed and increasingly in the developing world: that they share our tastes for such polluting luxuries as fast transport, holidays in the sun, and out-of-season produce." (Cripps, 2013, p. 76). Given the media coverage on the issue of climate change, the still persistent political awareness (e.g. expressed in the conclusion of the Paris agreement) and public figures who increasingly stress the status of the Polluters lifestyle (e.g. Leonardo DiCaprio, Emma Thompson or Robert Redford), awareness of others acting alike is probably hardest to deny.

Chapter 6.5 will highlight the rapidly increasing level of air and marine travel for tourist purposes, in industrialised countries in particular. That given, the Polluters' lifestyle itself is probably one of the best reasons to assume discovery of the fact that there are millions of others who act alike. Spending a holiday on a cruise with thousands of others, being on a plane with hundreds of others or transiting with thousands of others in one of the large flight hubs around the world is a sufficient ground to assume that members of the Polluters are aware of the existence of other people acting in the same polluting way as they do themselves.

Avoidability in the Case of Climate Change

To condition (3), Cripps very briefly clarifies:

> Finally, on the avoidability condition, these [polluting actions such as fast transport, holidays in the sun, and so on] are luxuries. [...], if all or most of us changed our consumption patterns (and at rather less effort if we did so collectively), the harm could be prevented. (Cripps, 2013, p. 76).

Though I very much agree with Cripps that there is good reason to assume avoidablity on a general level, I believe she is too quick with condition (3). Chapter 3.2 highlighted the demandingness-objection to be a serious one and stressed that it is a thin line between demanding too little and demanding too much. Chapter 4.4 stressed several fields of tension with regard to reasonable costs. The avoidability condition (3) is located right in the middle between them. Though many polluting actions are certainly actions linked to convenience and prosperity, which makes them avoidable, labelling all elements of the Polluters' lifestyle as luxuries is simply too quick. Taking air travel for tourist purposes as an example, chapter 4.4 argued that there is good reason to assume a *prima facie* duty to limit many emission-intensive actions of the Polluters. But resulting economic costs have a moral dimension too if they violate peoples' generic rights – in particular with regard to non-members. Considering air travel, many people globally work in aircraft industry ranging from air traffic control and airport logistics to airline staff and flight catering. Limiting air travel for tourist purposes will cause thousands of people to lose their jobs and with it the possibility to secure their own living. The ethical discourse must take these aspects into account too.

To provide a different example, that chapter 6 will take up, consider an individual agent who is aware of human-induced climate change, that enough others act alike and who now tries to consider possible, less-polluting alternatives for her grocery shopping at the nearest discount store. Assuming that the individual agent knows what kind of information to look for it probably takes her a lot of time, flexibility and patience to figure out relevant information on origin, transport, seasonal nature, and so on. Critics may oppose that the individual agent may simply choose an organic food store or the local farmer's market. But that is too simple. Organic products are usually more expensive than conventional products, which is actually a good sign. These products often reflect the true cost of sustainable consumer goods. But purchasing more expensive consumer goods is a matter of financial ability. That way, grocery shopping is not only about time, flexibility and patience but also about financial ability. Is that enough reason to reject avoidablity as a condition? First of all, this is a reason to underline the importance of promotional action. With regard to consumer goods, chapter 6.5 will stress that individuals are currently at the very uninformed end of a long, non-transparent production chain: life cycle information is often not accessible and, in some cases, not even available. With regard to consumption, increasing accessibility and availability of climate-relevant information could be a key target for promotional action of the Polluters. But furthermore, the grocery example demonstrates that avoidance of individual actions often comes at costs that are easy to overlook or underestimate. So, what does that tell us about the avoidability condition with regard to the Polluters' weakly collective responsibility?

My first point is that changing consumption patterns in practice may not be as easy as Cripps assumes. I have made that point with regard to tourist air travel and I have touched upon it with regard to consumer goods. When Cripps writes that individual changes to consumption patterns would be "less effort if we did so collectively" (Cripps, 2013, p. 76), reference to the collective level is the necessary addition. In fact, if individual agents are to perform changes as members of the Polluters, promotional action is required first to guarantee that individuals will not to be overburdened in practice. A profound answer to the fulfilment of the avoidance condition can probably only be given in a case-related analysis as part of the ethical discourse. I have tried to provide a more profound answer with regard to limiting tourist air travel (chapter 4.3).

My second point, on the other hand, is that the existence of alternatives may not be underestimated either. An alternative to convenient and fast private car travel is taking a train, bike or bus. An alternative to spending the summer holidays on the Galapagos Islands or a weekend in London is traveling to neighbouring countries or to a national weekend-destination by train and hiring a car on location. An alternative to buying raspberries during winter is to buy regional fruits or growing one's own fruits and vegetables. Sustainable alternatives that do not interfere with peoples' generic rights exist, as I see it, not only in many cases but in the majority of cases. Changing consumption patterns must not be confused with giving up the system of driving, the system of travel or the variety of consumer choices. But if it does, as in the case of air travel, compensatory effort will have to balance out the negative consequences for those whose generic rights are violated. Therefore, I suggest assuming that condition (3) is fulfilled on a theoretical level and that promotional action should aim to ensure that it is fulfilled in practice.

5.2.2 *Responsibility as a Matter of Individual Contribution?*

The previous chapter argued that predictability and avoidance (on a theoretical level) are fulfilled. Cripps' collectivized no-harm principle states that the Polluters' moral duty is the result of "the predictable aggregation of their avoidable individual acts" (Cripps, 2013, p. 74). Given that climate change is the collective, aggregated result of individual contributions, why should I not be obliged to stop contributing? This is consistent with the widely held intuition that individuals should refrain polluting actions individually. Hohl (2017) who was mentioned earlier (chapter 5.1) has followed this line of thought. She defends a version of a contribution theory that aims to hold individuals directly responsible for contributing as members of a subset.[93] The following will first introduce Hohl's approach in close connection to Cripps' considerations. Next, Cripps' concept of weakly collective responsibility is defended against Hohl's argument.

Hohl's Defence for an Individual Duty to Refrain from Contribution

Major parts of Hohl's argument are concerned with responsibility in a backward-looking sense. Hohl argues that existing theories of individual responsibility, though modified, are prepared to justify direct individual responsibility to refrain from polluting actions that contribute to causing collective harm if there is a relevant connection between the individual action and the collective harm as well as an acceptable option to act otherwise. In distinction to Christopher Kutz (2000), who suggests considering complicity in order to bypass causal inefficiency, Hohl argues that causal contribution is a necessary condition to establish moral responsibility (Hohl, 2017, pp. 53–68). Just as Cripps does, Hohl identifies climate change as a case of overdetermination where no individual action is necessary to cause or prevent relevant harm (Cripps, 2013, p. 69; Hohl, 2017, p. 36). Both Hohl and Cripps agree: In cases of overdetermination where no individual agent makes a difference individually,

93 Hohl refers to climate change throughout her book but is concerned with cases of collective harm in a more general way.

it is still no option to consider the harm to be causeless (Cripps, 2013, pp. 70–71; Hohl, 2017, p. 34).

While Cripps moves to the level of weak collectivities, Hohl sticks with individual moral duties to refrain from contributing and defends her theory as a plausible alternative to requiring individuals to make a difference (Hohl, 2017, pp. 94–98). Hohl's objection, discussed in chapter 5.1, with regard to a version of the weak bridge from Cripps' earlier contribution must be considered in this context: Three individuals approach the bridge, all step onto it at the same time despite warning signs – two would have been enough to cause the bridge to collapse and there is a person (Fred) swimming underneath the bridge who gets seriously harmed (Cripps, 2011a, p. 180).[94] Hohl writes:

> The duty not to step onto the bridge is among other aspects justified by other present agents' expected acting – it is only because others can be expected to step onto the bridge that I should not step onto the bridge in the given situation. If no other agent steps onto the bridge, my stepping onto the bridge would not cause any problem. That way my individual action cannot be considered without relation to other agents' acting. However, that does not imply morally addressing the group instead of the individual agent. (Hohl, 2017, p. 34, own translation)

Building upon the so-called NESS-theory (Braham & van Hees, 2009, 2012; Wright, 1985, 1988), Hohl claims an individual action to be causally contributing to collective harm if it is a necessary element of a sufficient set of actions (Hohl, 2017, pp. 98–99).[95] The crucial point is that an individual agent is morally responsible because she could have avoided contributing to severe harm but not because she could have produced a different outcome (Hohl, 2017, p. 99). Turning to prospective responsibility, Hohl claims climate change to be a textbook example where individual actions pass the NESS-test (Hohl, 2017, pp. 103–104). Assuming that moral claims are violated collectively and that emission thresholds are likely to be reached, she argues that individuals contribute prospectively and are therefore individually responsible (Hohl, 2017, p. 104). According to Hohl, passing the NESS-test is only the first of five conditions to attribute moral responsibility in cases of collective harm (Hohl, 2017, pp. 123–125):

(1) The individual action passes the NESS-test.
(2) Performing the action is within the individual's control.

---

94 In the updated version, Cripps refers to a "three-legged race" to ensure that all agents step onto the bridge at the same time (Cripps, 2013, p. 70).

95 The emphasis indicates the abbreviation given: NESS = necessary element of a sufficient set.

(3) The individual contribution was predictable or at least the individual agent could have reasonably been expected to predict its contribution to the collective outcome.
(4) There is no moral claim to perform the individual action.
(5) There is at least one morally acceptable alternative to act otherwise.

Hohl defends her approach against a number of objections: Starting off with Julia Nefsky (2012) who believes it to be trivial to claim individual actions to be part of collective harm. Others may claim that statements on causal contributions as causal factors remain unclear as they always require normative presuppositions on what actually counts as a causal factor (see e.g. Mackie, 1965). And still others criticize theories such as the NESS-test, which are concerned with individual contributions to include too many cases (Budolfson, 2009). Indeed, the NESS-test does not provide a criterion that differentiates between indigenous people who contribute by sparking off a fire to keep themselves warm and people living in industrialized countries who use their tilted fireplace to create a cosy atmosphere while using a renewable heating option at the same time. The NESS-theory as such may have many weak points, which I do not intend to discuss in detail. Here, I would like to focus on whether the NESS-theory on a very general level makes sense in connection to climate change. Is the NESS-theory a viable alternative to Cripps approach with regard to the Polluters? Why must the Polluters be weakly collectively responsible if all individuals could also be responsible for contributing individually?

Defending Cripps' Concept of Weakly Collective Responsibility against Hohl's Argument

By all appearances, the extended NESS-theory defended by Hohl seems to explain the widespread intuition that I should not contribute to climate change even though it does not make any difference to the overall outcome. Assuming the existence of sufficient sets of acts, the crucial point is that reference to sets of acts is only the first step. I stressed elsewhere (chapter 3.1) that Cripps claims an additional and "ineliminable" reference to the collectivity as a whole (Cripps, 2013, p. 34). To see the difference, consider the swimming teenagers again: Suppose there are twenty teenagers who cause turbulences leading the child to drown and suppose that ten of them would have been sufficient to stop major turbulences. Any teenager can correctly claim that the child could have been saved by ten or more others without her having to refrain from splashing and swimming. In fact, however, based on her membership in the respective potential collectivity, any teenager is a member of a subset that is able to avoid harm. There are many different subsets, which could have avoided the drowning the child. In fact, in the present case there are 184,756 possible combinations for a sufficient subset.[96] To apprehend them all in one statement, the weak collectivity of twenty teenagers remains the locus of responsibility. Assuming that predictability and avoidability are fulfilled, there is a weakly collective duty to the group of twenty not to harm the child. Given that the collectivity is not yet prepared to act, the primary corresponding individual duty is of a promotional nature in order to organize the collectivity in a way that prevents the child from downing.

From the fact that the Polluters comprise a few hundred million people, the size of a sufficient subset in the case of climate change is an indefinite number of people. We can only tell it is by all means very large and covers at least a major part of the Polluters.[97] This leads to an indefinite number of combinations. Here too, reference to the collective level is essential to capture all relevant subsets in one statement. The relation to others in the form of subsets is the first step. However, relation to others is insufficient to fully capture the significance of individual actions. The only way to fully grasp the significance is by reference to others. Considering sets of individuals with reference to the collective level is the necessary second step.

How is it, that Hohl admits the relationship but not the reference to others? Why does she insist on identifying the individual agent against all odds instead of referring to the collectivity? A serious conceptual problem right at the beginning of Hohl's theory is that she overlooks the normative addressability of collectivities as weak collectivities. In the tradition of intentional collective accounts, there is indeed no group of agents that intentionally causes climate change. Likewise, there is no group of agents that is responsible in a strong sense. The Polluters' mutual dependence and the predictability and avoidance of their individual actions allows for holding them weakly collectively responsible. Assuming that collectivities as groups cannot be held responsible, leads Hohl to identify alternative reasoning to capture moral responsibility in cases of overdetermined collective harm. She writes:

> It would be odd to assume that a group of agents causes collective harm, but no individual member is responsible ["zuständig"] for preventing the harm (except

---

96 Considering combinatory analysis, the definition of a sufficient subset for the case of the swimming teenagers is like choosing 10 balls from a basket of 20. Choosing combinations is of an unordered type with no repeats. The result follows the formula: $\binom{n}{k} = \frac{n!}{(n-k)! \times k!}$ with n = 20 and k = 10.

97 Successful mitigation requires carbon neutrality as soon as possible. That given, the sufficient subset probably covers the whole collectivity.

> the group could be regarded to be responsible ["zuständig"] as a group but often, preconditions are not fulfilled [...]). (Hohl, 2017, p. 128, own translation).[98]

The point is, however, the Polluters can in fact be regarded to be weakly collectively responsible as a group. Fulfilling their moral duty requires cooperation – they are mutually dependent to achieve their goal of preventing harm to others.

Reconsidering Hohl's five conditions for moral responsibility and Cripps' three-part sufficient condition for weakly collective responsibility, Hohl and Cripps do not seem to be too far apart: both require predictablity and avoidance.[99] Is it, after all, that the crux of the matter is whether or not one accepts mutually dependence as a criterion? The answer is 'no'. The reasons are bound to effectiveness. With regard to climate change, the overall goal is to successfully mitigate further emissions. I am sure that Hohl is aware that, prospectively, climate change requires large-scale action and not a single Polluter's action. The NESS-theory in connection to climate change requires the intention for mimicking under full compliance. Full compliance of a sufficient subset. The challenge with mitigation is that there is no sufficient subset of motivated, like-minded, conscientious people dedicated to take responsibility. For reasons of effectiveness, that have been mentioned in chapter 3.1.4, the collective scheme has to be initiated in the first step. That requires promotion. Promotion of large-scale action falls back to individual members of the collectivity. To clarify the point, recall the swimming teenagers one more time: Twenty teenagers create major turbulences, which causes the child to drown. A subset of ten teenagers who refrain from causing turbulences would be enough to save the child. One teenager feels responsible and leaves the water. She mimics what would be required under full compliance. Still, there are nineteen teenagers

98 As indicated, Hohl uses the German word 'zuständig', which may either express a *competence* or a *responsibility* to do something. On the one hand, the quote is connected to Hohl's discussion of a version of the competence-objection against individual moral responsibility, which suggests the translation of a *competence*. On the other hand, right before the quote Hohl refers to D. Miller (2011) who speaks about *responsibility* and she also connects to her own analysis of *group-responsibility*. I have defended individual moral duties against no-competence-objections myself (chapter 3.2.1). Thus, the relevant matter is to distinguish the locus of responsibility: Is it the collectivity or the individual? That given, I suggest using *responsibility* and reading Hohl as follows: If the harm is caused collectively but the group is not collectively addressable it would be wrong to assume that no individual is responsible because it would leave the harm causeless.

99 I do not intend to talk down the differences: the perspective (backward-/forward-looking), the NESS-condition and the identification of the locus of responsibility are, without a doubt, major differences.

causing turbulences. The child drowns. Suppose the child does not drown but is 'only' severely harmed. After the recovery, it will legitimately claim that the single teenager, who left the water, could have done more. It might ask: If you felt the urgency to leave the water because you saw me drowning, why did you not raise awareness, motivate others and ask them to leave the water too?[100]

Chapter 1.2 highlighted that theories in defence of isolated individual duties usually account for promotion too. Hohl does not remain silent on this matter either. The final section of her book reconsiders cooperation and promotion in a very weak sense and suggests prioritizing mimicking over promotion.[101] She writes: "It makes more sense for individuals as loose collectivities to cooperate to prevent collective harm if they can assume each individual to have an impact already." (Hohl, 2017, p. 156, own translation). In fact, precisely the opposite is the case. The explanations above defended reference to the collective level as a necessary step to capture the significance of individual actions. For reasons of effectiveness, promotional duties are required first to initiate the collective scheme. The priority of promotion over mimicking rests upon the assumption that there is no collective scheme willing, organized and prepared to act. Implicitly, this means if the relevant collectivity was organized and ready to act already, 'mimicking' as doing one's part may indeed be primary. Yet, in that case, individuals would no longer mimic what would be required under a collective scheme. Instead, they would simply be required to do their part of collective climate action. Given that the collective scheme is required to reduce individual costs to a bearable level (e.g. by initiating structural changes), doing one's part under a collective scheme is unlikely to include reducing personal emission only. But the key point with climate change is that there is no sufficient set of agents ready to act collectively. As long as there is no such effective collective scheme, there is every reason to assume the primacy of promotional duties over mimicking.

This section's headline asks whether climate change is a matter of individual contribution. Climate change is indeed a matter of tiny, insignificant, individual contributions that do not make a difference in isolation, which is why it is a matter that cannot forgo reference to the collective level. Because action on the collective level does not exist yet, promotion is a priority.

100 Cripps emphasises this point in the version of the weak bridge from (2011a, pp. 180–181).

101 Cooperation in this very weak sense is limited to willingness to communicate, to use available cooperation-based products such as carsharing or to support political consumerism (Hohl, 2017, p. 156).

5.2.3 *A Limited Case for Mimicking as a Matter of Integrity?*

Theories in defence for individual mimicking have a point: if every member of the Polluters or at least a sufficient subset refrained from their extremely polluting lifestyle, there would probably be no climate-induced harm. Following Cripps, I held that there is no sufficient set of members of the Polluters willing or prepared to act. But let us assume for a moment that large-scale promotional action already took place, has shown effect and the collective scheme was actually on its way. Though subordinated and very limited, mimicking is part of Cripps' theory too. The following will first explain Cripps position towards mimicking in a bit more detail before discussing her suggestion to make a limited case for mimicking action alongside promotional duties.

Cripps claims that individual action as part of a collective scheme includes more than reducing individual emissions (Cripps, 2013, p. 130). Individual action as part of collective action of the Polluters also includes adaptation and compensation though mitigation is of primary importance. But mitigation too, requires more than reducing emissions from individual agents. Chapter 3.1.4 clarified already that individual agents' actions amount for a substantial, but not for a major part, of global emissions. Effective mitigation certainly requires large-scale technological and structural changes across sectors. But effective mitigation includes using the Polluters' potential to reduce emissions. Indeed, as I have stressed earlier, the Polluters have individual corresponding duties on two levels: (1) acting promotionally in the process of initiating collective action and (2) acting as part of the collective scheme that secures the Polluters' common goal (Cripps, 2013, pp. 71, 79).

Though it is reasonable to assume that doing one's part of the collective scheme is *not limited* to reducing personal emissions, doing one's part of the collective scheme *includes* refraining from polluting actions to some extent in order not to increase harm even further. Promotional duties are primary – I do not intend to develop an objection here. However, promoting large-scale collective action in order to avoid climate-induced harm requires at some point changing one's own lifestyle as part of the Polluters' collective action.

To illustrate the point, considering the swimming teenagers again: Twenty of them swim and splash and the child is about to drown. Regardless of the size of the sufficient set, they, as a weak collectivity, face a weakly collective duty to organize themselves as fast as possible to prevent the child's death. Suppose, two of them detect the collective duty to act. They start promoting collective action. Before taking political measures to generally limit the risk of drowning in the long-run, for example by capping the number of swimmers or recruiting a lifeguard who takes care of weaker or non-swimmers, the two teenagers will probably raise awareness within their own collectivity. Now imagine, they shout out to others asking them to be more considerate, to help the child stay on the water's surface or to leave the water. Suppose others pay attention and detect the need for collective action too. If, however, the two teenagers happily continue to swim and splash while promoting collective action, others are likely to act in one of the following ways: Others will either legitimately say that the two teenagers do not take the situation seriously so that the risk of drowning cannot be too high after all. Or others will conclude that the child's drowning requires at least some action and they find themselves in the respective position. But then, others will legitimately accuse the two teenagers of being liable if the child drowns. Both ways, the two teenagers contradict themselves. They promote action to prevent the child's harm but continue to contribute.

The same point is true for climate change. Suppose, after the publication of this book, readers may do some research and find my social media profile filled with last year's holiday pictures from Seychelles, Australia and Canada, a weekend trip to London and a video of my latest speed-drive in a powerful SUV. If readers are already well aware of climate change, they will probably not doubt my argument content-wise, but they will legitimately doubt me to be serious. Given that I intend to convince readers unfamiliar with the topic that climate change is a matter that requires immediate, large-scale collective action, I will probably lose credibility with regard to the strength of my argument. Promotional action and mimicking action do not coexist independently from each other. Individuals who detect that promotional action is necessary should mimic what is required under a collective scheme alongside promotion even if others neither promote nor mimic.

The matter has been raised as loss of integrity by Hourdequin (2010) in response to Sinnott-Armstrong (2005):

> A person who is truly concerned about climate change and is committed to alleviating it to the best of her ability must make some effort to effect social change [...]. However, a person of integrity who has this commitment will act also on a personal level to reduce her own emission and will, in general, avoid frivolous emissions of greenhouse gases: her actions at the political level will be integrated with those at the personal level. (Hourdequin, 2010, pp. 448–449)

In the context of the swimming teenagers I claimed the two promoting teenagers to contradict themselves. Hourdequin would probably not go that far. She writes:

> It may be too strong to say that someone who is working on the political level to reduce collective greenhouse gas emissions, but is not doing anything as an individual to reduce her emissions, is practically inconsistent; yet it certainly seems that an individual who worked for emissions limiting policies while steadily and frivolously increasing her own emissions would be working at cross purposes. The kind of unity that integrity recommends requires that an individual work to harmonise her commitments at various levels and achieve a life in which her commitments are embodied not only in a single sphere, but in the various spheres she inhabits. (Hourdequin, 2010, p. 449).

Recalling Gewirth's argument, individuals and state institutions do not act in separate spheres. Individuals have moral duties with regard to political institutions. Maybe Hourdequin is right and regular individual agents and governmental institutions are too far apart to refer to practical inconsistency. However, weak collectivities are in between the individual and the institutional level. The two teenagers are members of a weak collectivity whose moral duty consists of organising themselves to be able to act collectively. Changing the collectivity's behaviour is an essential part of the solution and the two teenagers are inherently connected to the collectivity as members.

Cripps has also recognized the point of consistency: She admits that there is "a limited consequentialist case for some mimicking actions [...]. The case for mimicking action is derivative of promotional duties: not a defence of mimicking as a duty in itself." (Cripps, 2013, pp. 151–152). She differentiates two versions of the consistency-claim: First, the "mild version", which states promotional duties to be best fulfilled by living a sustainable lifestyle oneself and second, the "stronger version", which holds that mimicking is a precondition for promotion (Cripps, 2013, p. 152). Cripps locates Hourdequin's position in the stronger version and transfers her integrity-argument to an extreme position. She argues that if it was "psychologically impossible for the individual to campaign for collective action on climate change whilst continuing to live a high-emitting lifestyle herself" then "mimicking would be a universal prerequisite for fulfilling promotional duties" (Cripps, 2013, p. 152). She admits that such inconsistency may be "psychologically problematic", but with reference to Al Gore she stresses that it is quite possible for people to promote climate action while travelling around the globe (Cripps, 2013, p. 152).

Because Al Gore is a living public figure, I would like to refrain from discussing Al Gore's personal consistency here. But the point of consistency may be discussed without reference to real life persons: Suppose there is a public figure who finds herself in the position to promote collective action against climate change and environmental pollution. Suppose she has a secret weakness for meat dishes, long and hot showers, excessive heating and cooling, and leaving electric devices needlessly running. Though there is always a risk of public detection, it is likely that no person outside her family will ever notice. Though consistency from the internal perspective is desirable (and, as I will argue in the following, from a rights-based perspective perhaps even more required than Cripps assumes), it is not necessarily required as a psychological precondition for promotion.

Lack of consistency is not limited to public figures. Taking a look at reality, we see regular people who constantly contradict themselves: Doctors might tell their patients that smoking is likely to shorten their life-expectancy while smoking themselves. Sport scientists doing university research know that little physical activity is likely to cause back pain, cardiovascular or metabolism diseases and still, scientists naturally spend too much time sitting at the computer. Parents might tell their children not to have too many sweets while not missing out on any piece of cake or chocolate themselves. In practice, individuals do not always act as rational agents. Integrity, as consistency from the internal perspective, is desirable but not necessarily required in practice. That being said, assumptions in ethics are certainly affected by individual psychological effects but they are not guided by them. The morally right thing to claim in theory is not always limited by what is implementable or performed in practice.

In contrast, having defended Gewirth's rights-based moral theory as a normative foundation to weakly collective duties, I build on a theory that follows the principles of logic. As a matter of consistency every agent must accept the sequence of necessary judgments leading to the PGC. The PGC is not a recommendation for action. The PGC is unchallenged and categorically obligatory as a supreme moral principle. With regard to Gewirth, I have discussed the role of "virtues of character" in connection to additional well-being (Gewirth, 1978, p. 242). Even though the PGC primarily prescribes the performance or avoidance of certain actions, the individual agent has an indirect obligation to exercise moral and prudential virtues on her own "because of the assistance such development provides for his fulfillment of the various moral duties of action prescribed by the PGC." (Gewirth, 1978, p. 244). Considering integrity as a virtue, agents ought to act in ways that help to develop integrity because integrity may help to meet their moral duties towards others. Thus, even though the sequence of judgements is a method that only rational agents would pursue, it probably allows for a stronger case for integrity, as internal consistency, than Cripps' defends.

Cripps gives the consistency-argument another twist by appealing to the external perspective: the key word is "hypocrisy" (Cripps, 2013, p. 152). This is what I mainly have in mind with regard to the swimming teenagers above. The external, negative impact of promoting one thing while doing the opposite does nothing for an individual's credibility. Cripps admits "publicly promoting systematic change whilst continuing to maximize your position within the old system tends not to improve public opinion of you or of the worthiness of your goals" (Cripps, 2013, p. 152). Hypocrisy is the reason for Cripps to make mimicking a limited, derivative case of promotional duties. Again, with regard to Al Gore she holds:

> Those more in the public eye (those individuals of whom [...] more promotional success might reasonably be expected), who are likely to have their actions overhauled by an often unsympathetic press, can least afford to ignore the consideration of perceived hypocrisy. (Cripps, 2013, pp. 152–153).

I continue to refuse pillorying Al Gore as a person. Instead, I will transfer Cripps' point to an anonymous prominent climate scientist who gives international lectures, provides political consultancy on international scale or attends climate conferences abroad. Her promotional actions might on the one hand have a positive climate impact. On the other hand, international activity and a tight schedule are likely to involve international air travel. Because climate scientists may be members of the Able and members of the Polluters at the same time, polluting activities – within very strict limits – might be legitimate in order to enable large-scale promotional activities. The point is that for some persons, promotional impacts as members of the Able might outweigh their limited negative climate impacts as members of the Polluters.[102] Such a position is in risk of being overused as an excuse for any polluting activity, which makes it mandatory to highlight strict limits. From the theoretical perspective, strict limits can only be determined with difficulty. There is always a risk that the public will assess any polluting action as an act of hypocrisy by the public figure herself and a reason against the matter of fighting climate change. Any kind of inconsistency is likely to damage the collective concern. That is true for hypocrisy, but it is just as true for inconsistency from the internal perspective, that is publicly discovered. Therefore, I generally agree with Cripps that

102 Note that this does not account for regular individuals. They do not have a social reach by living a polluting lifestyle. Chapter 6.5.2 will develop a model of social roles that makes sense of these differences. It will show that some social roles – and climate scientists are not the only ones – inherently incorporate air travel to some limited extent.

those in the public eye should avoid inconsistency as far as possible. However, avoidance of inconsistency is not limited to public figures. It applies to regular individuals too. Even though their inconsistency might not do any large-scale damage to the collective concern, persons in the private surroundings might doubt the seriousness of climate change.

With regard to the "mild version", which stresses promotion through mimicking, Cripps differentiates between private and public mimicking (Cripps, 2013, pp. 152, 153). She correctly concludes that refraining from air travel in silence is unlikely to initiate others to rethink their own lifestyle (Cripps, 2013, p. 153). This way, mimicking may only guarantee consistency from the internal perspective. Promoting one's own mimicking by promoting one's reasons in public at any occasion, however, may have a decisive impact together with others acting alike (Cripps, 2013, p. 153). The premise 'do good and talk about it' does not only prevent others from doubting one's personal commitment, but public mimicking is also a necessary requirement of the collective scheme in the longer run. Promotional action of the Polluters is all about advancing measures that initiate and secure lifestyle changes but also collective changes in lifestyle themselves. Nevertheless, public mimicking in isolation will not have any relevant impact in the first place. For reasons of effectiveness, public mimicking also requires promotion of the collective solution first.

Kingston and Sinnott-Armstrong (2018) believe it to be "risky to move from the political to the personal" level. Not only because some individuals might get the impression they have done enough by refraining from polluting actions which "makes them less inclined to do the harder work of politics" but also because "[p]ersonal freedom needs to be protected against the encroachment of political obligations in order to allow personal expressions and experimentation" (Kingston & Sinnott-Armstrong, 2018, p. 185). They seem to regard polluting actions such as fun-driving as an expression of identity. Following Gewirth, persons are free to perform any action they wish as long as they do not interfere with others' freedom and well-being. Just as in the case of air travel (chapter 4.4), the problem is that performing polluting actions collectively does in fact interfere with other persons' rights. It is therefore risky not to move from the political to the personal level as it overlooks what actions of personal expression may do to others.

The appeal to consistency does not make mimicking a priority. Neither does it replace additive measures that decrease individual costs and ensure zero emissions in the long run. This leads Cripps to conclude a limited case for mimicking *action* instead of mimicking *duties*. Having presented Gewirth's rights-based approach as an alternative justification, I rely on an approach that is built around the avoidance of self-contradiction. Against this background, integrity, as consistency from the internal perspective, is probably more important than Cripps assumes. Given that there is still no collective scheme and given that promotional action is required to initiate collective action, the transition from promotional duties on the one hand to promotional duties alongside with mimicking on the other hand remains unclear. The point where promotion requires supplementary mimicking is hard – if not impossible – to specify beforehand. Without collective action, which decreases individual costs to a bearable level and sustains the impact of individual mimicking, individual mimicking is hardly justifiable. Yet, as a matter of internal and external consistency it is required at some point. Therefore, I suggest speaking of a limited, secondary individual duty to mimic alongside with promotion.

5.2.4 *Who belongs to the Polluters? The Need for a Threshold-level*

Up to this point the Polluters were largely treated as a group of people who contribute to climate change at all levels. Most people living in industrialised countries contribute huge amounts of carbon. They certainly belong to the Polluters. But what about those who contribute only very small amounts? Consider indigenous people for example or people in the West who live self-sufficiently in a sustainable way. They can hardly count as members of the Polluters. But then, the approach for the Polluters is in urgent need of an account of membership. Given that climate change is a matter of contribution with an "ineliminable" reference to the collectivity as a whole (Cripps, 2013, p. 34), who contributes in a normatively relevant way? Who belongs to the weak collectivity of the Polluters?

Cripps suggests that an individual becomes a member of the Polluters once her individual contribution exceeds a level such that were everyone else only contributing to that level, there would be no harm (Cripps, 2013, pp. 59, 72–73). With reference to Joel Feinberg (1970), Cripps stresses that responsibility is about those actions that stand out, that are "unusual" and deviate from a situation without harm (Cripps, 2013, p. 73). She exemplifies the given definition with reference to a version of the weak bridge: Four people with an average weight of 70 kg can safely cross the bridge. Seeing the others approaching, each individual can predict that all are average weight and are likely to be below critical weight. Now, one person carries a bag of 20 kg, which causes the bridge to collapse. Cripps argues that there is something normatively relevant about the additional 20 kg: this bag of 20 kg is the "unusual cause" – the element that deviates from the situation without harm (Cripps, 2013, p. 73). Bringing the example closer to collectivities, Cripps assumes two people with a bag with 20 kg each. Here, she argues, the "unusual cause" is the combination of two people (Cripps, 2013, p. 73). In a third version, three of the four people have a bag of 15 kg each. Two people form a sufficient set to cause the bridge to collapse. And yet, as defended in the second section of this chapter, in order to cover all relevant subsets, reference to the collectivity is required. According to Cripps, the collectivity, this time, consists of three people carrying a bag but not of four people (Cripps, 2013, pp. 73–74).

In the case of the weak bridge, the individual contributions deviate between stepping onto the bridge and avoiding stepping onto the bridge. Individual contributions in the case of climate change can hardly be compared with that. Though individuals decide to either drive the car or refrain from it, to take the plane or travel differently, to buy off-season products or avoid them, etc. performing these actions in isolation does not cause climate change. Cripps stresses contributions to climate change do not to involve some "all-or-nothing action" – contributions are rather "a matter of doing too much or too little of something" (Cripps, 2013, p. 72). The swimming teenagers are better equipped to depict individual contributions to climate change. The teenagers' normatively relevant contribution is not their being in the water. Their relevant contribution to the child's drowning is their swimming *too much* and splashing *too much* that causes the child to drown (Cripps, 2013, p. 73). Suppose one of the teenagers only refreshes her feet in the shallow water, another is floating in the water and yet another is doing steady hydrotherapy stretches.[103] These actions do not actually contribute to the child's drowning. According to Cripps, people below the threshold-level can hardly count as members of the weak collectivity (Cripps, 2013, p. 73).

That given, who, in fact, belongs to the Polluters? With reference to Singer (2002) and statistics of the Global Footprint Network, Cripps points out that people living in Western countries as well as the US and Japan already crossed this threshold several times since the beginning of the 21st century (Cripps, 2013, p. 220, footnote 34). In the absence of statistically recorded personal carbon footprints, determining the cut-off point for the Polluters relies on national average numbers.[104] The Global Footprint Network (2018)

---

103 The last example is taken from Cripps (2013, p. 73).

104 Given that I focus on individual emissions, it is somehow unsatisfying to use average numbers derived as a share from the total on state-level. However, I consider it helpful to get a feeling who actually is a member of the Polluters – at least on average.

estimates Earth Overshoot Day[105] as well as so-called Country Overshoot Days. These are country specific dates that the Earth Overshoot Day would be if all people emitted as much as people of the specific country. Earth Overshoot Day 2018 fell on the 1st of August. Country Overshoot Days in the first quarter of 2018 were for example those of Qatar (9th of February), Luxembourg (19th of February), United Arab Emirates (UAE) (4th of March), the US (15th of March) and Canada (18th of March). Many others fell before the 1st of August: Most of the European countries including Sweden, Finland, Austria (all in April 2018), Germany, France, Switzerland, the UK, Greece, Italy (all in May 2018), Spain, Portugal and Hungary (all in June 2018) but also non-European countries like Japan, Israel, Malaysia, Chile, Fiji, Argentina and Brazil (all in May to July 2018). Country Overshoot Days of Mexico, Thailand, several nations from Central America, Egypt and Vietnam fell after the 1st of August 2018.[106] These insights suggest that residents of many different countries – not only industrialized nations – actually belong to the Polluters.

Average numbers provided by the Global Footprint Network contain not only CO_2 emissions but also factors for general over-exploitation of natural ecosystems such as overfishing and overharvesting. Narrowing factors down to carbon emissions, numbers provided by the World Bank (2018a) may be considered. Their open data shows that global average per capita CO_2 emissions are almost 5 t annually. Considering that global emissions already produced about 1 °C of warming, 5 t annually with a rising tendency are definitely too much. A selected number of countries that produce more than 5 t are: China (7.5), Germany (8.9), Gibraltar, Australia, Oman and Canada (all about 15), the US (16.5), Luxemburg and Bahrain (both 23.4) and the top emitters are Kuwait (25.2), Trinidad and Tobago (34.2), Curacao (37.7) and Qatar (45.4). Critics might argue that these numbers only tell a lot about economic growth, existence of high-tech industry in relation to population size in the first place. Luxemburg, Trinidad and Tobago as well as Curacao are first of all fast-growing countries with a small population.

In order to stay within 2 °C warming, Arnold Tukker et al. (2016) estimate a required range of 2 to 2.5 t carbon dioxide equivalent (CO_2eq) per capita

105 The date when humanity used up more natural resources than Earth can renew in one year (including not only emitted amounts of carbon that exceed the Earth's absorbing capacity but also overfishing and overharvesting).

106 This is only a fraction of Country Overshoot Days. For a complete presentation see Global Footprint Network (2018).

in 2050 (Tukker et al., 2016, p. 174).[107] CO_2eq is a measure that indicates the warming potential of a certain amount of GHG (e.g. methane) taking CO_2 as a reference. Accordingly, the per capita carbon budget limited to carbon or CO_2 is even lower. As chapter 1.1 stressed, projections involve a great deal of uncertainty. The Mercator Research Institute on Global Commons and Climate Change (MCC) visualizes the remaining carbon budget for 2 °C warming and 1.5 °C warming with regard to CO_2 emissions. Depending on the scenario, they estimate the remaining carbon budget for 2 °C warming to be 900 Gt CO_2 in an optimistic scenario, 720 Gt CO_2 in a neutral scenario and 350 Gt CO_2 in a pessimistic scenario (as of October 2018).[108] Based on data provided by the UN's World Urbanization Prospects 2018, I calculated an average world population of 8.5 billion people between 2010 and 2050.[109] That given, the remaining per capita budget per year is 2.58 t CO_2 (optimistic scenario), 2.07 t CO_2 (neutral scenario) and 1,00 t CO_2 (pessimistic scenario).[110] Correspondingly, per capita budgets are generally lower for 1.5 °C warming. In November 2018, the MCC updated their data for 1.5 °C warming with reference to the IPCC's SR5. Accordingly, the new carbon budget is 420 Gt CO_2.[111] That given, the annual per capita carbon budget for 1.5 °C warming is 1.2 t CO_2.[112] Assuming a neutral scenario and following the open data provided by the World Bank (2018a) (reference year 2014), the following countries emit below 2.07 t CO_2 per capita (those countries that emit 1.2 t CO_2 or below are indicated with an additional mark (*):

107 Tukker et al. (2016) assume a carbon budget of 18–25 Gt CO_2eq range by 2050 and a global population of 9–10 billion people.

108 Assuming that 1 Gt C equals 3.67 Gt CO_2, the remaining carbon budget measured in carbon is 245 Gt C, 196 Gt C and 95 Gt C (for conversion factors see chapter 1.1).

109 The data sheet provided by the United Nations' Department of Economic and Social Affairs (DESA) is available for download: (UN DESA, 2018).

110 Using updated data as indicated above, I follow the calculation used for the carbon footprint calculator provided by Atmosfair, the leading German organization for voluntary carbon offsetting.

111 According to the SR15, data ranges from 420 Gt CO_2 to 580 Gt CO_2 using global mean air temperature and 570 Gt CO_2 to 770 Gt CO_2 using GMST [both with medium confidence] (Allen et al., 2018, p. 16). In order to provide a range from a pessimistic scenario under 1.5 °C warming to a neutral scenario for 2 °C warming, I continue to use a budget of 420 Gt CO_2 as suggested by the MCC.

112 I continue to assume an average world population of 8.5 billion people.

Africa		
Burundi*	Burkina Faso*	Nigeria*
Somalia*	Sierra Leone*	Sao Tome and Principe*
Chad*	Comoros*	Senegal*
Congo, Dem. Rep.*	Guinea*	Benin*
Central African Republic*	Liberia*	Congo, Rep.*
Rwanda*	Tanzania*	Mauritania*
Malawi*	Gambia, The*	Zimbabwe*
Mali*	Zambia*	Djibouti*
Niger*	Mozambique*	Eswatini*
Ethiopia*	Kenya*	Cabo Verde*
South Sudan*	Cameroon*	Lesotho*
Madagascar*	Togo*	Angola
Uganda*	Cote d'Ivoire*	Namibia
Guinea-Bissau*	Ghana*	Morocco
Asia		
Nepal*	Tajikistan*	Korea, Dem. People's Rep.
Lao PDR*	Yemen, Rep.*	Kyrgyz Republic
Afghanistan*	Sri Lanka*	India
Timor-Leste*	Pakistan*	Vietnam
Myanmar*	Philippines*	Indonesia
Cambodia*	Bhutan	Armenia
Bangladesh*	Syrian Arab Republic	
Oceania		
Solomon Islands*	Tuvalu*	Fiji
Kiribati*	Samoa*	Micronesia, Fed. Sts.
Vanuatu*	Tonga*	Marshall Islands
Papua New Guinea*		
Europe		
Liechtenstein*	Moldova	Albania
North America		
Haiti*	St. Vincent and the Grenadines	Dominican Republic
Dominica		
Central America		
Nicaragua*	Honduras*	Belize
El Salvador*	Guatemala*	Costa Rica
South America		
Paraguay*	Bolivia	Peru
Colombia	Uruguay	

Irrespective of using 2 °C or 1.5 °C as a target, numbers support the claims that mainly the world's poorest do not belong to the Polluters and that sustainable development requires significantly less than current contributions.

The purpose of a threshold level is to exclude those individuals from the weak collectivity who do not contribute to a normatively relevant amount. Using the average numbers provided above, indigenous people who light a fire to keep themselves warm during night are excluded from the Polluters. What about individuals from typical polluting nations who adopt a green, sustainable lifestyle below the threshold? Though I doubt that, right now without collective measures, a lifestyle below the threshold is actually feasible in industrialized nations without incurring major individual costs – suppose a German citizen who lives self-sustained, on a vegan diet and exclusively travels by foot or bike. Using average numbers, she is a member of the Polluters (because Germans emit way more on average) though she does not contribute in any relevant way. Thus, indicating personal contributions from national average numbers may be too inclusive.

My first comment is that the green German may use one of the online carbon footprint calculators to calculate her personal carbon emissions.[113] If she finds her carbon footprint is indeed significantly below the threshold, she might regard herself to be free from promotional duties that are based on being a member of the Polluters. My second comment is that reference to single, isolated people below the threshold does not invalidate the need for a threshold-level altogether. In fact, the green German actually demonstrates that assuming a threshold is a suitable criterion for membership to the Polluters. If everyone only emitted to her level, climate change would probably not exist.

Before moving on, I would like to clarify one point: Defining a threshold may create the impression that there is an individual right to emit a certain amount. Caney (2012) has forcefully rejected this view even if emitting an equal amount of a total would allow sustainable development. Cripps' concern is not about a fair-share distribution of the carbon budget. Rather, her goal is to respond to the need of a criterion that decides on membership of the Polluters.

Though I agree with Cripps' proposal to hold only those who contribute above the threshold to be weakly collectively responsible, I see one major

113 Given the variety of calculators, their accuracy may legitimately be put into question. The variables used such as diet, use of transport modes, shopping patterns, home size, use of electronic products, heating patterns, and so on use estimated average values. Personal calculations are therefore filled with uncertainties. Given the use of calculators based on peer-reviewed data, such as the ones provided by the Global Footprint Network or the CoolClimate Network (University of California, Berkley), personal calculations may provide legitimate reason to exclude oneself from the Polluters.

challenge: why should the green German be free from promotional duties only because she emits below the threshold? To see the challenge, consider the swimming teenagers once more: Given a number of swimming teenagers, one does her stretching in the shallow water, contributes under the critical threshold but recognizes the danger of the drowning child. Why would this teenager not be obliged to promote the collective action of the others? The child may legitimately claim that the relevant teenager could have promoted the others' collective action. The essential point is the following: From the victims' perspective, all individuals situated with others who contribute in a relevant way are required to act promotionally. That is true for the green German, but it may also be true for indigenous people, though unlikely, if they had the chance to promote collective action, they should be obliged to do so.

How can people who emit below the critical threshold legitimately face promotional duties? My suggestion is that their promotional duty is not grounded on their responsibility for contributing to climate change. They have a duty based on their ability to promote. That, however, requires a broadening of the definition of the Able. On most accounts with regard to states, the Able are the financially able. Cripps writes:

> By 'The Able' I mean the global affluent, or those who could contribute to action on climate change mitigation and adaptation at less than significant cost to themselves. This includes many (but not all) of those in the developed world, as well as the rich minority in the developing countries. (Cripps, 2013, p. 59).

According to Cripps, this reading considers the green German as a "'green rich' individual whose emissions were too low to be in the Polluters but who nonetheless met the conditions for duties of beneficence to apply." (Cripps, 2013, p. 77). In contrast, an indigenous person from Papua New Guinea hardly qualifies as a global affluent. Consider the case of a middle-aged person who counts as poor (e.g. from Papua New Guinea) and who emits below the threshold. Suppose she has the opportunity to raise awareness for climate change. Why would she not have a duty to promote collective action? In the context of climate change, ability is not always tied to financial means. Assuming that others cannot protect themselves and individual agents can do so at no comparable cost, my claim is that if a person can do something – in the broadest sense – to protect others from interference with their rights to freedom and well-being, individuals, as members of the Able, ought to do what she is able to contribute. This broader understanding of the Able, gives rise to a group of individuals who can do something about climate change regardless of their financial ability: Scientists. Chapter 6.4 elaborates the promotional duties of scientists in further detail.

Thus, my key claim with regard to the Able it to use ability in a broad understanding. This is an important finding, in particular in preparation of chapter 6. Yet, I assume that ability does not require a lot more discussion as positive duties based on ability have been standardly applied to groups or collectivities in the philosophical debate (e.g. with regard to overcoming poverty). I therefore chose to focus on the two critical weak collectivities. Having discussed and defended the weakly collective duty of the Polluters above, the following section critically discusses the second controversial weak collectivity: the Young.

5.3 The Case of the Young

Given that climate change will increasingly and globally impact peoples' rights in the coming decades, the Young have a weakly collective duty to cooperate to enable mitigation in order to protect their own rights. The relevant sets of people according to Cripps are "young generations of humans across the globe" (Cripps, 2013, p. 27). Speaking of young generations, I suggest thinking of people under forty.[114] With regard to the Young, I would like to focus on the weakly collective duty of a subgroup: young parents. The majority of people whose personal life planning includes parenthood, will naturally consider having children before turning forty.[115] Next to their weakly collective duty regarding their own interest, do young parents have climate duties with regard to their offspring? The first part of this chapter focuses on present mitigation. It analyses possible justifications of why parents owe mitigation to their descendants. In that context, I will consider children and future descendants alike. The second part of this chapter concentrates on future mitigation and asks whether young parents have a duty to ensure zero emissions in the long-run by raising their children to be climate-aware and responsible agents. In the years following the publication of her theoretical approach to individual climate duties, Cripps has herself published on parental duties with regard to both,

114 The way I approach the Young is to differentiate their age from the middle aged as the period going beyond adolescence. Depending on the source, middle age is often referred to as 40 or 45 to 60 or 65 (see e.g. the Oxford Dictionaries, 2019; Erikson and Erikson, 1998; Lee Phillips, 2011).

115 I am well aware that parenthood is generally possible for men of any age. Also, considering innovations from reproductive medicine (such as egg freezing), postponement of childbearing is increasingly an option for women too. Given that the majority of people still experiences parenthood well before middle-age, I believe that it is reasonable to focus on young parents as a particular group of persons of the Young.

mitigation and raising children. Her contributions will be part of the discussion too.

5.3.1 *The Duty of Young Parents to Mitigate*

Impacts of climate change are predicted to be fully felt latest by the end of the century. Depending on how ambitiously climate change is mitigated, the most severe threats will occur roughly in the second half of this century. Thus, approximately thirty to eighty years from now. That given, children of young parents today, will be forced to live most of their adult life suffering from climate threats. That holds even more true for future descendants of today's young parents. Climate change will progressively violate children's generic rights across generations. Current generations, and young parents as a subset, are the last people given the opportunity to limit climate change to a manageable extent. This section asks: Do young parents, and current young parents in particular, have a weakly collective duty to mitigate with regard to their children and future descendants? The first point to elaborate will be the relationship between parents and their children. Against this background, I continue by defending a duty of young parents to mitigate, that is directly owed to their children. Next, I elaborate in which ways and how far this duty extends to future descendants.

To be clear, this chapter is not concerned with the general moral duty present people have to future people as a matter of intergenerational justice. There is little doubt that present generations have moral climate duties with respect to future peoples' rights.[116] Once they come into existence, future people will have the same rights that present persons have. Because these rights can be fully guaranteed only if present generations mitigate now, future peoples' rights create duties for presently living generations. In contrast, the duty elaborated in this chapter is a special, weakly collective duty. The duty is special because it is a duty that parents have to their children only in virtue of parenthood. And, as this part of my book advances Cripps' idea of weakly collective duties, the duty I am concerned with is one that parents can fulfil only in cooperation with one other.

Parental duties are probably the least questioned special duties. But what justifies the stringent duties parents have towards their children? What is so unique about the relationship between parents and children that it serves as a sufficient reason to assign moral duties beyond the more general duty

116 For a defence see e.g. Shue (2014, pp. 230–234, 2015a) or Steigleder (2016a). The Non-Identity Problem is one of the challenges (Parfit, 1984, pp. 351–379). For discussion on its relevance see e.g. Caney (2005, pp. 757–758), Nolt (2017, pp. 344–349) or Shue (2010, p. 159).

towards children qua persons. In the face of changing family concepts (e.g. single or same sex parents, adoption) and innovative reproductive medicine, the ethical debate has been dealing with the following question: What feature makes an agent a child's parent (Brake & Millum, 2018)? Elizabeth Brake and Joseph Millum (2018) distinguish four key positions: One position argues that parenthood is grounded in the genetic relation between parents and their offspring (see e.g. Hall, 1999). Another position assumes that parents, and women in particular, have put substantial effort, work or investment into their children, which justifies their parenthood regardless of a genetic relation (see e.g. Gheaus, 2018; Millum, 2010). A third type are intentional or voluntarist accounts (Brake, 2005, 2010; Hill, 1991; O'Neill, 1979; van Zyl, 2002). They refer to the agency-aspect of parenthood: the person who agrees to care for a child is her parent. Parenthood on this line of thought is of a social rather than of a biological nature. A fourth line of thought is based on causation: parents cause their children to live, which establishes parenthood regardless of their intention or voluntary agreement to be a parent (see e.g. Archard, 2010; Blustein, 1997).[117] I remain neutral with regard to the ground of parenthood. Building on the respective nature of parenthood, the claim is that parents have special duties to their children because they either intentionally or voluntarily agreed to be a parent, because they caused their children into existence, have put effort into their children or because they are genetically connected. Accordingly, the primary addressees of parental duties may not only be biological but also social parents (such as adoptive parents).

The relationship is unique not only from the parents' side. Children are not just any kind of agent. Broadly following Harry Brighhouse and Adam Swift (2014), children are unique for at least four characteristics: they (1) are dependent, (2) are vulnerable, (3) lack foresight with regard to what is valuable or required to live a flourishing life and (4) still have the capacity to develop into independent, less vulnerable adult agents responsible for leading their own lives (Brighouse & Swift, 2014, p. 62). Because children still develop into autonomous agents, children require special protection. As parents, adult persons are personally responsible for meeting their children's physical and emotional needs. In addition, however, they are required to protect their children from infringements from third parties or, more generally, from outside

117 For further discussion on genetic view see e.g. Kolers & Bayne (2001), on effort- or work-based accounts see e.g. Bayne & Kolers (2003), Fuscaldo (2006), Gheaus (2018) or Millum (2010), on intentional or voluntarist accounts see e.g. Fuscaldo (2006) or Prusak (2011) and on causation-based accounts see e.g. Brake (2005, 2010) or Prusak (2011). Note that accounts that combine several approaches (see e.g. Bayne & Kolers, 2003) are possible as well.

the parent-child-relationship. From a rights-based perspective, parents ought to protect their children's generic rights.

Connecting to climate change, the following limits the relevant group of persons to persons who are both parents and members of the Young: young parents. Readers may ask why focusing on young parents is important. Parents in their fifties or sixties are likely to have grown-up children, who are independent and able to responsibly lead their lives. Grown-up children should be able to stand up for their own rights. In contrast to middle-aged parents, young parents, and current young parents in particular, are an especially crucial group of agents. Thinking of young parents as a subset of the Young, young parents have a weakly collective duty based on their collective self-interest. I have defined the Young as people under forty leaving a threshold age open. Children may be defined either as a separate group of persons such as 'the future Young' or as members of the Young who are not yet mature – as 'members in progress' so to say.[118] Because I defend a parental duty to accompany the process of becoming a member of the Young in the next chapter, I adopt the notion of members in progress but the argument works equally well for either definition. The point is that children will also have a weakly collective duty once they become autonomous members of the Young too. Children and young parents are connected by a unique relationship as well as (potential) membership of the Young. Talking about climate change, time is a highly relevant factor. If the level of carbon emissions remains largely unchanged, the period of time that allows for the limitation of warming to anything close to 2 °C is likely to end roughly around 2035 (see chapter 1.1). Assuming that the legal age of eighteen is the decisive factor for being an autonomous agent, children of young parents today will not be given the opportunity to choose an emission pathway that restricts warming to a manageable extend of 2 °C. More generally, the relevant difference between young parents' being members of the Young and children's future membership is the loss of opportunity for effective action.[119] In view of rapid climate change, it is impossible to do later what is possible right now. Once children of today's young parents are grown up, they will legitimately ask

118 During research for the next chapter, I came across Burtt (2003) using a similar description for children. In distinction to liberal theories that exclusively focus on children's limited rational capacity he focuses on children's needs and writes: "[…] children are best understood as comprehensively needy adult "works in progress," […]." (Burtt, 2003, p. 258).

119 Considering warming scenarios, children as members of the Young, (or even future young parents) will still be required to act in order to limit warming to at least below 3 °C even if their parents fail to act. However, as I will emphasise in the following, 3 °C warming is anything but a manageable scenario. The relevant point here is that mitigation is immediately needed.

their parents: 'Given that you and I have a unique relationship that you initiated, why did you fail to mitigate for me while there was still time?'. My point is that the relationship between parents and their children requires action in the interest of one's own child if that child otherwise loses the opportunity to do so. Because limiting warming to below 2 °C is so centrally important, I discuss this point with a special focus on current young parents. But the argument applies to future young parents too. With regard to rapid climate change, each young parent can only fulfil her duty in cooperation with other young parents. Only if others mitigate too, will young parents be able to protect their children's generic rights. From a rights-based perspective, the claim is: Regardless of the general duty to children, young parents have a special, weakly collective duty to mitigate to protect their children's generic rights in virtue of parenthood and membership in the Young.[120]

Given that climate predictions are subject to uncertainties, readers may point to some likelihood that both predictions on probability and magnitude may be overestimated. Accordingly, critics might emphasize that if things turn out to be not as bad as expected, children will have enough time to mitigate to secure their rights themselves. First of all, there is absolutely no reason to assume that predictions are too pessimistic. In fact, rather the opposite tends to be the case.[121] Though in relation to climate duties present generations have to future persons, Shue (2010) has commented on uncertainties in the following way:

> That something is uncertain in the technical sense, that is, has no calculable probability, in no way suggests that its objective probability, if known, would be small. [...] There is grand illusion here: if we cannot see what the probability is, it must be small. (Shue, 2010, p. 148).

---

120 Making this claim, I assume for now that costs are reasonable. Recall that, in the absence of collective action, the primary duty of individual members is promotion. Making a profound statement on individual costs in the case of parental duties certainly requires further analysis (for an elaborate discussion on individual costs with regard to air travel see chapter 4.4). However, given the unique relationship between parents and their children, I assume it can hardly be too costly for parents to promote action that will spare their children tremendous suffering.

121 An example is the East Antarctic ice sheet. The West Antarctic ice sheet is one of Earth's tipping elements and has long been a matter of concern. Due to its geologic structure, the considerable larger East Antarctic Ice Sheet was generally thought to be less sensitive to warming than the Western part of Antarctica. Recently, an internationally conducted study has gained worldwide attention: The East Antarctic Ice Sheet is not nearly as stable as expected. Considering total loss of Antarctic Ice, the East Antarctic Ice Sheet has contributed way more to overall loss than expected (Rignot et al., 2019). The trouble is: The East Antarctic Ice Sheet holds way more ice than the Western Ice Sheet. If it melts, consequences will be devastating.

Further, he emphasizes that probabilities play no role whatsoever because the magnitude of possible harms and losses due to climate change is disastrous (Shue, 2010, pp. 148–150). He concludes: "Plainly, delay will not make the necessary transition [that is required for effective mitigation] less painful – it will only shift it off us and onto others." (Shue, 2010, p. 150). But that is not all: Failure to mitigate climate change now will not simply shift circumstances on to the next generation. Rather, it creates further climate threats, adversely endangers future, personally unknown generations, creates and enhances climate feedbacks and is likely to tip climate elements beyond controllability and return (Shue, 2010, pp. 150–153). The same point is true for young parents and their children: Rejecting young parents' duty to mitigate because of uncertainties associated with climate predictions will shift an increased, less manageable burden to the next generation of the Young. A generation that owes its existence exclusively to these young parents. Therefore, I claim that uncertainties may safely be ignored.

Mitigation with regard to one's own children is probably the most intuitive argument. It is therefore surprising that Cripps (2017a) strongly disagrees. Referring to it as the "children's interests argument", Cripps believes the claim fails because "it ignores the possibility that many parents could protect their own children more directly" (Cripps, 2017a, p. 312). Large-scale adaptation could secure children's interests during childhood as well as the "prospect for a flourishing adulthood" without further mitigation (Cripps, 2017a, p. 312). Because adaptation is required to avoid "significant risks [...] over a more immediate timescale", parents may be obliged to cooperate "to organize for state- or local-level adaptation" (Cripps, 2017a, p. 313). But even with regard to adaptation, Cripps emphasizes that some parents might be able to protect their children individually with less effort than she had to put into promoting collective adaptation (Cripps, 2017a, p. 313). Taking Cripps' claim seriously, parents may individually install air-conditioning, rely on independent vehicle cooling or move to regions of higher altitude in order to escape summer heat. Facing local food and water shortages, they could import grocery goods themselves, or build up greenhouse facilities with just the right climate conditions to grow their own food. In the longer run, young parents could pay for their children to move elsewhere when climate conditions no longer allow them to live in their home countries.

First of all, such an argument, if at all, applies to affluent young parents only. What about young, non-affluent parents in least developed countries? Second, with regard to climate change, there is no reason why the implementation of adaptation measures supersedes mitigation. With regard to Cripps (2013) and Caney (2010), I stressed elsewhere that, from a moral perspective, neither adaptation nor compensation outweigh the failure to mitigate in the first place (chapter 3.1.1). Also, performed collectively, many individual adaptation measures directly or indirectly contribute to further climate change.[122] What about the general duty not to cause foreseeable and avoidable harm – especially when it comes to one's own children? Referring to arguments of beneficence (the Able) or no-harm (the Polluters), Cripps can easily be proved wrong. Affluent young parents still ought to mitigate to enable non-affluent young parents to protect their children too (beneficence). Performed collectively, many individual measures ought to be avoided because they indirectly or directly contribute to further climate change (air conditioning, vehicle cooling, shipping or air trade, etc.) (no-harm). Connecting the Polluters, the Able and the Young, the case is very clear: Affluent, polluting, young parents have a duty to mitigate because they are affluent and polluting. But do young parents also have a duty to mitigate as parents and members of the Young only? My claim is still that young parents have a special, weakly collective duty to mitigate in virtue of parenthood and membership in the Young.

To defend such a duty, Cripps declares that "mitigation would have to be necessary to secure each child such fundamental interests as continued life, health and community" and finds that "[t]his is questionable." (Cripps, 2017a, p. 312). With reference to the AR5's Summary for Policymakers Cripps further writes: "In fact, adaptation could keep serious health impacts in Europe and North America within or below 'medium risk' levels until well into the second half of the century." (Cripps, 2017a, p. 312). And with reference to the Stern Report she explains that "it would be several decades before mitigation efforts now would take effect." (Cripps, 2017a, p. 312).

The IPCC assess all scientific findings related to climate change. Attempting to objectively evaluate the extensive amount of publications, I assume that the nature of the IPCC's reports is mediating and rather cautious. And there is nothing wrong with that. Providing objective overviews is the IPCC's purpose. However, the drawback of objectivity is the losing sight of possible devastating scenarios. This may lead people to underestimate the *possible* effects of warming. In fact, I claim that Cripps greatly underestimates the effects of warming towards the middle of the century as well as the effectiveness of adaptation measures. Reconsidering chapter 1.1, the IPCC predicts mean average surface temperature changes between 1.0 °C and 2.0 °C by the middle of the century.

122 As an example, consider air conditioning. Air conditioning contributes indirectly via an increased energy demand, which still relies on fossil fuels. But more than that, air conditioning contributes directly via leaking of hydrofluorocarbon coolants, which are many times more polluting than carbon (see e.g. Campbell, Kalanki, & Sachar, 2018).

At first sight, that may not seem too tragic although even 2 °C warming are far from being unproblematic (see chapter 1.1). In a high emission scenario (RCP 8.5), however, the projected change may rise up to a maximum of 2.6 °C between 2046 and 2065 (likely range 1.4 to 2.6 °C, mean 2.0 °C) relative to the period between 1986 and 2005 (Stocker et al., 2013, p. 90). Taking pre-industrial level as a reference, warming is likely to cross a maximum of 3 °C by the middle of the century (Field et al., 2014, p. 13).

Mark Lynas (2008) has made an effort to outline how the world will look with warming of varying degrees. Because affluent young parents seem to be the critical group of agents, I limit his findings to effects in developed countries and I will enter the most pessimistic scenario for the middle of the century: 3 °C warming. Because possible El Niño's activity increases up to permanent effect, the risk of "super El Niños" rises, leading to freak weather around the globe (Lynas, 2008, p. 115). Europe will experience drier winters (Lynas, 2008, p. 115), Australia will be the driest continent on the planet facing temperatures of up to 40 °C (Lynas, 2008, pp. 122–123), the West Coast of the US will suffer dramatic droughts, wild fires and water shortages (Lynas, 2008, pp. 143–144) while New York City will experience surrounding water to rise between twenty-five centimetres to one meter (Lynas, 2008, p. 147). Extreme cyclones will haunt the UK, Spain, France, Germany and, in particular, the Netherlands (Lynas, 2008, p. 149). Norway, Finland and Canada will experience longer growing seasons and yet, benefits do not outweigh global losses (Lynas, 2008, p. 131). With severe droughts and, resulting from it, famines across the African continent, Europe will face millions of refugees leading to severe conflicts (Lynas, 2008, p. 159). The huge problem with 3 °C warming is that the risk of tipping elements being tipped beyond their threshold increases tremendously (see chapter 1.1). With 3 °C warming, the Amazon rain forest basin will dry out completely, which would lead to wildfires and thereby to an even higher release of CO_2 that is otherwise stored in trees and wood (Lynas, 2008, pp. 116–120). In view of various tipping elements, Lynas concludes: "[...] if we reach three degrees, that could lead inexorably to four degrees, which in turn could lead inexorably to five" (Lynas, 2008, p. 190).

Leaving such a worst-case tipping-scenario aside for another moment, even at 3 °C, the possibilities of withstanding the strength of natural forces is limited. Heavy freak weather, droughts with water shortages across a vast country like Australia and extreme cyclones should not be taken lightly. If ambitious collective mitigation throughout the next two or three decades fails, warming will cause climate impacts that adaptation – even at the national or international level – will be unable to cope with. That is true for parents in the world's poorest countries in particular, but it is just as true for young parents living in industrialized countries. Facing warming effects of anything beyond 2 °C, potential small-scale adaptation of affluent young parents will reach its limits rather quickly.

Nevertheless, the matter of unknown likelihoods is persistent. It is true that prediction always has a great amount of uncertainties attached. However, recalling Shue (2010), only because probabilities cannot be fully specified, it does not follow that they are small. In fact, quite the contrary is the case. Facing reinforcing effects that possibly lead to unstoppable, irreversible warming, young parents ought to ignore the degree of uncertainties attached to climate predictions. Also, they ought to ignore the chances of protecting their children individually or by adaptation only. In a note, Cripps admits:

> An alternative children's interest argument might appeal to the danger of runaway climate change. This would have more rapid and far worse implications than those currently considered 'likely' [...]. If parental duties extend to a requirement to insure one's children against even relatively small risks of catastrophe, the shared duty would follow. This has potential, although I do not pursue it here.

In fact, this alternative argument is the only relevant argument with regard to devastating climate-induced harm.

Risks are commonly defined as "the product of magnitude and probability" (Shue, 2010, p. 147). I think the seriousness of loss with regard to climate change has become clear by now. So, the magnitude is not the matter in question. What about probabilities? As discussed above, predictions have a great amount of uncertainties attached, which makes it hard to specify probabilities. The beginning of this chapter stressed Shue's claim to ignore the matter of probabilities with regard to climate change because the magnitude of possible harms is disastrous (Shue, 2010, pp. 148–150). Going into detail, the severity of possible losses is only one of three features that requires to ignore probabilities. According to Shue, cases require the disregard of the matter of probability if they fulfil three characteristics: (1) "massive loss", (2) "threshold-likelihood" and (3) "non-excessive costs" (Shue, 2010, p. 148). Climate change is such as case. Regarding (1), Shue refers to resettlements due to rising sea levels and food shortages causing "volatile food prices" (Shue, 2010, p. 150). With Lynas' analysis in mind, it is safe to say that potential losses of climate change will be massive – even for 3 °C temperature change and as early as the middle of the century. Concerning (2): If the mechanism leading to severe losses is sufficiently known and circumstances push the mechanism close to its threshold for functioning, "likelihood of the losses is significant, even if no precise probability can be specified" (Shue, 2010, p. 148). Shue argues that the mechanism relating carbon emissions to climate change as well climate change to human suffering is well understood (Shue, 2010, p. 150). Indeed, the scientific consensus on climate change and climate-induced harm has never been clearer. Climate change is already happening and up to 3 °C warming by the middle of the century is a realistic (though pessimistic) scenario. Given the findings on cumulated carbon, almost any amount of CO_2, no matter how small, adds to the total amount of atmospheric carbon leading to further climate change (see chapter 1.1). Considering the risk of tipping central elements of the climate system beyond their threshold and the risk of passing points of no return, the likelihood for severe suffering is certainly significant even if its probability cannot be fully specified. With respect to (3): From a rather general perspective on mitigation, Shue claims: "The costs of prevention are moderate, although far from insignificant." (Shue, 2010, p. 150). Costs of mitigation are indeed a serious issue and they are the constant challenge of this book. Connecting to young parents, costs should be bearable on the collective level. Promotion is the primary individual duty. As I stated above (footnote 120), given the unique relationship between parents and their children, I assume it can hardly be too costly for parents to promote action that will spare their children tremendous suffering. Facing climate change as one of the biggest challenges of humanity, probabilities ought to play no role.

Now, recall Cripps' note: "If parental duties extend to a requirement to insure one's children against even relatively small risks of catastrophe, the shared duty would follow." (Cripps, 2017a, p. 321). In fact, the risk of catastrophe is anything but small. For risks to be small, either magnitude or probability must be small. The magnitude of potential harms and losses due to climate change is severe. It may be difficult to determine probabilities in detail but that does not imply they are small. Cripps has disproved the misconception that Shue refers to when he emphasizes: "There is grand illusion here: if we cannot see what the probability is, it must be small." (Shue, 2010, p. 148). The risk of catastrophe may be uncertain, but it is no way small. The risk of catastrophe will increase with every year of inaction, endangering the next generation with even more harm and leaving them with less potential to act. With respect to their children, young parents ought to ignore uncertainties and mitigate right now.

Given that current young parents are essentially the last people to mitigate effectively, I wish to broaden the perspective to future children: young parent's grandchildren, great-grandchildren, great-great-grandchildren, and so on. Do young parents have a duty to mitigate with regard to future descendants too?

Distance in time or space can be challenging for climate ethics. Critics may object that duties cannot extend to forever. Given that the most severe climate impacts are predicted to occur by the end of the 21st century in most scenarios and assuming that duties to future people may be most stringent if individuals are somehow connected, I claim that duties of current young parents extend to the relevant time frame with regard to climate change – namely to the end of the century. Assuming an average parental age of 30 years across countries and generations, children born today will have children themselves by 2049. They are grandchildren of today's young parents. By 2079 these grandchildren will experience parenthood themselves (great-grandchildren of today's young parents). The next generation (great-great-grandchildren of today's young parents) will then be born in 2109. Given life expectancy in industrialized countries of almost 80 years for children born recently, current young parents have to assume that their own children will standardly know their grandchildren (born in 2079) but they cannot exclude that their children would know *their* great-grandchildren (born in 2109) in person either.[123] This is only a very rough estimate of evolution for the 21st century. But it is sufficient to make my point: Assuming that the relationship between parents and their children is unique and imposes a special duty on parents, the scope of this relationship extends indirectly from, at least, young parents to their great-grandchildren and even their great-great-grandchildren. The argument even gains strength when linking parenthood with membership in the Young. By the time future children across generations become members of the Young, they will have a weakly collective duty to cooperate to secure their entitlements themselves. Because it will truly be impossible to effectively secure this interest towards the end of the century, young parents ought to mitigate now. So, at first sight, there seems to be a strong case for a duty of young parents to mitigate with regard to their future descendants.

But, taking one step at a time, I would first like to take a quick glance at living conditions in the second half of the century if emissions do not undergo major changes. What happens if carbon continues to rise? Like earlier, I focus on the most pessimistic scenario. In a high-emission scenario (RCP8.5), the IPCC predicts average surface warming of 3.7 °C between 2081 and 2100 relative to the period between 1986–2005. The likely range of RCP8.5, however, extends to a maximum of 4.8 °C (Stocker et al., 2013, pp. 89–90). And, taking preindustrial times as a reference, extreme predictions are likely to extend to more than 5 °C. Taking Europe as an example, the IPCC's AR5 refers to extreme predictions, estimating sea level rise of up to 1.05 meters (in the Netherlands)

123 For data on life expectancy at birth see World Bank (2018b). Data shows an average of 80 years life expectancy at birth in 2016 for 'OECD members' as well as 'High income'.

(Katsman et al., 2011) and 1.9 meters in the UK by the end of the century (Lowe et al., 2009). The IPCC further predicts that extreme sea level events will be followed by storm surges along the European coastlines. The frequency and intensity of heat waves will increase dramatically with large parts of Southern Europe experiencing extreme dry spells during summer. As early as 2050, inland water transport (e.g. on rivers) as well as rail infrastructure are projected to experience unnavigability and damage due to high temperatures and drought. Also, infections are projected to severely affect human health across Europe (Kovats et al., 2015, pp. 1278–1281, 1291).

In view of these findings, there seems to be good reason to consider Lynas' analysis once more and to draw an even more sinister perspective of 4 °C and 5 °C warming. As previously mentioned, I focus on climate-induced effects in the wealthy regions of the planet. For 4 °C warming Lynas found that sea level rise will be irreversible leading cities such as Shanghai, Boston, New York, London and Venice to be under water (Lynas, 2008, p. 165). Australia (except for Tasmania) will no longer be able to support any kind of agriculture (Lynas, 2008, p. 173). Severe dry spells will be commonplace in the Southwest of the US, Australia and the Mediterranean region (Lynas, 2008, pp. 173–174). Heat waves would last much longer causing wildfires, even in the Alps (Lynas, 2008, p. 177). UK summer temperatures could reach 45 °C (Lynas, 2008, p. 178). Europe would experience 80% less snowfall resulting in water shortages through the year (Lynas, 2008, p. 178). Melting of permafrost (Greenland, Siberia, Arctic, Antarctic) will lead to massive release of carbon and methane, which is likely to double global emissions (Lynas, 2008, pp. 186–190). A rise in temperature of 4 °C will lead to chaos and conflicts linked to climate migration across Europe in particular (Lynas, 2008, p. 179).

Now, 5 °C warming: First of all, 5 °C warming will leave us to a completely different planet (Lynas, 2008, p. 193). Further, Lynas expects the following effects: ice sheets, just like rain forests, will no longer exist (Lynas, 2008, pp. 203, 209). Droughts, floods, desertification and the new sea level will leave habitable areas only towards the poles up in the far North of Europe, Canada, Russia and possibly also New Zealand and Tasmania (Lynas, 2008, pp. 196, 209–211). Methane hydrates, inclusion of methane that only exist under certain ambient conditions (deep in the sea with high water pressure, low ocean temperature), will release further quantities of methane if ocean temperature rises (Lynas, 2008, pp. 204–205). The results will be submarine landslides, tsunamis and the tipping of major tipping elements of the climate system beyond their thresholds (Lynas, 2008, pp. 205–206, 208). And I would like to add one more piece of information: in contrast to carbon, methane is flammable even at a concentration of 5%.[124] Thus, air will become increasingly explosive. Or, as Lynas puts it: At 5 °C warming, "[h]umans would sit powerless to intervene as their planet began to turn into Venus" (Lynas, 2008, p. 205).

Readers will probably insist this to be a too pessimistic even apocalyptic view. On the one hand 4 °C and 5 °C warming until the end of the century is indeed pessimistic and I, personally, wish it was pure fiction. On the other hand, given the IPCC's likely range for RCP8.5, given the facts that the trend of global emissions in 2018 was actually upwards instead of downwards and even leading parties of international climate policy (such as the EU) will tragically fail to meet their intended climate goals for 2020, there is very little reason to be optimistic. If nothing changes, 5 °C warming by 2100 is anything but pure fiction. As I have stressed earlier with reference to Shue (2010), it is impossible to do later what is possible now. Failing to act will not *shift* an increased, less manageable burden to the next generation as in the case of young parent's children. Failing to act now and throughout the next decades will *leave* a huge, non-manageable burden to future children. Considering the end of the century, failure to mitigate is not about reducing opportunities, it is about giving up on opportunities. Given these insights, given the urgency to act and assuming that future descendants matter to parents via an indirect relationship that expands from young parents over children to their children and again over to their children, etc., current young parents have a duty to mitigate with regard to further descendants. Because they can fulfil this duty only in cooperation with one another, the parental duty with regard to future descendants is a weakly collective duty.

Cripps (2017a) has discussed parental duties to future descendants too. Again, with regard to parents in general, she elaborates two arguments that could justify an obligation to mitigate with respect to further descendants: Cripps (2017a) calls the first the "indirect further descendants argument" (Cripps, 2017a, p. 313) and the second the argument "direct duties to further descendants" (Cripps, 2017a, p. 316). The following will briefly introduce each argument before discussing her reasoning as well as possible objections.

On the first argument, Cripps assumes that climate impacts will become more and more severe towards 2100, which will decreased the likelihood that future

124 With regard to vapours and gases such as methane, scientists specify certain flammability characteristics. Gases have a certain minimum concentration, the Lower Explosive Limit (LEL), and a certain maximum level, the Upper Explosive Limit (UEL), for flammability and explosiveness. At room temperature and atmospheric pressure, methane is flammable in a range between 5% (LEL) and 15% (UEL) (ILO & WHO, 2017).

parents will be able to protect their grandchildren and great-grandchildren on their own (Cripps, 2017a, p. 313). Protection of further descendants requires "at least community or state-level adaptation" but also global mitigation across mainly the affluent parents (Cripps, 2017a, p. 313). Because it may be in the interest of one's child to be able to protect their own children, parents today are obliged to mitigate in order to protect their own children's interest (Cripps, 2017a, p. 313). According to Cripps, this duty of parents is still owed to children and only *indirectly* to further descendants. I will therefore simply refer to this as the indirect version. Even though parents today cannot know if their children will have children themselves, they cannot exclude it either. Because parenthood is a central human interest, parents owe it to their children to be able to choose to have children themselves (Cripps, 2017a, p. 314). Further, Cripps highlights it to be part of the parental duty to spare one's own children the choice of leaving *their* own children with severe climate-induced harm or not experiencing parenthood at all (Cripps, 2017a, p. 314).

So, inability to protect children and preserve opportunities becomes a matter for Cripps too. Yet, much later than there is reason for worry when entering scenarios other than average or optimistic pathways. In more detail, Cripps assumes the effects of 4 °C warming between 2080 and 2100 to "impose serious health risks which adaptation, though necessary, could only curb to a limited extend" (Cripps, 2017a, p. 313). She believes that these are the effects children of current children will experience. In fact, however, 4 °C warming is a major human tragedy that is to be avoided under all circumstances. And, as described above, even 3 °C warming is far from being easy to deal with. Also, because 3 °C warming increases the risk of passing points of no return, it may unavoidably lead to 4 °C and 5 °C warming.

Warming only occurs with a significant time lag. The huge problem is that once humanity starts to suffer from climate impacts that adaptation is largely unable to cope with, it will be too late for effective mitigation that spares humans severe harm. Accordingly, the critical years are way earlier than 2080 to 2100. Here, it becomes evident that Cripps underestimates the effects of warming towards the middle of the century that were described above. To be fair, though, the critical temperature level that adaptation is able to cope with is indeed largely uncertain. Concerning this matter, the IPCC concludes that (with very high confidence) adaptation efforts required for 4 °C warming "may be significantly greater than adapting to lower magnitudes of change" (Klein et al., 2014, p. 913). Considering scenarios for three, 4 °C and 5 °C warming, it is reasonable that the critical temperature that adaptation at any level is able to cope with is rather around 2 °C than around 3°C but definitively not at 4 °C warming. Critics may argue that the difference between my view and Cripps' position is a matter of risk perception. Referring to Shue (2010) again, I wish to stress once more that uncertainty does not mean that probabilities are small. And when stakes are high – as in the case of climate change – the question of probabilities ought to be ignored as far as possible.

Cripps argues that parents today ought to spare their children the tragic choice of not being able to experience parenthood at all or seeing one's own children suffering (Cripps, 2017a, p. 314). Narrowing parents down to young parents, the tragic choice is bitter truth already. Current young parents may not be aware of it. But, considering their children's lifetime, young parents have already chosen to see their children experiencing severe harm when becoming a parent. In order to spare themselves this kind of fundamental regret of having caused their children to live in a warming world, young parents certainly ought to promote collective mitigation. Given that I have defended a duty of young parents with regard to their own children, Cripps' indirect version may supplement the duty of young parents to mitigate.

Cripps' second, and probably more intuitive, argument connects to my remarks regarding future descendants above: young parents are obliged to mitigate because they owe it to their future descendants directly. Cripps writes:

> In reproducing, parents bring not only their child into the world but the prospect of a whole line of descendants, all also vulnerable in central ways. If parents know this and know, too, that they (individually or collectively) could protect the fundamental interests of their grandchildren, great-grandchildren, and so on, the same initial responsibility that grounds their specific duties to their children may ground a special duty to do this [the protection of further descendants directly]. (Cripps, 2017a, pp. 316–317)

Cripps stresses that this argument assumes a causal view of parenthood (Cripps, 2017a, p. 316). Only because parents have caused their children to exist, they have also caused the preconditions to create further children. The direct duty to further descendants is an argument that builds upon the duty to one's own children with regard to their children (indirect version), but with "further reaching consequences" (Cripps, 2017a, p. 317). Cripps points out that a direct duty to grandchildren and great-grandchildren applies to procreative parents only (Cripps, 2017a, p. 316). Further, she argues that the duty of parents extends to descendants "with whom their children could build a relationship" (Cripps, 2017a, p. 318). Cripps believes that the probability for having children is essential and concludes:

> Depending on factors such as the number of one's children, the number of children they can be expected to have (factoring in choice and chances of fertility problems) and so on, and the possibility of external events rendering whole

> families, communities or populations extinct, there must be some limit to the number of generations unto which parents have special duties to their descendants. However, my argument would apply for at least several generations. (Cripps, 2017a, p. 313)

I have three comments: The first comment is with regard to the time-related extent of direct duties to further descendants. Cripps assumes that her argument "would apply for at least several generations" (Cripps, 2017a, p. 313). Narrowing parents down to young parents, it is possible, to roughly specify the relevant generations by considering the relevant timeframe. I have shown above that children of young parents today will almost certainly know their grandchildren (born in 2079) but they cannot exclude that their children would know *their* great-grandchildren (born in 2109) in person either. Assuming that a relationship is either directly or indirectly necessary to cause stringent duties, the moral duty of current young parents certainly extends to the last quarter of this century.

My second comment is about Cripps' limitation to a causal view of parenthood with regard to direct duties to further descendants. I admit that the ground for duties to further descendants is easiest to see with regard to causing a line of children to exist. However, I see no reason to limit the weakly collective duty to procreative, biological parents only. Why would adoptive parents, who are connected by social parenthood, not be responsible for the protection of further descendants' entitlements? Even though they did not bring their adopted children into existence, they agreed to take care of them. Thereby adoptive parents also created the prospect of a line of descendants. People become grandparents roughly after two third of their lifespan.[125] From the sociological point of view, one could add that seeing one's own roots, values, family history, and so on being continued towards the end of one's own life is a value that motivates persons with regard to their grandchildren in particular. Witnessing the passing on of one's own history is not limited to procreative parents. Family bonds extend to social parenthood too.

My third comment is on the relation between mitigation and adaptation. In the summary of her defence of a parental duty, Cripps writes that she has defended a parental duty "to organize to act on climate change, through both

125 Depending on method and observation period, the demographics of grandparenthood vary substantially between countries (Arpino, Gumà, & Albert, 2017; Leopold & Skopek, 2015; Margolis, 2016). Comparing twenty-four European countries, Leopold and Skopek (2015) for example found the median age to range between 46 (Ukraine) and 57 (Switzerland). To make the point, the qualitative statement about the timing of grandparenthood above suffices.

adaptation and mitigation" (Cripps, 2017a, p. 319). Though I generally agree with Cripps that the duty of young parents includes both adaptation and mitigation, I am not convinced that adaptation and mitigation should be assigned or only appear to be assigned equal status. The devastating effects of three, 4 °C and even 5 °C warming have been elaborated above. The point is simply that without large-scale mitigation even the most effective adaptation measures will not be able to protect descendants in thirty to eighty years' time. With that in mind, I claim that collective mitigation is strictly a priority over adaptation.

A first objection, that applies to my remarks too, could be why the obligation is a duty of current parents to one's grandchild or even great-grandchild and not a duty of one's child to her child. Basically, why does the duty based on the parent-child relationship extend to further parent-child relationships? Cripps clarifies that the primary locus of responsibility with regard to further descendants may indeed be with one's children. But because grandparents or ancestors in general "cause something to happen which will lead to some other event unless another person makes a very great sacrifice, it seems reasonable to consider me partially responsible for that further event" (Cripps, 2017a, p. 317). In contrast to Cripps, I have defended a duty of young parents to their own children. In my account, current young children will certainly have a duty to mitigate once they experience parenthood. The problem is that once young children experience parenthood (recall this will be roughly around 2049), humanity will already be right in the middle of severe climate change if collective action today fails. In this context, Cripps surprisingly emphasizes that generations living right now are the only (in terms of time) generations who are in the position to fight climate change (Cripps, 2017a, p. 318). In view of this, I hardly understand why Cripps believes that the moral duty of parents today is nothing owed to their own children directly.

A further, well-known objection may again be raised with regard to reasonable costs. I wish to keep the objection as short as possible. In the summary of her contribution, Cripps (2017a) responds to demandingness-objections by stressing that the moral duty to future descendants is a duty to all present and future parents, which distributes burdens across generations and therefore leaves current parents with a reasonable burden (Cripps, 2017a, p. 319). Recall that costs must be bearable on the collective level. The duty of young parents to children and future descendants alike, is a collective duty – collective in the weak sense. Thus, costs apply to the collectivity in the first place. Not only do costs distribute across generations as Cripps stresses, but they also distribute within the generation of young parents. With regard to their financial ability, young parents from developed countries may certainly be able to carry more of a burden than young parents from developing regions. Just as any other relevant weak collectivity, young parents are probably not yet organized to act collectively. Because the collective duty would otherwise be left unfulfilled, individuals, as members, have a promotional duty derived from the weakly collective duty. As soon as there is collective mitigation, parents have a duty to do their part of the collective scheme. I provided guidance on reasonable costs elsewhere (chapter 4.4). Given the range of promotional actions, I suggested that promotional duties are probably not the biggest challenge. With regard to promotion, chapter 6.3 will argue that the Young (and young parents as a subset) are already well connected – mainly online. This leads me to the very brief statement that the required effort of young parents to promote collective action is probably lower than for any other relevant collectivity. And yet, one should avoid jumping to conclusions: More discussion is certainly needed to make a profound statement on reasonable costs with regard to both promotion and doing one's part of collective parental mitigation.

This chapter aimed to show that young parents have a moral duty to mitigate. The duty to mitigate is owed to children directly but also to future descendants. For young parents, climate change will become reality in the course of their life. Children of young parents will be the first generation living the largest part of their adult life suffering from serious climate-induced harm. In virtue of parenthood and common membership in the Young, young parents have a special duty to their children. Because no parent can fulfil this duty on her own, the duty is of a weakly collective nature. I have stressed that Cripps probably underestimates the suffering towards the middle of the century while overestimating the potential of adaptation measures. Given that inaction will severely endanger the next generation with even more harm, leaving them with less potential to act, I have stressed with reference to Shue (2010) and in contrast to Cripps (2017a), that young parents ought to ignore uncertainties and mitigate immediately. Focusing on young parents, sparing tragic choices between seeing one's own children suffering and not experiencing parenthood at all is already a matter of young parents regarding their children and not, as Cripps assumes, just a matter of children and their future children. The duty of young parent to their own children may therefore be supplemented by Cripps' indirect version of the further-descendants-argument. Taking future descendants directly into account, young parents (together with all other presently living generations) are the last people to mitigate climate change to a manageable extent. Taking it seriously that the indirect relationship to future descendants matters to young parents today, young parents have a weakly collective duty to mitigate. In contrast to Cripps, I have defended this as a duty to biological and social parents alike.

5.3.2 *The Duty of Young Parents to Ensure Mitigation in the Long-run*

The previous chapter focused on mitigation as a weakly collective duty of young parents with regard to their children and future descendants. Present mitigation is very important and the decisive factor for limiting climate change to a manageable amount. However, in order to sustainably secure present mitigation, future generations will have to be part of the solution too. Only if generations from now onwards succeed in collectively living a less polluting, climate-friendly lifestyle, climate neutrality will be guaranteed in the long run. The previous chapter emphasized the central role of young parents when it comes to climate change: They are the last generation who is able to make a change towards 2 °C warming. This section continues to explore moral duties of young parents. It asks: In order to ensure mitigation in the long run, do young parents have a duty to raise their children to be responsible, climate-aware agents?

The previous chapter developed several arguments that will be taken as premises for this chapter. I therefore consider it to be helpful to recall them first:

- The relationship between parents and their children is unique from both the parents' and the children's side.[126]
- This special relationship justifies stringent parental duties.
- In view of rapid climate change, current young parents are a particular important group of agents.
- Young parents and their children have a central common feature: the (present and future) weakly collective duty as members of the Young.
- With regard to common membership, the relevant difference is loss of opportunity for effective action.

To specify what this chapter deals with, recall that children were highlighted to be unique for at least four characteristics: (1) vulnerability, (2) dependence, (3) lack of foresight and (4) their capability to develop into independent, less vulnerable, responsible future adult agents. The previous chapter focused on the first three characteristics to justify parental duties to mitigate. This chapter focuses on the fourth characteristic and on parental duties linked to children's capacity to grow into adults. And, connecting to climate change, the focus is on parental duties linked to children's becoming adult agents in a rapidly warming world.[127]

---

126 I continue to include social and biological parenthood.

127 This emphasis on parental duties is not meant to deny that institutions and governments in particular should also be concerned with children becoming adult agents. However, as I will emphasize in the following, parents' influence on their children is unique. Since it

Children's becoming future adult agents certainly requires many different things. As a first approach, one may consider Harry Brighouse and Adam Swift (2014). They distinguish "five categories of children's interest" (Brighouse & Swift, 2014, p. 63). These are physical, cognitive, emotional and moral development as well as the appropriate conditions to enjoy childhood (Brighouse & Swift, 2014, p. 64).[128] Accordingly, parental duties may be established regarding each interest. Enabling physical development by providing food, shelter, clothing, and so on is for sure the most important parental duty. An appropriate setting to enjoy childhood is certainly beneficial for adulthood too. According to Brighouse and Swift (2014), cognitive, emotional and moral development require "education and upbringing" with regard to different aspect such as "critical reflection" (cognitive development), the ability to regulate emotions and to build emotional bonds to others (emotional development) and the ability "to understand the basic demands of morality" and the ability to act in accord with these demands (moral development) (Brighouse & Swift, 2014, p. 64).

Turning to climate change, what do children require to become fully responsible, independent, less vulnerable adult agents in a warming world? They certainly require cognitive development to understand basic science of the climate system and the nature of climate change. In order to initiate change, children also require emotional development to identify themselves as part of a collectivity. But most of all, children require moral development. Understanding one's role and duties as individual moral agents and behaving accordingly is very important. As members of the Young, children will have a weakly collective duty. In order to fulfil this duty, children require moral preparation. Though a focus on moral development suggests itself when talking about moral duties, I do not exclude other kinds of development required to fully grasp the nature of climate change. Thus, in general terms, accompanying children in their becoming future adult agents in the age of climate change is all about climate education. Accordingly, I frame parental duties mainly as obligations connected to children's climate education. The general claim that I will develop below is that children require parental climate education for their adulthood. Just to be clear, I do not ask what children require as *children* in the

is almost impossible for institutions to make up for parents' failure, I consider parental duties with regard to children's becoming adult agents to be especially important.

128 Just as Cripps, Brighouse and Swift are strongly influenced by Nussbaum's capability approach (see Brighouse & Swift, 2014, pp. 59–62), which is also evident from the kind of categorization they suggest. Even though I have distanced myself from the capability approach with regard to the normative foundation, the five-fold categorization suggested by Brighouse and Swift helps to locate the kind of parental duty I am concerned with in this section.

process of becoming adult agents but rather what they need to be fully prepared to master adulthood in the age of a rapidly changing climate. To avoid further misunderstandings, I therefore define the duty of young parents to provide parental climate education as a duty to raise one's child to be a climate-aware and responsible agent.

If young parents have a duty to their children to provide climate education, the duty must be owed to them in virtue of parenthood and membership in the Young. To provide proof, recall that young parents and their children (will) have a weakly collective duty as members of the Young. Recall also that individual corresponding duties are generally assigned on two levels: promotion (in the absence of collective action) and doing one's part (in the presence of collective action) (see chapter 3.2.3). Currently the Young are in the phase of overcoming collective inaction, the primarily individual corresponding duty of members is therefore promotion. I have defined young parents as members of the Young. Thus, young parents' current primary duty is to promote collective action. As stressed before, by the time children have grown into adult agents, the window to mitigate climate change to anything close to 2 °C will have closed unless global emissions decrease fast and immediately. Thus, there will hardly be time left for promotion. Children of young parents will have to live up to what their parents promote: collective action. Correspondingly, their primary individual duty will be to do their part of the collective scheme. However, in order to be set up for effective collective action, children require preparation in terms of climate education. Failure to provide children with relevant awareness, knowledge and a sense of responsibility regarding climate change will leave them unable to fulfil their duties as members of the Young.

The importance of changing the dominant moral narrative of the climate discourse is the overarching plea of this book. The duty on the collective level as well as the duty to act promotionally instead of changing one's own behaviour in isolation is a fundamentally new thought to many. Moral development and with it the development of persons' fundamental moral convictions begins in childhood. Being prepared to live up to this new moral narrative is best achieved by being taught as a child and not just as an adolescent or adult. Changing moral convictions during adulthood is more difficult and probably even psychologically painful. On the contrary, being raised with a conviction of climate awareness and responsibility is easier. Suppose a child enters school and has never actually heard about climate change, her role as an individual agent or that individual duties actually exist. Suppose teachers as representatives of public education try to teach children about the importance of climate change and the imperative of mitigation. At best, the child will come home asking their parents about climate change. At worst, the child will simply deny what her teacher said. My point is that parents in general have a unique and life-sharpening influence on their children's education. And if parents fail to exert this influence in a way that sustainably secures mitigation action, it is hard to make up for parents' failure further in the process of children becoming adults. Current young parents are not just any generation of parents. They are the last to effectively initiate change, which is in their own but also in their children's interest. Assuming that children require climate education to fully meet their duties as adult agents, young parents have a duty to provide such education due to their unique relation to their children. Thus, young parents ought to raise their children as climate-aware and responsible adult agents.

Now consider a young parent who believes that climate education will put a needless burden on her child. After all, climate change is about massive potential loss as well as devastating human suffering and children ought to enjoy a jaunty, care- and anxiety-free childhood. Against this background, critics may oppose that young parents act sufficiently if they fulfil their promotional duty, which will leave the child with a collective setting in which her child only has to step in and fulfil her own duty. However, first of all, raising children as climate aware and responsible adult agents does not mean raising them in fear and anxiety. It should also not be confused with children having to experience a childhood of burdens and frustration. In a similar context, Cripps (2017b) highlights "age-sensitivity" and children's "emotional and cognitive readiness" as "crucial" (Cripps, 2017b, pp. 42, 43). I hope it is superfluous to mention that very young children should not be left with a nightmarish view of starvation, drowning people, extensive fires and climate-induced conflicts. Here, conveying the general nature of climate change or the feeling of commitment is more appropriate and also sufficient. Climate education is about raising awareness and empowering children to be prepared to meet their moral duties as adults. For children, climate change will probably be the greatest challenge of their lifetime and a constant threat to their environmental, social and political conditions as a generation. In view of such a life-shaping challenge, I doubt that children are sufficiently prepared by simply leaving them with a situation that may be in their best interest but of which they know nothing about.

To prove such a position wrong, Cripps has provided an additional reason. She explores whether parents (in general, not young parents) have a duty to raise children to be "good global climate citizens" (Cripps, 2017b, p. 41). Her key argument to support such a duty is the so-called "moral community/relationship argument" (Cripps, 2017b, p. 48).[129] Starting off with the assumptions that

129 In order to simplify reading, I will refer to Cripps' argument as the community-relationship-argument.

"parents have climate justice duties" and that parents and their children are connected by a "life-shaping relationship" she argues that parents have a duty to provide "a certain quality of relationship" but also "not to undermine the relationship in a central way, for example, by destroying trust" (Cripps, 2017b, p. 48). I consider the moral community-relationship-argument to basically consist of two parts: the first part locates the parent and the child as members of the same *moral community* across time. The second establishes the parent's acknowledgement to be an essential element for the *relationship* between the parent and her child. With regard to the first part (moral community), Cripps explains:

> In fulfilling her moral duties, the parent acknowledges her place in the community of moral agents. In fulfilling climate justice duties, she identifies herself as a member of a global and intergenerational moral community: those entitled to at least basic moral consideration and required to show it to one another, by virtue of common humanity. The parent also knows that she is in an asymmetrically influential relationship with her child. Given this, someone who brings her child up as a good global climate citizen – teaches him about climate change, discusses her own response with him, and enables and encourages him to respond – acknowledges that child as a future member of the same global moral community and prepares him to act as such. (Cripps, 2017b, p. 48)

Recognizing the child as a future fellow adult agent is a "necessary component of a truly human relationship" (Cripps, 2017b, p. 48). This leads Cripps to the second part of her argument (relationship):

> In failing to help her child to understand and enable him to pursue the goals that she pursues as a moral agent, the parent is failing fully to acknowledge and value him as both her child, on whose development she has a perhaps unequalled influence, and a future member of the same moral community. (Cripps, 2017b, pp. 48–49)

Simply put, Cripps' key claim is that parents cannot (a) fulfil their own individual duties, (b) recognize the importance of her relationship to her child and (c) recognize the importance of her child as a future adult member of the same community and (d) at the same time refuse to raise her as a "good global climate citizen" (Cripps, 2017b, pp. 41, 49, note 13). Based on that, Cripps draws the following comparison: If a parent recognizes her child as a future adult belonging to the same community, "parental failure to prepare him to respond morally to climate change is akin to not teaching him the language of the county in which they both live." (Cripps, 2017b, p. 49). Cripps adds that if a parent fails to raise her child as a climate aware person this might have "undermined the relationship in the child's eyes" (Cripps, 2017b, p. 49).

I understand Cripps' reasoning as an argument of consistency. If parents agree to do (a), (b) and (c) than, as a matter of consistency, they ought to do (d) too. As such, Cripps' argument may be helpful in the context of young parents: If young parents fulfil their own promotional duties, recognize that their children will have a weakly collective duty as members of the Young too and acknowledge their unique influence on their becoming adult agents, then, as a matter of consistency, they ought to provide sufficient climate education.

Against the background of weakly collective duties elaborated in this book, Cripps' terminology such as 'good global climate citizens' or 'global and intergenerational moral community' probably causes confusion. Since I do not consider reference to either justice or community as particularly helpful, I wish to make a few more comments regarding her terminology. Different passages suggest that Cripps' shows openness to other ways of justifying individual duties regarding climate change. In the first paragraph of her article on raising children to be climate citizens, she still refers to weakly collective duties as defended in her book from (2013). With reference to Caney (2014), she notes that individual moral duties in the absence of action on the collective level may be labelled as "climate justice duties" (Cripps, 2017b, p. 42). Opening up her own defence of the primacy of promotion, Cripps states that individual duties may be fulfilled "by promoting collective action, minimizing individual emissions or aiding victims directly" (Cripps, 2017b, p. 42). And with reference to scholars contributing to the debate on causal inefficiency, she mentions the prioritisation of the three to be open for discussion.[130] In the course of her analysis, she distances herself from accounts of intergenerational justice when it comes to parental duties (Cripps, 2017b, p. 43) but at the same time, she assumes that the parent and her child belong to the same community across time and space in virtue of common humanity (Cripps, 2017b, p. 48).

Perhaps, supplementation with regard to global justice is possible. Maybe a global cosmopolitan approach even increases intuitiveness – climate change is a challenge to humanity after all. Cripps may also aim to increase compatibility with leading climate ethicist or accessibility to a broader audience. Her change in terminology may also be due to limited space that does not allow for a full reproduction of her approach from (2013) in depth. Or, Cripps is simply increasingly sympathetic with global justice and cosmopolitanism. In contrast, this chapter is part of defending and advancing the idea of weakly collective duties. Having explored Cripps' approach on weakly collective duties and corresponding individual duties in detail, I have already briefly shown above how

130 Cripps names her own contribution from (2013) as well as Sinnott-Armstrong (2005), Broome (2012), Baatz (2014) and Garvey (2011).

Cripps' community-relationship-argument as an argument of consistency may be applied to young parents. The following will briefly demonstrate that it is perfectly possible to make Cripps' argument above without having to rely on either global justice or cosmopolitanism. Building on thoughts developed in the first sections of this chapter it is basically a defence of Cripps' earlier work on weakly collective duties and corresponding individual duties as well as an advancement of parental duties assigned to young parents.

Following Cripps' terminology for now, I suggest reading her key argument with four key premises leading to the conclusion (C).

(1) Parents have "climate justice duties", which they ought to fulfil as citizens of the global community in virtue of common humanity (Cripps, 2017b, p. 48).

(2) Parents acknowledge the "asymmetrically influential relationship" connecting them with their children (Cripps, 2017b, p. 48).

(3) As a necessary element of this relationship, parents acknowledge their children as future members or citizens of the same global community (Cripps, 2017b, p. 48).

(4) Acknowledging children as *their* children and fellow future citizens requires parents to help their children "to understand and enable" how "to pursue the goals that she pursues as a [future] moral agent" (Cripps, 2017b, p. 49).

(C) Parents have a duty to raise their children as "good global climate citizens" (Cripps, 2017b, p. 41).

Following Cripps (2013), I have made a strong case for weakly collective and corresponding individual duties throughout this book. For current young parents, I continue to assume the primacy of promotion. Yet, the argument applies to future young parents and their children too. Since their corresponding individual duties may consist in doing one's part (in the presence of collective action), the first premise, in rather general terms, may be put this way: (1*) Young parents, as members of the Young, have corresponding individual duties with regard to climate change themselves.[131]

The previous chapter as well as my thoughts in this section take the special relationship between parents and their children as the foundation for parental

131 Generally, this first premise could be justified on parents' membership in either one of the relevant weak collectivities. Based on either their pollution (collectivized no-harm principle, the Polluters), their ability (principle of beneficence, the Able) or their youth (self-interest, the Young). Because this chapter focuses on young parents as a subset of the Young, I limit this first and all further premises to membership in the Young.

duties. Only because of this unique relationship, do parents have stringent duties in virtue of parenthood at all. So, the second premise may simply be: (2*) Young parents acknowledge their unique parent-child-relationship.

Given that climate change is a threat to humanity, it initially seems reasonable to assume a duty of all human beings to fellow human beings. I have stressed elsewhere that humanity is insufficiently discriminative to establish membership and assign duties accordingly (chapter 2). With regard to climate change one aspect has particular importance: the potential ability to act. In order to initiate change, collectivities ought to be formed in a way that leaves them with the capability to act effectively. Humanity is barely able to act as humanity. One of the strongest points of Cripps' theory on individual duties is the suggestion of three specific collectivities, their weakly collective duties and the promotion of collective action. That way, action is not likely to come to nothing. Having weak collectivities in mind, reference to humanity as a global community is not only unnecessary but it is also not helpful. Young parents and their children are in fact connected by membership in the same collectivity: the Young. However, the relevant premise is not only that children will be members of the Young too. Following my assumptions drawn in the previous chapter, the relevant premise with regard to this membership is that children will have a weakly collective duty too. Thus, in my view, acknowledgement of the common weakly collective duty in virtue of membership in the Young is the third relevant premise: (3*) Young parents acknowledge their and their children's common weakly collective duty as members of the Young.[132]

With regard to preparation for adulthood (Cripps' fourth premise), I suggest defining three further premises: (4i*) Children will have corresponding individual duties as members of the Young too. (4ii*) Fulfilling these duties requires preparation by way of climate education. Regarding parents' unique influence on their children: (4iii*) Climate education is a necessary component of education owed to children in virtue of parenthood.

In support of Cripps' comment on undermining the parent-child-relation from the child's perspective, one could even add a fifth premise: (5*) Failure to provide essential parts of education is a way of undermining the unique relationship from the child's perspective.[133]

132 One may add here that children are members of the Young 'in progress'. For reasons of clarity, I forego further additions.

133 Because demandingness is a huge challenge here too one could add: Unless provision was too demanding for example because it causes unreasonable costs or exceeds the parent's own capacity.

Based on these modifications, it is reasonable to conclude more broadly that young parents have a duty to provide parental climate education. Or, as I prefer to put it: (C*) Young parents have a duty to raise their children to be climate aware and climate responsible adult agents.

Now, I would like to turn to the question of what this parental duty actually incorporates. Following a distinction proposed by Brighouse (2002), Cripps defines the duty to raise "good global climate citizens" to have two components: Facilitation and promotion (Cripps, 2017b, p. 41). Facilitation is limited to educating one's child with regard to climatic science, collective failure to act, individuals' role and duties as well as "not rendering him [the child] over-dependent on fossil fuels" (Cripps, 2017b, p. 41). Promotion goes beyond facilitation and includes being a constant and active source of motivation by involving one's child in one's own climate action (Cripps, 2017b, p. 41). Accordingly, Cripps presents her community-relationship-argument in two versions: The "milder version" includes a duty to facilitate while the "stronger version" requires promotion (Cripps, 2017b, p. 49). In earlier section of her article, Cripps writes that promotion is likely to be a side-effect of facilitation and that both will probably be inseparable in practice: "[...] in educating her child about the moral failures of climate change, the parent is likely to motivate him to respond" (Cripps, 2017b, p. 47). Cripps' community-relationship-argument, however, is supposed to defend the strong version that involves facilitation and promotion alike.

In my view, facilitation is inherently connected with Cripps' definition of promotion here. In the context of weakly collective duties and corresponding individual duties, I believe that the facilitation-promotion distinction is misleading. Therefore, it is probably more appropriate to refer to active encouragement or motivation. As mentioned above, Cripps compares failure to provide parental preparation for adulthood with refusing to teach that child the language of the country of residence. Picking up this comparison, one could say that facilitation without promotion is just like providing learning material for grammar and vocabulary but not actively motivating one's child to actually learn and speak the langue. As an adult agent the child could blame her parent for not being able to speak the langue of the country they both live in. The parent could unreasonably reply: 'You know, I never stood in your way. I actually provided you with any information you could have needed. This language is something very important to me, but I did not see the need to encourage you to speak it.' Active encouragement or motivation is not just a side-effect. With regard to climate change, it is an essential part of the preparation required for adulthood – an essential part of climate education.

In fact, I consider Cripps' new definition of promotion to be misleading for a further reason: Overcoming the phase of collective inaction by promotional activity was emphasized as the primary derivative duty of current young parents as members of the Young. If current mitigation attempts are to be sustainably secured in the long-run, the Young ought to be organized to act collectively by the time children of current young parents become members of the Young themselves. The key driver of collective mitigation is promotion. Ensuring mitigation in the long run by raising a new generation of the Young who are better prepared to act is a weakly collective duty of young parents – in their children's interest. Right now, in view of collective inaction, I therefore regard raising one's child to be prepared to do her part of future collective mitigation as a promotional duty itself.

The reason leading Cripps to assume that parental duties may not always involve both, facilitation *and* actively encouragement is the vulnerability to criticism regarding the child's autonomy (Cripps, 2017b, p. 47). Some scholars seem to perceive active parental motivation as a greatly exceeding authority that violates the child's autonomy. The critical question may be formulated this way: Why do parents have a right to actively shape their children's values, opinions, attitudes and basically all of their moral development during childhood with a special regard to climate change? Such a criticism connects to a heated philosophical debate on education and libertarianism. Due to limited space, I have no intention to engage into the discussion here in detail. Rather, I limit myself to a brief comment.

I consider raising a child to be a climate-aware and responsible agent as enabling climate education. Climate education is part of preparation for adulthood. And preparation in the form of education is something owed to children. It is surprising that something like climate education, which is in the child's own very fundamental interest attracts criticism with regard to the child's autonomy. In fact, talking about autonomy, I regard climate education as a precondition for autonomy.[134] Climate education is not meant to indoctrinate children with a fundamental, deep green ideology. Climate education is also not meant to make children the property of their parents. They will be separate persons. They will have their own lives and decisions and it is still a matter of their own judgement to refuse to meet their duties as adult agents even if

134 This point is inspired by Brighouse and Swift (2014). In contrast to an extreme liberal perspective claiming that parents ought to enable only the child's cognitive development, Brighouse and Swift, who both work on liberalism themselves, have stressed that children's development in areas other than the cognitive are further "preconditions of autonomy" (Brighouse & Swift, 2014, p. 15).

they morally ought to fulfil them. The basic point is that climate change is the biggest challenge of the 21st century. Facing this challenge, climate education is simply meant to prepare children for their life in a warming world. Failing to prepare children with respect to climate change will leave them vulnerable and unable to effectively respond to the single biggest life-shaping threat that they will experience. Instead of being the authors of their own lives, their lives will be determined by droughts, floods, water shortages, wild fires and severe human suffering across the planet.

With regard to raising one's child to be an autonomous agent, Cripps writes: "[P]arents must enable their children to think and reflect rationally, revise even their own considered views, and engage critically but constructive with those of others." (Cripps, 2017b, p. 52). Following Cripps, I have made a strong case for promotional duties over mimicking. However, defending the primacy of promotion does not mean keeping other peoples' opinion on the primacy of mimicking or direct duties a secret.[135] Likewise, defending the primacy of promotion as a parent does not mean forcing this position on one's child. Parents should put their own views up for discussion, they should encourage their children to critically reflect on promotion, they should provide access to scholars making strong cases for other accounts such as political, virtue-based or direct moral duties. Why not use these accounts to elaborate and reinforce one's own position and to enable one's child to find theirs? Critical discussion will enable children to differentiate themselves from others to become autonomous agents who are ready to act – for reasons of efficiency, ideally on the collective level.

Is a parental duty to raise one's child to be a climate-aware and responsible agent too demanding? I regard this duty as a form of promotion. As a matter of promotion, parents might fulfil their duty drawing on a wide range of promotional activities. Climate education might incorporate integrating children into one's own promotional duties such as joining a climate march, discussing what to call for when contacting a local politician, and so on. It may include encouraging discussions at home, conveying a feeling of commitment, promoting the general willingness to accept and even to support mitigation measures even if they make life less comfortable. Climate education might be as simple as teaching the impacts of climate change or watching documentaries about climate change. That given, it is reasonable to assume that climate education will hardly be too demanding for young parents.

135 Note that while promotion is important, I have also defended a secondary duty to mimic with regard to integrity myself.

Focusing on raising one's child in practice, critics may further object that it is too demanding as long as young parents do not acknowledge their own duties. However, first of all, that does not invalidate the argument above. Even if young parents do not acknowledge or refuse to fulfil their duties, they still have these duties assigned and they ought to acknowledge and fulfil them. Yet, when it comes to practice, inability may be a problem. If young parents are unable to recognize their own duties it is difficult to ask them to provide climate education as a further duty. However, in my view, the central matter is not acknowledgement or recognition. In view of widespread climate awareness (chapter 5.2.1) and the dominance of the current moral narrative, young parents can be expected to generally recognize some kind of individual climate duty. Assuming that parents are generally interested in preserving their children's generic right and doing what benefits them in the long-run, I believe the problem is rather lack of empowerment and effectiveness. There is no way to effectively fight climate change in isolation. Likewise, there is no way to protect one's own child in isolation throughout their lifetime. This leaves young parents with a feeling of doubt, denial and the tendency to primarily regard others to be in the position to do something. I dare to hope that, in practice, promotion and raising one's child to be a climate-aware and responsible adult agent are duties that leave young parents with a feeling of empowerment and optimism to finally be able to do something that will effectively protect their children.

CHAPTER 6

Promotional Duties

Next to weakly collective duties, the idea of promotional duties is the key concept of Cripps' theory. I have emphasised elsewhere that Cripps is very open minded about the nature of promotional action. Individuals carry the burden to initiate change in the first place. In this respect, my concern is that leaving individual agents with a duty to act promotionally without providing at least some profound and concrete idea of what promotional action entails will result in inaction in practice.

This chapter elaborates the idea of promotional duties in two directions: one is conceptual, the other regards the content of promotional action. The first two sections substantiate the concept of promotional duties. The first section suggests distinguishing between promotion through the social channel and promotion through the political channel (chapter 6.1). The second attempts to conceptualise the transformation from individual agents who find themselves in the position to have a promotional duty, towards effective collective action (chapter 6.2).

Regarding the content of promotional action, sections three to five examine what individual agents ought to claim when promoting collective action. Yet, I am afraid, readers who expect a list of instructions may be disappointed, which is why I chose to stress the limited scope of this chapter right at the beginning. The last three sections of this chapter are a basis for specifying promotional duties in more detail. In contrast to fully elaborating the nature of promotional action, my aim is to demonstrate different factettes and areas of promotional activity in order to accelerate collective action.

Section three focuses on promotional duties of the Young and argues that their primary duty is to target the social channel defined in the first section as well as to promote large-scale climate education. The latter is not limited to their position as young parents (chapter 6.3). With regard to promotional duties of the Able, I will focus on a particular group of agents: scientists. The central claim of the fourth section of this chapter is that the road to zero-emissions is far from clear and that scientists have a promotional duty to ambitiously figure out a way to achieve zero-emissions (chapter 6.4). The last section of this chapter considers the promotional duties of the Polluters. Acknowledging that members of the Polluters basically promote against their own lifestyle, my general claim is that they ought to call for a framework that allows for mainstreaming a green, sustainable lifestyle across their own collectivity (chapter 6.5).

 | DOI:10.30965/9783957437426_007

6.1 Two Channels of Promotional Action

Given that none of the three relevant weak collectivities is in fact able to act as a group yet, individual members ought to promote their cooperation and organization. What exactly ought individuals do in this regard? In response to this question, Cripps has provided a variety of examples. Chapter 4.3 classified them as promotional activities with regard to top-down or bottom-up change. Regarding top-down change, individual action aims to promote regulations such as laws or rules forced upon people. Possible examples of promotional action are to vote for committed political candidates, to write e-mails to leading politicians, sign petitions or organise summits and climate marches (Cripps, 2013, p. 143). Regarding, bottom-up change, individual action aims to promote social change. In contrast to the kind of virtuous citizen movement that Cripps, following Jamieson, has in mind, my claim was that effective climate action contains both top-down and bottom-up change.

In view of this, I suggest referring to two different channels of promotional action: Promotional action through the social channel and promotional action through the political channel. I assume that the idea of promotion through the political channel has become clear by now. It involves any action that drives forward political action on any level and aims for carbon regulations being put into place. In contrast, the idea of promotion through the social channel is primarily about raising social awareness for climate change in general and for common weakly collective duties in particular. Promotion through the social channel is any action that calls fellow individual members into action.

Fragnière believes that "in the absence of a just climate regime [...] the aim of promotional duties is not so much to influence fellow citizens, but to foster the creation of new regulations and, if necessary, new institutions more adapted to the challenge" (Fragnière, 2016, p. 808). I disagree for at least four reasons that have been evident in previous parts of this book already: First, I have stressed elsewhere that the weakly collective duty as well as the duty to act promotionally instead of changing one's own behaviour in isolation is a fundamentally new thought to many (chapter 5.3.2). Changing the dominant moral narrative without people supporting and spreading its very idea does not sound very promising to me. The new narrative of promotion requires promotion itself – promotion through the social channel. The second reason is connected to the first. Suppose that I, after submitting this book, wish to promote collective action. I start by writing e-mails and voting for committed politicians. Next, I organise a climate march in my hometown. Organising a march is sufficient to see the need for influencing fellow members of the collectivity. How could I organize a march without raising awareness and without addressing at least some others first? Political activism without social activism may be exposed to the same kind of criticism as mimicking: causal inefficiency. Nevertheless, in contrast to mimicking, promotion can lead to more efficient and more effective results than mimicking in isolation. The point is that promotion necessarily has a public component towards society. Third, predictability is one of the two necessary conditions for holding the Polluters responsible and assigning moral duties. In view of growing shares of climate scepticism, I claimed that promotion through the social channel ought to ensure that fulfilment of the predictability condition can legitimately be assumed in the long-run (chapter 5.2.1). Fourth, any regulation aiming for effective mitigation right now, in a stage without major technological means, will require the Polluters to bear restrictions with regard to their lifestyle as a collectivity. Even if restrictions do not demand too much of individual agents, they will be felt, and they will be hurtful. Without major backing from society, political regulations will hardly come into force.

The idea of promotion through the social channel is the following: Right now, it seems that strong political will fades with social acceptance. Suppose there is a highly ambitious politician who takes climate change seriously and aims to implement effective regulations that apply to all members of the Polluters collectively. With regard to mobility, she might call for imposing traffic bans in larger cities, increasing airport charges or taxes on jet fuel and petrol or she might claim to arrange large investments in rail transportation. Regarding electricity production, she might claim to implement expansions of the electricity grid through rather densely populated areas or she might try to push the energy transition to its limits by decreasing reliance on fossil fuels. I am afraid that, right now, a politician who tries to implement extensive regulation of the kind just described will have to fear not being re-elected. Promotional action through the social channel is supposed to trigger a kind of moral revolution that aims to a different status quo: Ideally, politicians who fail to implement extensive regulations would have to fear not to having an additional term.

Yet, I believe, with regard to the social channel there is another conceptual gap. Individual agents ought to promote collective action in order for their own collectivity to cooperate and to organise themselves. But how do members connect with each other? In practice, what exactly can I do to start targeting the social channel? The following chapter attempts to give an answer.

6.2 From Individualism to "Networked Individualism" to Collective Action

Given that there are probably millions of people connected through membership in either one of the weak collectives, how does a single individual transform her individual action into collective action? My impression is that Cripps is too quick when she leaves individual agents with a promotional duty to initiate change without providing further and clearer guidance. Assuming that targeting the social channel is essentially important, this chapter aims to conceptualize the transition from individuals to connected individuals to collective action. Thereby, it supplements the nature of the social channel.

Given that an individual agent attempts to promote collective action against climate change, the first step is, I assume, to connect to other likeminded people who are aware of climate change and who wish to make a change for the better. The idea that I wish to examine with regard to climate action is Lee Rainie's and Barry Wellman's concept of "networked individualism" from social science (Rainie & Wellmann, 2012, p. 12). In light of technological innovations, individuals are provided with completely new networking opportunities. Rainie and Wellman consider three kinds of "revolutions" that open up ways to apply network theories to different social settings: (1) the rise of social networking, (2) the advent of the internet and (3) permanent availability via mobile devices (Rainie & Wellmann, 2012, pp. 21–58, 59–80, 81–108). Rainie and Wellman write:

> [T]he Social Network, Internet, and Mobile Revolutions are coming together to shift people's social lives away from densely knit family, neighbourhood, and group relationships towards more far-flung, less tight, more diverse personal networks. [...] we can see the changing the realities of this new social operating system. (Rainie & Wellmann, 2012, p. 11)

The "new social operating system" transforms into the concept of "networked individualism" that Rainie and Wellmann examine (Rainie & Wellmann, 2012, p. 12). The key characteristic of networked individualism is that "people function more as connected individuals and less as embedded group members" (Rainie & Wellmann, 2012, p. 12). Though Rainie and Wellmann aim to critically show how everyday life in social settings such as family, relationships or work has been influenced by these innovations, I believe that the very idea of networked individualism can be fruitful for weak collectivities too. Applied to climate change, the main point is that people in the 21st century are able to connect, communicate and exchange information with completely unrelated persons via online social media or via mobile telecommunication. Even if their close personal network is not aware of climate change or unwilling to promote collective action, individuals can initiate change far beyond their own social community. By being the centre of action and motivation, individual agents are empowered to take action within their digital community.

As a second step, a network of likeminded people can initiate social change more broadly, which will help to form a collectivity that is truly able to push collective action against climate change. It may still not cover all members of the Polluters, all of the Able or all of the Young. However, if enough people support the idea of climate action while showing willingness to back it up with even uncomfortable, hurtful and coercive measures against climate change, those who persistently show resistance against fulfilling their moral duty will not have any other chance than joining collective action.

To demonstrate the idea and potential power of networked individualism with regard to social change, I wish to provide two practical examples. The first is Pulse of Europe. As an independent non-party movement, Pulse of Europe aimed for setting a sign for the European idea and against the rise of anti-European sentiments across European countries. In face of the upcoming elections in the Netherlands, France and Germany in 2017, which were crucial for the political direction and political unity of Europe, Pulse of Europe called for politicians as well as citizens to put stronger emphasis on the need for a united Europe. Initially, a German couple initiated the network through their own social media network. In order to show society's support for the European idea, they soon organized weekly demonstrations starting in Frankfurt am Main (Germany), in the end of 2016. In the following weeks, the network increased its scope, visibility and public recognition. As a result, the initiative grew across borders to other European countries. According to the initiative's own information, in summer 2017 assemblies were held in 131 cities across 21 European countries (Pulse of Europe, 2017). By using social media reach, Pulse of Europe mobilised at least parts of the silent majority of pro-European citizens.

Irrespective of the great deal of public recognition by political leaders and the media, rather critical comments came from social and political science.[136] Criticism mainly questioned the lasting effect as well as the lack of concrete proposals providing a solution. Sandra Eckert for example expressed doubts that Pulse of Europe could sustain its commitment after the national elections in 2017 (Andreae, 2017). Dieter Rucht claimed that the aims and purposes were

136 In contrast to this, positive political recognition came for example from the former co-chairman of the German Green Party, Özdemir, who appreciated the initiative's commitment as a good example for citizens' taking collective responsibility for the European future (German Parliament, 2017). Further, German federal president Steinmeier honored Pulse of Europe's commitment in a speech given after his swearing-in (Steinmeier, 2017). For positive media attention see e.g. Bittner et al. (2017), Connolly & Oltermann (2017) or McGarne (2017).

not specific enough leaving a great deal of the potential that political activism could have untouched. Further, he described Pulse of Europe as a "heterogeneous crowd of people" connected by sweeping statements and demands that anyone could agree with (Gubernator, 2017 own translation). In view of initial diversity with regard to personal interests, political opinions and social ideas, Rucht predicted that Pulse of Europe would scatter as soon as it was to propose concrete solutions (Gubernator, 2017; Röös, 2017). According to Pulse of Europe's web presentation, the movement understood itself rather as a platform for discussion and sharing ideas than as a think tank aiming to propose concrete solutions. Choosing the 'right' means for sustaining the strength and unity of Europe was supposed to remain a private decision made through elections (Pulse of Europe, 2017). In 2018 public attention decreased significantly, which suggests that Eckert and Rucht were correct. Yet, in view of upcoming European elections in May 2019, it remains to be seen if Pulse of Europe succeeds in mobilizing widespread pro-European support among society.

The second practical example is Pantsuit Nation. It probably is the most shining example for networked individualism of the past few years. A couple of weeks before the last presidential elections in the US in November 2016, Libby Chamberlain, the founder of Pantsuit Nation, created a closed, invite-only Facebook group, invited a few dozen friends and asked them to wear pantsuits to the polls on Election Day in order to demonstrate support for Hillary Clinton. According to Pantsuit Nation's web presentation, the group grew to 24,000 members within twenty-four hours and to more than 3 million by the end of the elections. Today, Pantsuit Nation still organizes itself as a closed Facebook group – yet, a group with more than 3.8 million people around the world (Pantsuit Nation, 2019). Financial Times journalist Simon Kuper describes Pantsuit Nation as "an informal national support group for those who oppose Trump" (Kuper, 2018). Indeed, the group started with the intention to provide space for personal stories of those supporting Clinton. Pantsuit Nation now understands itself as a group that advances "equitable and engaged democracy" by supporting underrepresented groups of society, using personal stories to show how positions of privilege shape oppression – privileges connected to being white, male or heterosexual (Pantsuit Nation, 2019). In October 2018, Pantsuit Nation was mobilising society in view of the midterm Congressional elections in November 2018. By that time, Kuper presumed that "[i]f Democrats take one or just possibly both houses of Congress, it will be partly thanks to Chamberlain" (Kuper, 2018). Given that Democrats gained control of the House, Pantsuit Nation at least helped to shift the balance of powers.

Pulse of Europe, and Pantsuit Nation even more so, demonstrate the potential of social media reach with regard to the social channel. The central characteristic of both movements is probably that they both have political opponents or at least politically opposed tendencies that are worth distancing from. Neither Pulse of Europe nor Pantsuit Nation had to come up with concrete political solutions, and that distinguishes both social movements from the kind of collective action of individual agents that I have in mind with regard to climate change. Mobilizing 'the silent' is a great achievement and it is the first step of promotional activity with regard to climate change too. Yet, I conceptualised promotion through the social channel as a preparation stage and substitute for promotional action through the political channel. In order to sustain social change with regard to climate change, social movements need to propose and call for concrete solutions. Assuming that climate action is a matter of acceptance and means, the first task is to increase acceptance. But the second is to identify the means. Readers will probably ask what regular individuals can do with regard to the means to fight climate change. Broadly, the means demonstrate ways of transformation to zero emissions. They clarify what collective action ought to look like or at least what it ought to aim for.

Reflecting on the nature of promotional action, Cripps writes: "[T]here is another stage in determining which promotional actions would be most effective. This stage requires us to answer a question not tackled here: that of what form of collective action is most promising." (Cripps, 2013, p. 165). The much more important question that I wish to add is: What kind of technical and economic means are most promising for collective action? To emphasise the point, consider an agent who promoted through the social channel and successfully mobilised society to raise its voice in order to push for climate action. Now, it comes to promotion through the political channel. The network of individuals attempts to organise a climate march, write emails and open letters, create and sign petitions, and so on. What exactly are they, as a networked collectivity, supposed to be calling for? Overly general demands such as 'fight climate change', 'secure our future' or 'reduce carbon emissions' suit for the social channel, but they will hardly have an impact on the political side. Assuming that politicians are generally willing to accelerate the global fight against climate change, promotion through the political channel will fade before it even started. Not only because local politicians themselves probably do not know the answer either but also because members of the networked community probably have very different ideas about successful mitigation. With regard to the political channel, the second side of promotional action is figuring out the means of collective action. Now, readers will probably claim I demand far too much of regular individual agents. My argument indeed applies only to a certain group of individual agents. Having conceptualised the Able as those who have the ability to do something, the Able incorporate a certain group of individual agents who, regardless of their financial means, can do something about climate change: scientists. Part of my analysis in the following chapters will be to draw a picture of possible facets and solutions to climate change, claiming scientists globally to have a promotional duty to figure out a way to zero emissions.

Nevertheless, when Cripps claims that elaborating the nature of promotional action first requires the determination of what form of collective action is most effective, she hints to a central point. Even though I have limited collective action to the kind of action that regular individual agents can do together, networked individuals can hardly perform controlled collective action. Chapter 4.3 introduced necessary associations as the rights-based equivalent to weak collectivities. Given that necessary associations arise in the same line of thought as state institutions being organised collectivities, I believe that, inspired by Gewirth, some kind of specification of the nature of collective action performed by individual agents is possible. My claim is that the collective action of regular individual agents requires some kind of structure or organisation. That is the third and final step, which completes the transition from individualism to networked individualism to collective action.

Perhaps, my point becomes clearer by distinguishing the kind of collective action I have in mind from those that I do not consider to be effective. In view of some kinds of promotional action such as organising marches and international summits, what may come into readers' minds is a crowd of people or even an angry mob that blames political leaders for political inaction, that denunciates individuals who refuse to participate in promotional action, and so on. If they remain peaceful, demonstrations, marches and summits are part of collective action. Yet, in order to use synergy effects between individuals as collectivities and existing institutions, collectivities require some kind of structure in a very broad sense. Returning to the two practical examples, I assume that Pantsuit Nation has been more successful than Pulse of Europe so far because they, in addition to the Facebook group with 3.7 million members, organise themselves in regional and city groups. They have some kind of structure that allows them to turn intensive discussions into proposals. Just to be clear, my claim is not that individual agents ought to organise themselves in order to become a collectivity in the strong sense that is able to adopt green collective intention. My point unfolds with regard to effectiveness. If individual agents intend to push action through the political channel, they require some kind of standardisation.

The last argument to substantiate my claim is with regard to structural changes. Following Cripps, I have stressed elsewhere that the key task of collective action is to initiate structural changes in order to decrease individual costs (chapter 3.1.4, 5.2.2, and 5.2.3). Given that I limited collective action to the action of regular individuals, how can individuals together initiate structural changes if they have no structure or organisation themselves at all? My point is simply that in order to create long-lasting action with effective social and political impacts, individuals may start by connecting to each other online, raising awareness, setting signs for more ambitious climate action and leaving room for heated discussions. Yet, in the long-run, action of individual agents as collectivities requires some kind of structure. In order to avoid confusion with existing institutional structures and strong collectivities, I have tried to avoid the following terminology up to this point, but what individuals require as a means of becoming an effective part of climate action is institutionalisation.

The following three chapters attempt to elaborate the nature of promotional duties in more detail. Generally, they respond to the question what members of the Young, the Able and the Polluters are supposed to be calling for content-wise when taking promotional action. I will take up on action through the social channel and action through the political channel.

6.3 The Young: Being an Inconvenient Friend and a Promoter of Climate Education

For members of the Young, I see two major tasks of promotional action: The first is linked to targeting the social channel, the second is linked to education.

Promotional Action through the Social Channel

The previous two chapters highlighted the importance of promotional action through the social channel in order to organize collectivities. The idea is to influence fellow citizens and to raise social awareness for the collective solution. Though action through the social channel is not limited to promotion via social media, the potential and extended reach of social media platforms such as Facebook, Instagram, YouTube, Twitter, and so on is a decisive factor. Given that a large majority of the Young is already well connected through the social media, I claim that the Young's primary promotional task is to promote climate action through the social channel.

Reflecting on personal evidence, I assume that there is little doubt that young people in particular use social media. Reliable data is hardly available. The following is simply supposed to add quantitative weight. According to a cross-country survey by GlobalWebIndex, a market company that provides data for consumer profiling for a number of countries worldwide, about 90% of users aged 16 to 24 and 25 to 34 have at least one social media account (Mander, 2016). Data for the US shows that 88% of those between 18 and 29 who generally use online services use Facebook, 59% of them use Instagram and abound 35% use Twitter, LinkedIn or Pinterest (Greenwood, Perrin, & Duggan, 2016). Depending on the source, the average mean number of 'friends' on the most dominant (in terms of engagement per day and month) platform Facebook ranges from 155 (Dunbar, 2016) to 338 (Smith, 2014). Users of Facebook under 29 tend to have larger networks: 27% of them have networks larger than 500 contacts (Smith, 2014). Sources for other social media platforms are either too diverse or very poor on quality. My point is simply that members of the Young are already well connected and all they need to do in order to initiate action in practice is to use their existing digital networks.

The idea that those who have an extended social reach, such as public figures, ought to use their voice to raise awareness for climate change is nothing new to climate ethics (see e.g. B. Johnson, 2011; Sandler, 2010). Likewise, the very idea of using social reach has long been part of business strategy: Influencer-marketing is a key pillar of communication policy – not only in the consumer goods industry (Kietzmann, Hermkens, McCarthy, & Silvestre, 2011; Kumar & Mirchandani, 2013). However, having an audience and being influential is no longer limited to celebrities, actors, athletes or politicians. Due to their close relation to target groups, businesses consider regular citizens to be particularly appealing as social media influencers (Hellenkemper, 2018; Lammers, 2018). So-called micro-influencer with rather small communities but very high engagement rates are increasingly assumed to be important to reach certain target groups (Kirchmeier, 2018; Lin, Bruning, & Swarna, 2018).[137] Regardless of the follower base, social media influencers are in fact shaping their followers' attitudes and perceptions with regard to products, brands and businesses (Freberg, Graham, McGaughey, & Freberg, 2011). There is hardly any reason why these insights should be limited to corporate action. Regular citizens of the Young who actually have a reach either with regard to their own social media network or even as social media influencers ought to use their voice to promote collective action against climate change. Given the number of webpages obtaining numerous tips for becoming a social media figure, anyone

137 Though there is no single definition of micro-influencer, their criterion of definition seems to be the relatively smaller number of followers (as compared to e.g. celebrities or public figures), which is, in most definitions, limited to below 100,000 people (Audunsson, 2017; Clifton, 2016; Kirchmeier, 2018; Wolfson, 2017).

seems to be able to become an influencer. Why not become an influencer who promotes collective action against climate change?

Consider the thematic focus of top influencers, there is a huge challenge: The world's most successful social media accounts focus on travel, fitness, gaming, fashion, beauty, food, pets, parenting or home and interior.[138] They seem to represent a desirable lifestyle – a lifestyle that is directly or indirectly linked to facets of the Polluters' lifestyle. There is considerable doubt that using social media influencers for the promotion of collective mitigation will be successful in practice. Yet, that does not change anything from the moral perspective. Maybe promoting mitigation is not what young people wish to be confronted with in the first place. But as soon as they realise that collective mitigation is the necessary means to protect their rights, this might change. In the meantime, those members of the Young who already detected that mitigation is in their own self-interest ought to address their social and digital network by raising awareness. Inspired by Al Gore's film *An Inconvenient truth*, members of the Young will probably have to accept being the inconvenient friend. One who raises awareness for climate change as a matter that is not as comfortable or appealing as travel, fashion, fitness or interior design.

Enabling Mitigation through Education

Chapter 5.3 examined the duty of young parents with regard to their descendants. I claimed that current young parents owe mitigation to their children and to future descendants alike. Moreover, I provided a defence for a duty that ensures future mitigation by raising a new generation of climate-aware and responsible agents. The latter duty aims to provide parental climate education. Because I considered raising one's child that way to be a promotional duty itself, the analysis in chapter 5.3 already provided one aspect of the promotional duty that young parents as a subgroup of the Young have. Yet, parental climate education is only one aspect of climate education in a rather general understanding.

The weakly collective duty of the Young is a duty of all members towards each member. In conceptual accordance with the parental duty to children, members of the Young (regardless of parenthood) have a duty to each future member of the Young too. In order to back up the parental duty with regard to climate education, the Young could call for appropriate framework conditions for enabling climate education. What I have in mind is the more institutional

138 To get a feeling for the numbers: Forbes Media LLC annually rates the world's top ten influencer in twelve categories. In 2017 their total reach ranged from 13 million people to 228 million people (O'Connor, 2017).

aspect of education. The Young could for example call for schools to focus on communicating the personal dimension of climate change in order to empower students to act against climate change. They could call for a particular focus on climate change with regard to educating teachers – young teachers in particular. Young teachers could raise awareness among teachers in general by addressing for example the middle-aged teachers. Those of the Young who are in the position to strengthen collaboration between universities and schools, could do so for example by participating in established projects such as Children's University or by organising school-related projects themselves. Members of the Young could engage into voluntary environmental work with regard to education or organise social events such as workshops, public talks, penal discussions, and so on themselves. In view of recent demonstrations of pupils and students who refused to attend school on Fridays in order to raise awareness and accelerate political action with regard to the Paris Agreement, known as Fridays for Future or Youth for Climate, members of the Young who are beyond schooling age could simply join national school strikes in order to encourage pupils' promotional effort.

Yet, climate education is not only about future or younger members of the Young. In order to raise awareness across their own collectivity, members of the Young could call for wide-spread public climate education and engage into putting climate education provided by public institutions into practice. International policy has long recognised the role of public environmental education. Action for Climate Empowerment (ACE) has been formalised under Article 6 of the UNFCCC, Article 10 (e) of the Kyoto Protocol as well as Article 12 of the Paris Agreement. ACE encourages national governments to foster programs aiming to ensure climate education, increase public awareness as well as capacity building with regard to climate change (Paas & Goodman, 2016, p. 2). The United Nations Educational, Scientific and Cultural Organization (UNESCO), as the leading body for promoting public climate education, aims to address stakeholders in the framework of a large-scale education concept referred to as Global Action Program on Education for Sustainable Development (GAP on ESC) (Paas & Goodman, 2016, p. 8). In order to support and sustain national strategies on education in the framework of GAP on ESC, the UNESCO provides online training materials for teachers on the implementation of climate education in the classroom, directly fosters student education and training programmes across the globe and supports initiatives that aim to raise general public awareness (Paas & Goodman, 2016, pp. 26–27). That given, I assume that even the most ambitious education programs installed at the international level will fail to have an impact if efforts are not sustainably backed up by society. In that regard, members of the Young could promote the existence of established programs, tools and efforts as well as their fellow members' participation. Climate education is another example demonstrating that the political channel and the social channel are inherently connected.

6.4 The Able: Figuring Out a Path to Zero-Emissions

In view of rapid climate change, the political direction is clear and aims for zero carbon emissions as soon, as fast, and as comprehensively as possible. With regard to decarbonisation, the remaining question is: What does the path to zero carbon emissions look like? In view of existing political will, I claimed that climate action is rather a matter of means and acceptance. While the previous chapters mainly examined acceptance in more detail, the following chapters focus on the means in order to answer the more general question – what individuals actually ought to call for when promoting collective action. Having conceptualised the Able as those who have the ability to do something, I claimed earlier that the Able incorporate a certain group of individual agents who, regardless of their financial means, can do something about climate change. This group is scientists globally. The key claim of this chapter is that scientists globally have a promotional duty to figure out the way to zero emissions. So, the focus of this chapter is rather on the nature of promotional action through the political channel than through the social channel. Since my expertise is neither in natural science nor engineering, I limit myself to elaborating possible facets and proposed solutions to decarbonisation. In the context of decarbonisation, the IPCC's AR5 states:

> The energy supply sector is the largest contributor to global GHG emissions (robust evidence, high agreement). Annual GHG emissions from the global energy supply sector grew more rapidly between 2000 and 2010 than in the previous decade; [...]. The main contributors to this trend are an increasing demand for energy services and a growing share of coal in the global fuel mix. (Edenhofer et al., 2013, p. 69)

Further, the IPCC concludes: "Decarbonizing (i.e., reducing the carbon intensity of) electric generation is a key component of cost-effective mitigation strategies in achieving low-stabilization levels" (Edenhofer et al., 2013, p. 69). I will therefore limit my analysis to proposed solutions to decarbonising global electricity supply.

Electricity demand has been growing for a long time, but at the latest since the 1970s when electricity found its way into almost all markets (IEA, 2017b,

p. 41). Yet, in view of new communication technologies, ever accelerating digitalisation, robots and artificial intelligence it is reasonable to expect energy demand to grow intensively in the coming years – after all, sending data is inherently connected to electricity. So, following Dieter Helm (2017), the first observation is: "The future is electric" (Helm, 2017, pp. 64–65). And, broadening the perspective from the industrialised countries to the development of less developed countries, the future is even more electric, with millions of people from India, Africa and South America waiting to grow out of poverty using the same technological means as the industrialised countries.[139]

For decades, peoples' electricity demand has been satisfied by burning a combination of fossil fuels. Helm writes: "Oil played a role where it was abundant and cheap, most notably in the Middle East, but globally the world of electricity relied overwhelmingly on coal. It still does. It makes up over 40% of electricity generation." (Helm, 2017, p. 66). In contrast to oil and coal, natural gas only relatively recently became a powerful option. Burning coal was and still is the backbone of the world's economy. The problem with coal is, it is extremely polluting, and it is even more polluting than any other fossil option. About 45% of global carbon emissions are from coal, followed by oil with 35% and gas with about 20% as the 'cleanest' of all three (IEA, 2017b, p. 54). Scientist have made huge efforts to develop new and in particular green alternatives. So, the good news is: Options exist. However, the persistent disagreement is about the question whether any of these options can compete with fossil energy forms.

Anthony Patt (2015) has provided quite a profound analysis on alternative strategies and technologies to move to zero-carbon emissions. Connecting to competitiveness, he distinguishes three parameters: (1) Cost, (2) reliability, and (3) potential capacity (Patt, 2015, p. 178). The three aspects are the advantages of fossil energy: Oil, coal and gas are cheap, reliable and able to meet energy demands. Accordingly, the biggest challenges for alternative electricity generation are:

(1) Is the alternative technology able to compete with costs of fossil electricity?
(2) Is it able to generate electricity when needed?
(3) Does it have the capacity to fully cover societies' energy demand and is it available on a large-scale?

---

139 The International Energy Agency (IEA) estimates that 1.1 billion people (14% of the world's population) have been without access to electricity. 95% of those without access to electricity live in sub-Saharan Africa and the least developed parts of Asia (e.g. India, Philippines, and Indonesia) (IEA, 2017a, pp. 40, 114–117).

In the face of climate change, public opinion on fossil energy is probably not enthusiastic but its acceptance is high. For some technologies it is precisely bad reputation, which prohibits the way into politics. I therefore suggest adding public opinion as a fourth parameter. Thus, the relevant fourth challenge is the following:

(4) Does the alternative technology have adequate public support?[140]

Findings of recent scientific analysis demonstrate great agreement: Today, none of the existing options are able to fully compete with fossil power. Regarding the predictions, the situation is different: Some scientists are clearly optimistic that advanced versions of renewables will guide the way to clean energy. Others, on the contrary, argue that current renewables are insufficient even if they are subject to further development.

Patt (2015) is an optimist. He stresses that "there is reason to believe that every one of the challenges we face with each [alternative] technology is one that we can overcome with some concerted strategic action" (Patt, 2015, p. 223). In his view, "[c]oncerted strategic action is synonymous with public policy" (Patt, 2015, p. 223). Alternative technologies will be able to compete with fossil fuels if governments successfully accompany the process by putting appropriate policies into force. According to Patt, there are "two separate elements of the technological pathway" that he explains as follows:

> The first element of the pathway is to get all of our electricity from renewable, carbon-neutral sources. The second element of that pathway is to switch as much of our energy consumption onto electricity as possible. Where electrification is not possible, we will need to use biofuels, and for this we need to consider the issue of agriculture and land use as well. (Patt, 2015, p. 244)

In contrast to Patt, Helm is rather pessimistic and regards gas as the only way to bridge the lack of truly competitive green technological innovations. He writes:

> Existing technologies have much to offer, but are nowhere near up the task. Gas is a *transitional* option that ought to be taken seriously. But beyond that, the gap between what needs to be done and what is on offer is just too great. The carbon problem will not be cracked with what is currently available. (Helm, 2015, p. 213)

The reason for such divergent assessment is with regard to parameter (1): Future cost competitiveness. A great deal of the analysis concerns predicting and comparing the development of alternative and fossil energy prices.

140 With regard to public support, a promotional action of the Able through the social channel could be to provide public education to increase reputation for suitable technological means. Yet, in this chapter I focus on promotional action through the political channel.

Over decades, scientists and politicians have built on peak-theories: The reasonable notion that fossil resources were limited, and that extraction potential was either close or even behind its peak. By applying market forces, limited supply would, so the assumption goes, lead fossil energy prices to continually rise. Assumed certainty about rising prices in view of technological advancements created the vision that cost competitiveness of alternative technologies was only a matter of time. Regarding oil, Helm refers to it as the "peak oil delusion" (Helm, 2017, p. 23). Instead of going up, Helm predicts that fossil energy prices will go down in the medium and long term. That is for two reasons: The first is that the use of new technologies will increase the accessibility of even more oil and gas. In particular with regard to shale gas technologies, Helm stresses that fracking was only the beginning (Helm, 2017, pp. 29–34). In combination with political measures such as carbon taxes or emission trading the extraction of new fossil resources gives rise to the second reason: Competing OPEC-countries, Iran and Iraq in particular, are likely to float the market (Helm, 2017, p. 34). Increased supply of fossil sources combined with a decreased demand caused by increased carbon prices means that economic prosperity in oil producing countries is no longer self-evident. Helm asks: "What do oil-producing countries do when the money runs out? The answer in the short run is: pump more." (Helm, 2017, p. 35). Obviously, Helm's predictions are in sharp contrast to peak theories and ever rising fossil energy prices. If Helm is correct, his assessment bodes badly for future cost competitiveness of alternative energy technologies but also for global warming altogether.

Next to uncertainties with regard to future cost competitiveness, each technological alternative has its individual challenges. Focussing on different renewables as well as nuclear power, the following will substantiate the claim that alternatives are insufficient to compete with fossil energy. I will also discuss whether shortcomings may be solved by power storage, super and smart grids. Most of the discussion below is limited to Europe.

Nuclear Power

Nuclear power plants have always had huge potential. Patt notes: "One thing that makes nuclear power really attractive, as a decarbonizing option, is simply the fact that nuclear power plants produce a ridiculously huge amount of electricity." (Patt, 2015, p. 182). And not only that, in producing this huge amount, they run carbon neutral.[141] In many countries, nuclear power generates a sub-

141 Life-cycle analysis shows that building nuclear power plants as well as production and processing uranium involves carbon emissions. A comprehensive analysis should consider these factors. But that applies to the production of renewable power plants too:

stantial amount of electricity generation. In France, for example, almost 72% of the electricity supply comes from nuclear power plants. Slovakia, Ukraine, Belgium and Hungary rely with roughly 50% on nuclear electricity and even Switzerland, Sweden or Finland, as countries that are generally known for rather taking the lead on renewable energy production, cover about one third with nuclear energy.[142] And still, nuclear electricity will probably not be the future. Nuclear power has a very bad public reputation and that is for at least four reasons: First, waste management for nuclear power plants is not at all a well-designed concept. Long-term storage prior to disposal, radioactive exposure to people and lack of disposal options are serious challenges. Second, nuclear accidents in Chernobyl and Fukushima affirm the public perception of danger that comes with running nuclear power plants.[143] Third, nuclear weapons and vulnerability to terrorism are further aspects that underline related dangers. Fourth, it is still unclear whether nuclear energy has or will ever have any cost advantage over fossil energy.[144] Nuclear power is perceived to be very dangerous compared to running renewables and even fossil power plants, it is expensive and produces a whole lot of waste that no one ultimately knows how to deal with. Given that nuclear power comes with a huge carbon-free electricity potential some countries such as China heavily rely on nuclear energy for the future. For other countries such as Germany, there is serious doubt that nuclear power will (ever again) find its way into either the market or politics any time soon.

---

quarrying rare earths for solar panels, cement production for hydropower plants, production of iron and steel as well as transportation of wind mills are only a few examples that demonstrate environmental costs of renewables. The analysis of this chapter is limited to emissions during the electricity production.

142 Statistics are taken from the database provided by International Atomic Energy Agency (IAEA) with 2017 data (IAEA, 2019).

143 Various authors have thoroughly investigated the central role of nuclear power for a green future and show that there is good reason for doubting that nuclear power really has a greater life-threatening risk than fossil power plants (coal in particular) (W. Allison, 2015; Fox, 2014; Lynas, 2013).

144 Various studies on historical (regional) nuclear cost suggest that there is no cost advantage (Cooper, 2014; Du & Parsons, 2009; Grubler, 2010; Hultman, Koomey, & Kammen, 2007). Yet, others trace historical cost disadvantages back to the cost of pioneering in nuclear energy (Lovering, Yip, & Nordhaus, 2016). It suggests that for countries, which are relatively new to nuclear power such as South Korea, nuclear electricity is cost competitive and may become cost competitive for countries like China or Russia. Helm (2017) stresses that safety standards in these countries may be low. Thus, cost competitiveness might unacceptably come at the cost of security (Helm, 2017, p. 51). For the purpose of this book, it is probably safe to say that there is no clear cost advantage.

The hope of some scientists is that future nuclear power might incorporate new technologies such as the widespread introduction of nuclear reactors that are able to handle core melting, usage of thorium (instead of plutonium or uranium) or small and modular nuclear reactors that might be mass-produced, less vulnerable to terrorism and are publicly perceived as less negative.[145] Helm stresses that with regard to the next generation of nuclear energy "horizons are very long" so that "[n]one [of the innovations] will make any serious impact until at least 2050" (Helm, 2017, p. 67). In addition, I doubt that small-scale reactors or different combustion material will receive any better public reputation. As long as waste management and security are not guaranteed to a minimum, perception of life-threatening danger is not going to change. But even if next-generation and small-scale reactors will turn out to be technologically, economically and socially feasible, the current electricity grid infrastructure in many countries will be unable to handle the continually rising potential of new nuclear power plants (Goldberg & Rosner, 2011, p. 21). With innovative waste management strategies, new safety and efficiency technologies and a thus enhanced change of public reputation, nuclear power might become a viable option for electricity generation. However, the gap between science on the one hand and market as well as political reality on the other will probably not make nuclear power an option any time soon.

Renewable Energy Sources

To date, renewable energy sources have a hard time counting as large-scale electricity sources. The numbers speak for themselves: Worldwide, renewable sources only make up a share of around 24.3% of the electricity generation only: 16.3% coming from hydro power (excluding electricity from pumped storage) and about 8% from non-hydro and waste (including e.g. geothermal, solar, wind, tidal, ocean and biofuels). An additional 10.4% of the global electricity production comes from nuclear power. The remaining two thirds (65.1%) are still produced by burning coal, gas and oil (IEA, 2018a, p. 14). The picture looks slightly better for the EU-28. Renewable electricity sources make up about 30.9% of the European electricity generation: Hydropower (excluding

145 For an analysis of the historic development of different nuclear power reactor generations see e.g. Goldberg & Rosner (2011). Goldberg and Rosner show that cost-effectiveness for small modular reactors as well as safety improvements for future generation IV reactors solve much of the nuclear barriers (Goldberg & Rosner, 2011, pp. 11–17). According to the IAEA, small modular reactors are currently under development or construction in a number of countries, among them Argentina, China, France and Russia (IAEA, 2018a, p. 9). Generation IV reactors are currently under development or construction and are supposed to achieve operational stage from 2030 onwards (IAEA, 2018b, p. 10).

electricity from pumped storage) (11.7%) and wind (9.3%) are the most important source for renewable electricity generation. Followed by all other (accumulated) non-hydro sources and waste (9.9%). Nuclear power has a share of 25.8% and 48.2% comes from fossil sources (IEA, 2018b). Thus, the two main findings are that, first, renewables contribute a slightly bigger share of electricity generation in the EU compared to worldwide and that, second, renewables hardly count as large-scale electricity sources.

Even though tidal energy is the most developed type of ocean energy, it is almost not worth the discussion.[146] Tides are certainly more predictable than wind or sun power, but the existing tidal power plants seem to be in a testing stage rather than being a future large-scale power option. Next to environmental impacts for example on the marine ecosystem (Frid et al., 2012; Sannino & Cavicchioli, 2013), tidal electricity production is very expensive due to its technological and supply chain barriers (Ernst & Young, 2010; Segura, Morales, Somolinos, & López, 2017), which makes it highly uncompetitive to fossil energy.[147]

Geothermal energy is more developed but available only in limited countries around the world. In electricity generation, geothermal produces less than 1% of the world's output (World Energy Council, 2016, p. 2). Geographically, generation capacity is located along tectonic plates and in particular along the Pacific Rim (the so-called Pacific Ring of Fire). In terms of heat and electricity generation from geothermal energy, Iceland is probably the most prominent example. But not only Iceland relies on geothermal electricity. In addition, New Zealand, China, Japan and Turkey cover parts of their electricity demand through relatively easily accessible geothermal energy, which unfortunately is not as easily accessible in other places. Right now, high installation cost, impacts of deep drill holes and complex extraction processes leave geothermal energy as a small-scale cost-effective option only.[148] Nevertheless, Earth's stored thermal energy potential is immense. Electrical potential is estimated to be sixteen times the currently installed generation capacity (World Energy Council, 2016, p. 3). Indonesia for instance is a high-potential country for geothermal electricity. Its general energy potential is estimated about 40% of the global geothermal energy potential. But only 4.5% is being used for electricity generation right now (Nasruddin et al., 2016). Maybe that is a case for globally

146 For an overview over types of ocean energy see e.g. the IPCC's Special Report Renewable Energy Sources and Climate Change Mitigation (Lewis et al., 2011).

147 For updated comprehensive reviews on ocean energy and tidal power in particular see e.g. Segura, Morales, Somolinos, & Lòpez (2017) or Uihlein & Magagna (2016).

148 For a detailed overview of the status of geothermal energy including geographic distribution and cost economics see World Energy Council (2016).

funded energy-oriented development projects. Together with technological progress, a green future might include a greater share of geothermal electricity, but the geographic location remains an issue.

Cost-ineffectiveness and limited availability hold also true for hydropower used in run-off-river or storage power plants. Power plants are extremely expensive to install. Economically, that only make sense for regions with extensive, reliable reservoirs and precipitation quantities, since energy prices are directly related to availability of water. Hydro power is the major source of renewable electricity and yet, appropriate conditions for hydro power are unevenly distributed (IEA, 2017b, p. 21). Half of the global hydroelectricity is generated in China, Canada, Brazil, the US and Russia. When it comes to covering electricity demand with hydro power, which is the decisive factor for being a reasonable alternative to fossil energy, Norway is the showcase example. Hydropower makes up almost 96% of the Norwegian national electricity generation. But also Brazil, Venezuela and Canada generate more than half of their national energy production through hydro power (IEA, 2017b, p. 21). Hydropower plants are in use where the descent of rivers, the size of existing reservoirs as well as the amount of precipitation is sufficient. Geographical and topographical differences as well as the intensity of droughts, will cause hydropower to remain a regional alternative for electricity generation. Yet, hydropower might gain a different status when it comes to electricity storage from wind or solar power, which will be discussed below.

Biomass can be either a heating or electricity source. However, biomass is widely criticised for competing with food production for cultivation areas leading to increased food and commodity prices (Ignaciuk, Vöhringer, Ruijs, & van Ierland, 2006). In order to provide a solution, some scientists examine ways of producing biomass using plants adapted to drier areas such as cacti or succulents (Mason et al., 2015). However, my impression is that the argument regarding the competition with food crops carries a lot of weight and cannot simply be brushed aside. Moreover, heavy use of pesticides and fertilizer may negatively impact the eco-system including groundwater. Further, scientists still do not agree on the emission footprint of biomass (Haberl et al., 2012; E. Johnson, 2009; Röder, Whittaker, & Thornley, 2015). Due to high moisture content, biomass requires an energy intensive preparation process of drying, grinding, and briquetting. Plus, the high content of alkaline and halogen elements leads to an increase of harmful substances and fine particulate emissions during the burning process (Vassilev, Vassileva, & Vassilev, 2015).[149] For

149 For three elaborate reviews on biomass and biofuels see Ioelovich (2015), Rosillo-Calle (2016) and Vassilev, Vassileva & Vassilev (2015).

transportation based on biofuels (such as bioethanol, biodiesel or biokerosene), the picture might be different, but for electricity production, biomass seems to fail at being a convincing large-scale energy source.

That leaves solar and wind power as possible options. The good news is that sun and wind energy are resources that, apart from the concentrating process, do not require much preparation and are essentially unlimited. The bad news is, however, that both solar and wind power come with major challenges concerning intermittences, which causes follow-up problems such as overloading the electricity grid and lack of reliability.

Sun energy generation relies on two forms: First, photovoltaic, which is electricity generation through moving electrons and second, concentrated solar power, which is energy generation by focusing sun rays and thus producing high temperatures (Patt, 2015, p. 193). With regard to concentrated solar power, heated water either directly serves to supply the heating system or water vapour is converted into kinetic energy by driving a generator to produce electricity. Naturally, the electricity production of both systems is directly connected to the amount of sun exposure. However, Patt clarifies a considerable difference:

> The amount of electricity that a PV [photovoltaic] panel generates, depends on the amount of total sunlight hitting it. This means that it generates no electricity at night, a fair amount when skies are cloudy, and a lot when skies are clear, thus providing a slight benefit to building it in places with generally sunny skies. (Patt, 2015, p. 194)

In contrast, energy production by concentrated solar power "depends on the total *direct* amount of sunlight hitting it, which means no clouds" (Patt, 2015, p. 194), or, more precisely, lots of electricity in sunny conditions, no electricity in cloudy or nighttime conditions.

The good thing about solar power is that public opinion, unlike opinion on nuclear power or other renewables seems to be rather positive (at least in Europe) (Bertsch, Hall, Weinhardt, & Fichtner, 2016; Ribeiro, Ferreira, Araújo, & Braga, 2014).

Cost effectiveness for photovoltaic energy, however, varies greatly across the planet due to variations in sunlight and differences in installing photovoltaic modules themselves (Patt, 2015, p. 194). For most photovoltaic systems, electricity is transmitted back into the grid connecting cost effectiveness closely to market prices. In most countries, market prices are determined by fossil electricity. Without state subsidies, photovoltaic power is not competitive to coal or gas – at least not right now. That may change if prices for fossil energy were to increase but, as I stressed above with reference to Helm (2017), its likelihood is rather low. In contrast to photovoltaic electricity, which is based on modular construction, concentrated solar power offers significant economies of scale (Patt, 2015, p. 194). With available space as well as dry and sunny conditions (*direct* sun exposure), deserts count as *the* place to produce large-scale solar energy from concentrated solar modules giving rise to projects like Desertec.[150] Patt describes Desertec as "the idea of producing solar power in the North African desert and transmitting part of that electricity to Europe" (Patt, 2015, p. 195). With decreasing costs of concentrated solar power panels, large-scale sun power could become a feasible option. For Desertec, however, things got more complicated with the Arab Spring that increased the risk of investments and the European debt crisis forcing Spain to stop their investments in building long distant electricity interconnections (Patt, 2015, p. 196).

The major problem with solar power lies somewhere else: Solar power is subject to intermittences. During summer, electricity supply peaks during midday while public electricity demand is very low by that time. During winter, demand peaks during the day due to heating, while electricity supply is rather limited due to the lack of sunlight.[151] At least during the night when there is no sunlight, electricity demand is low. As long as sun power is only a small share of the power system, it's volatility might be a serious but manageable problem for sensitive power grids. As soon as the share of solar power increases it is either not a competitive large-scale option or it requires reliable energy storage options or smart grids that I will be dealing with soon.

Helm (2017) points to a further challenge of current solar technology: Since material in solar panels is able to generate electricity only from 7% of the light spectrum, great solar potential is left unused. He calls for efficient, innovative materials and technologies that broaden the use of available sunlight and criticizes that subsidies for current solar power should rather be used in research and development. With technologies like solar films or use of graphene, ground-breaking innovations seem to be just around the corner, which is not true for all renewables, he argues. Helm introduces the idea of covering any surface with such a solar film as a kind of coating. Not only would that be indeed a large-scale solar power plant, Helm argues further that marginal costs would decrease to almost zero (Helm, 2017, p. 70). That would indeed increase cost-competitiveness and is therefore a tempting idea. A photovoltaic coating,

150 Installing solar projects in deserts has environmental impacts too (see e.g. Dessouky, 2013).

151 Patt (2015) states that this is no problem for hot and dry areas such as California with an extensive use of air conditioning during the day and intensive sunlight. Photovoltaic power might indeed be an option in these areas. However, sunny areas that are wealthy enough to install air conditioning *and* photovoltaic electricity are rather limited (Patt, 2015, pp. 196, 200–201).

however, would still require multiple electric connections and I find it hard to imagine that any surface with solar coating is equally equipped with electric connections. But that is subject to nanoscience and thus, research problems.

For the future, there might be reason to hold on to large scale sun power if costs further decrease. But also, further research and development is required. And especially for overseas power plants such as Desertec, the right international policy interventions need to be taken for example to financially support Spain to develop a transmitting grid with France and Morocco, to truly open up and integrate the European energy grit.

Focusing on wind energy, cost effectiveness has increased a lot. Patt (2015) provides three arguments for decreased cost of wind power: First, engineers were able to develop bigger windmills reaching high into the sky where wind is stronger. Second, the windmill's material resistance increased over the years, making them less costly to maintain. Third, most parts of windmills are now subject to mass production leading to economies of scale. Decreasing costs of wind power lead to prices that are almost compatible with fossil energy (Patt, 2015, pp. 190–191).

One problem with wind power is its low public acceptance. In many European countries but also in Brazil or the US, people generally favour renewable energy but do not like to have one close by (Patt, 2015, p. 193) – the so-called "not in my backyard phenomenon" (see e.g. Buffoli, Odone, & Signorelli, 2016). To keep them out of peoples' sight, an alternative to on-shore windmills are off-shore wind power plants. The problem with off-shore is that cost of installation and maintenance are much higher than for on-shore windmills, which makes wind power cost ineffective in comparison to fossil energy. For the future, innovations that might let prices for off-shore power plants drop, might turn out to be at least a cost-competitive solution for wind energy. In this regard, floating off-shore would be an option (see e.g. Cruz & Atcheson, 2016).

The remaining major problem with wind power is that it is – even more than sun power – highly intermittent. Periodic occurrence causes extreme volatilities in terms of electricity production pushing electricity grids to their limits while, during calm days, they lack the capacity to cover public energy demand. For large-scale energy supply, current wind power infrastructure does not count as a reliable source. In autumn 2017, Northern Germany – where wind power is fairly widespread – was extremely impacted through storm 'Herwart'. Herwart caused much wind, generated even more electricity, flooded the energy market and forced the market price down to minus 83 Euros per megawatt hour. Usually, prices are at about (plus) 37 Euros (Sommerfelt & Zschäpitz, 2017). For physical goods, over-production would simply drop prices down to zero forcing suppliers to dispose unused products. For electricity, however, there are no disposal options. Over-production necessarily has to find a buyer – even for negative prices.

Negative electricity prices are not only a German phenomenon. They occur in any country with high-level wind power coverage where electricity demand is either way smaller (e.g. due to a small population such as in Denmark) and thus unable to outbalance wind electricity supply or where regional electricity grids are not sufficiently integrated so that electricity demand is technically unable to meet electricity supply. What I said about sun power, holds true for wind power as well: As long as wind power is only a small share of the power system, it's volatility might be a serious but manageable problem for sensitive power grids. As soon as the share of wind power increases, it loses its competitiveness as a large-scale option or it requires long-distant interconnections, smart grids or reliable energy storage. That is what I discuss next.

Long-Distant Grids, Smart Grids and Electricity Storage

Electricity production close to the grid's capacity limit is a huge problem for comparatively small countries like Denmark or Germany with high-level wind electricity in particular. For Denmark, it is the vast number of on- and off-shore wind parks along the coastline and in the North Sea, which enables Denmark to get one third of their electricity from wind energy. Due to its geographic location, Denmark is usually a transit country for electricity trade across Europe – mainly between Germany, Norway and Sweden. But even interconnections through the North-Sea have large potential for development. Also, with regard to other large-scale renewable electricity projects such as Desertec, transmission lines between Morocco and Spain as well as Spain and France would be needed. Most of these European long distant interconnections, however, are still in the planning or development stage. Expansion of power grids is a very expensive matter and cost-intensity is certainly one of the main challenges.[152]

152 The idea of a European grid can actually be extended: Building super grids is one approach to dealing with intermittencies of sun or wind power. The idea is to balance intermittencies by connecting electricity grids over distances large enough to balance wind differences and even availability of sunlight. A study for wind energy suggests the idea of a super grid extending from Iceland to Morocco over to Russia (Czisch, 2005). Findings show that a super grid designed that way not only has the capacity to fully cover Europe's energy demand through wind power but is able to do so at a reasonable cost, which is competitive to the cost of fossil energy. In the framework of a super grid, a further option could be to connect wind and solar energy with hydropower in terms of power-storage plants for example in the North of Norway or in Switzerland.

What is true for long distant interconnections is even more true for national short distance interconnections. Germany's wind power production for example is concentrated in the North. For storms like Herwart, reliable electricity grids would have (at least partly) balanced over-production from wind power plants in the North and energy demands in the South. The problem is, these high-voltage transmission lines through Germany do not yet exist. Even nationally, grid integration is not advancing sufficiently enough. German national interconnections are planned to go through densely populated areas, which gives rise to the not-in-my-backyard phenomenon again. Underground cables would probably count as an option – an even more costly one that comes with huge ecologic impacts.

In the setting of underdeveloped national electricity grids, over-production even forces relatively large countries like Germany to export electricity to their neighbourhood. Neighbour countries either use the provided renewable power and reduce power production of their own conventional power stations (coal, gas or nuclear power) or store over-production in pumped-storage power plants – both ways are free of charge. The tragic moment is that for calm days with high power demand, electricity is imported. Electricity that is either produced by using pump-storage plants (backwards basically) or by additional use of conventional fossil power– and neither way is free of charge – while the latter is not even free of carbon. But leaving the tragic moment aside, extreme variations count as stress tests for the existing European power grid. It is no wonder that grid operators in European countries with less or no wind energy have taken action: Electricity produced in the north of Germany has caused the Czech Republic and Poland to install safety transformers to protect their national grid from high-altitude variations (see e.g. Kortschak, 2015; Wetzel, 2012).

With electricity overproduction, blackouts are no longer apocalyptic scenarios. If the strong expansion of renewable energy (solar and wind power in particular) goes hand in hand with strong depletion of fossil and nuclear energy, electricity supply becomes more and more volatile and thus unstable and unreliable. The number of regulatory interventions, the so-called redispatch, in countries with a high share in wind and solar energy such as Germany, has been growing rapidly. In 2015 and 2016 for example, frequency of intervention was eight to ten times higher than in 2010. Blackout-scenarios force nations to adopt safety measures and precaution. The recovery plan includes shutting down large-scale consumers such as industrial areas and re-starting independent fossil power plants that had officially been taken of the grid but remain in stand-by due to their systematic importance (Bundesnetzagentur, 2017, p. 33).

As long as there is no long-distant grid that reliably balances electricity variations, there might be different options: Building smart grids is another way of dealing with intermittencies (Patt, 2015, pp. 201–202). If the power grid would not only transmit electricity but also information about the current market price, technical systems and machines could respond to electric volatility. They could turn themselves off when prices are high to decrease overall electricity demand and could turn themselves on when prices are low. I consider smart grids to be an attractive option rather for private households than for industry. With industrial production being attached to delivery dates, the production process cannot simply be switched off for a few hours. In terms of that, private washing machines, refrigerators, air conditions or even electric cars are different. Smart grids are mainly subject to research and development. The European Commission (EC) and also several national ministries promote (very few) isolated pilot projects and fund research projects on legal limitations on data transfer, technical optimization for grid communication with demand and supply control in particular.[153]

Intermittencies suggest providing some source that occasionally takes electricity out of the power system when energy supply peaks and leads electricity back into the system when energy supply drops significantly below energy demand. Basically, sufficient regional power storage systems would reduce the need for a reliable super grid. Pumped-storage hydropower and batteries are two promising options with relatively high degree of efficiency.[154]

Hydropower storage already works well between France and Switzerland – sadly not for renewable energy but for nuclear power. During night-time Switzerland buys excess-supply from France, operates their own hydro power units by pumping water up the mountains and when electricity demand peaks in France, water flows back down, operating a turbine to feed stored nuclear power back into the electricity grid. Even though pump storage is very expensive to install and also comes with an electricity loss of about 20%, hydropower is still one of the best electricity storage options that is currently available (Rehman, Al-Hadhrami, & Alam, 2015). But as I said above, potential for hydropower is located in limited areas only. That holds even more true for hydropower storage, which necessarily requires sufficient difference in height to reach its maximum storage capacity. Switzerland, due

153 One prominent example is the European research project Energy to Smart Grids (EC, 2016). Several pilot projects, exploring the use of smart grids in order to balance feed-in capacity and network congestions, have been implemented in Salzburg, Austria, e.g. in connection to e-mobility (Smartgrids Salzburg, 2019).

154 For an overview of different electricity storage options see e.g. Palizban &Kauhaniemi (2016).

to its geodetic altitude, has huge potential for pump-storage. Germany with way fewer mountains or even a less mountainous Denmark will not be able to sufficiently cover their over-production by using hydropower storage. Norway in comparison, would indeed have sufficient potential for electricity storage. The hydropower-based system in Norway (in connection to wind power in particular) has received lots of attention as potentially being Europe's battery (Gullberg, Ohlhorst, & Schreurs, 2014; Lindberg, 2012; Pérez-Diaz, Chazarra, Garcia-González, Cavazzini, & Stoppato, 2015). For instance, Germany and Denmark with high level wind power must have a particular interest here. The idea of a European green battery in Norway still requires a grid system that enables transmitting electricity across Europe at least from wind-intensive parts in Germany or Denmark to the North of Norway. Connecting the North of Europe, however, does still not transmit green electricity to the South of Europe. Thus, hydropower as a storage option will only work in the framework of a developed European electricity grid. The message, I think, is that pumped hydropower electricity storage can help the issue of electricity reliability, but it will not be able to help the need for a reliable electricity grid across Europe.

Batteries are a further and a rather obvious storage technology. Batteries have – with 90% and more efficiency – even higher degrees of efficiency compared to hydropower storage (Patt, 2015, p. 207). Costly manufacturing processes, limited life-spans, high self-discharge and limited storage capacity in particular leave conventional batteries rather uncompetitive. Next to well-developed lithium-ion batteries, which have been used in electronics for decades, there are further types of batteries using different types of ions.[155] The majority has only been used in small-scale storage but creates high potential for integrating electric transportation in the framework of "vehicle-to-grid power" even today (see e.g. Kampton & Tomic, 2005).

When it comes to directly dealing with intermittencies from renewable power plants, conventional batteries might offer sufficient and maybe even economically feasible storage capacity for private households using photovoltaic electricity. Large-scale application in direct connection to power plants still fails to be cost-competitive (Helm, 2017, p. 72; Patt, 2015, p. 207). Nevertheless, conventional batteries are subject to various pilot projects connected to grid stabilization around the world. The world's largest battery storage project in terms of capacity was announced in early 2017 and will be built in South Australia. Its storage capacity is huge compared to regular battery storage. Compared to the world's average pumped-hydro storage power plant capacity,

155 For an overview on battery types see e.g. Ferreira, Garde, Fulli, Kling, & Lopes (2013) or Poullikkas (2013).

however, it is still only a tiny fraction of about 6 to 10%.[156] Nevertheless, these pilot projects raise hope for powerful and feasible electricity storage options in the future. The lines of development so far go to either improving the potential of conventional batteries by using different mixtures of substances or developing a completely different type of battery such as fuel cells. I think the question is not whether but when conventional battery storage will become a competitive option, but until then, large-scale storage is subject to research and development.

More broadly, electricity storage options such as hydropower storage and batteries are part of a pathway for balancing electricity that is referred to as power-to-X. The X refers to various uses or storage options. Examples are power-to-mobility, power-to-heat, power-to-fuels, or power-to-gas (Sternberg & Bardow, 2015). Strictly, hydropower storage is a power-to-power technology and using electric mobility to balance intermittencies is power-to-mobility. I wish to broaden the perspective here a bit more. The basic idea is twofold. First, power-to-X balances exceeding (renewable) electricity generation (e.g. by wind power plants during storms) by transforming electricity into other forms of energy or by storing electricity. Second, power-to-X refers to the connection of the electricity sector with other sectors such as transport or chemical industry. Both ideas are interlinked. Consider power-to-gas as an example. Using electrolysis processes, renewable electricity (or electricity in general) is split into hydrogen and oxygen, the components of water. Following, there are three basic options: First, hydrogen is used in transport (e.g. for fuel cell vehicles). Second, hydrogen is used in industries such as the chemical sector. Or third, hydrogen is directly fed into the regular grid for natural gas as a storage option (Götz et al., 2016).[157] The challenges of power-to-gas are manifold and considering other forms of power-to-x, there are numerous challenges to overcome. I limit myself to two challenges with regard to power-to-gas. The first concerns the transformation process of one energy form to another. Any transformation is subject to energy loss (Sternberg & Bardow, 2015). Given that the alternative option is to sell electricity to neighbour countries for low and even negative prices and given that technological innovations further improve efficiencies,

156 The world's largest hydro power plants range between 2000 and 3000 MW. Around the world they range between 1000 and 1500 MW (Rehman, Al-Hadhrami, & Alam, 2015). The battery storage plant in South Australia offers a capacity of 102 MW (Zhang, Wei, Cao, & Lin, 2018). For the estimation of 6 to 10%, I took average capacities of 1000 to 1500 MW as reference values.

157 Power-to-gas is an extremely rich scheme. I am sure that natural scientists and engineers will be able to specify many more options. For the purpose of this chapter, it probably suffices to limit power-to-X to a very rough idea of the scope such a pathway might have.

that is probably an acceptable drawback. The second challenge is directly concerned with power-to-gas as an option for electricity storage. The further up and the further left an element is in the periodic table, the less energy dense an element is. Hydrogen as an element is the first element on the upper left side of the periodic table. That given, hydrogen has a "low volumetric energy density" and storing hydrogen requires massive storage volumes (Sternberg & Bardow, 2015, p. 392). In order to ensure the contractually guaranteed energy density of gas to energy suppliers, the storage capacity of the natural gas system is limited. In Germany, for example, the limit is at 5 to 12 vol% (Sternberg & Bardow, 2015, p. 392).[158] Current research is therefore investigating further usage of hydrogen such as producing methane, methanol or syngas as gases with higher energy density, which could then be used in industry or transport (Foit, Vinke, de Haart, & Eichel, 2017; Sternberg & Bardow, 2015). Power-to-X technologies are particularly important in countries that greatly rely on renewable energy while attempting to fulfil ambitious mitigation targets. Given that Germany opted out of nuclear power a few years ago, it heavily relies on renewables and aims to take a global lead in climate mitigation, investing in power-to-X technologies is particularly important for the German electricity market. Thus, Germany, if successfully backed-up by further large-scale research, could become a showcase country for transiting to a zero-emissions economy and society.

Great potential might unfold if smart grids combined electricity supply and feed-in, electricity transformation, storage options and sector coupling either regionally or in the framework of a transnational, high-voltage and high-volume grid. If large-scale storage worked regionally, there might not even be the need for a transnational grid anymore. But again, that is subject to further research and I will leave the final assessment to those with expertise in energy science and engineering.

In the beginning of this chapter, I cited Patt (2015) who argues that strategic policy interventions will be able to move all electricity generation to renewables and then transform as much energy demand to electricity. I am afraid, things are not that easy. Sufficient, reliable and also cost competitive technologies do not seem to be around the corner. As long as alternative technologies are insufficient, increasing their share in generating electricity is rather a burden than a potential. At this point I have to say a few words about sectors other than electricity generation. Against the background just described, the claim

158 Percentage by volume (vol%) is a unit to express the composition of a solution. In the case above, a solution of 100 volumetric units consists of 5 to 12 volumetric units of hydrogen mixed with 88 to 95 volumetric units of natural gas.

to move industries such as transportation and travel to electric mobility must be considered in a different light. Increasing electricity demand right now will either destabilize the electricity system even more if demand is met with renewables, or it will increase the amount of fossil electricity generation in order to meet increased demand and to ensure reliability. Increasing fossil electricity generation bodes badly for climate change.

At the same time as I regard things to be not as easy as Patt outlines, I hope that Helm is too quick when he suggests using gas as the interim technology until completely reliable and potential technologies are found. That is for at least three reasons: The first is that extraction of shale gas releases significant quantities of methane – a much more climate damaging greenhouse gas than carbon. Increased extraction of shale gas will result in sharp increase of atmospheric methane concentration, which will amplify global warming several times over.[159] The third reason is that shifting to gas establishes new, persistent interdependencies that make a truly green future not very likely anytime soon. Fossil energy sources and oil in particular have persistent and great importance, causing heavy dependencies on the consumer and producer side. Therefore, turning our backs on fossil sources will have tragic impacts. Helm (2017) devotes almost two thirds of his book on these geopolitical and economic consequences. They become particularly evident for oil producing countries in the Middle East. The big players such as Saudi Arabia, Iraq and Iran but also the UAE and Kuwait have built their entire economic growth and social welfare on producing and exporting oil. Helm shows that through these heavy dependencies on the producer side, shifting away from oil will take away the basis for the livelihood of their citizens, will threaten the existence of their autocratic regimes and comes with the potential of giving rise to (even more) armed conflicts, terrorism and migration (Helm, 2017, pp. 107–125). New technologies certainly produce new winners. Turning to shale gas for example, will give the US a dominant position in electricity generation. Yet, the new winners of shale gas production are unlikely to be the biggest losers of increased fossil energy prices. Helm explains:

> Shale [technologies], […], solar films being widely applied, electric cars, new batteries solving the electricity storage problem, information technologies activating the demand side, new material like graphene, […] – these are the things OPEC, Middle Eastern monarchs and autocrats, Putin and oil executives should worry about. These developments could make the current falls in oil prices look like a picnic in comparison. They are what our energy future may be all about. (Helm, 2017, p. 9).

159 For analysis of environmental impacts of fracking see e.g. Jackson et al. (2014), Small et al. (2014) or Sovacool (2014).

Helm believes that shifting key energy generation from oil, coal and gas to mainly gas as the 'cleanest' fossil source will decrease carbon emissions and will leave more time for the development of reliable new technologies (Helm, 2015, 2017, pp. 53–61). But that is only one side, the other is that shifting to gas may also have the potential to weaken the negative impacts for those with heavy dependencies on oil if right measures and policies are taken (Helm, 2017, pp. 10, 194–203). And he is probably right. However, shifting electricity generation to gas does not mean slightly changing the system. Gas will cause similarly heavy dependencies as oil does. Replacing the status of oil with the importance of gas will not make the geopolitical, social or economic impacts any better. On the contrary, gas technologies will cause new dependencies. These technology shifts may best be compared with paradigm shifts, I think. It is not an easy switch – it is a fundamental transformation. The point is, shifting from oil to gas to renewables is in fact two fundamental transformations. I am sceptical that once producers and consumers adapt to gas, they will adapt to renewables any time soon. Considering that there are less then 16 years left to bring emissions down to zero (if emission levels remain unchanged), I am doubtful that these two fundamental changes will be carried out in time.

This rather brief analysis leaves a huge amount of aspects untouched: In any case, there are many more aspects with regard to alternative technologies. I am sure that there are plenty of strong arguments and concerns (such as country-specific aspects or environmental concerns) that I did not keep in mind. Also, I have limited the analysis to electricity generation only. Even though for good reasons (number one carbon source, net-growth), the road to zero emissions requires attention paid to carbon sources other than electricity generation. For example, I have left sectors such as industry or building and heating without a comment and said almost nothing about transportation.

Also, controversially discussed technologies of geoengineering, which either aim to prevent carbon from getting into the system or to take carbon out of the climate system, such as carbon capture and storage (CCS), carbon capture and utilization (CCU) and Carbon Dioxide Removal (CDR), or technologies that aim to intervene with the planet's radiative energy budget, such as planes distributing certain substances in the stratosphere in order to protect the planet from further solar radiation or even space mirrors, should be taken into account.[160] Yet, geoengineering comes with a whole lot of risks and impacts.[161]

160 Lawrence et al. (2018) provide an elaborate overview of geoengineering options in view of mitigation targets agreed upon under the Paris Agreement.

161 Gardiner (2017) provides an overview of several relevant ethical questions to geoengineering. For an earlier contribution see Gardiner (2010). Steigleder (2016c) presents a comprehensive risk-ethical evaluation of CCS.

In the face of these risks and impacts, geoengineering is not an option that spares mitigation and that allows to continue with fossil-reliant business as usual. However, regarding the urgency of climate action and continuous levels of carbon emissions, I suggest at least considering them as possibly complementary options.

In order to keep the discussion fairly short, I have limited the analysis primarily to a developed-countries context. In the context of electric transportation, Patt (2015) stresses that driving an electric car in Germany and driving an electric car in Kenya are two completely different things and that, for Kenya, it will take a long time until anyone will want to drive an electric car instead of a diesel engine (Patt, 2015, p. 222). This rather obvious example demonstrates the weakness that is true for almost all discussions about decarbonisation. Discussing the potentials of renewables in Germany or Denmark is a crucial thing, but the real challenge rises when populous countries like India, Pakistan, Indonesia, Nigeria or Ethiopia begin or continue to break out of poverty. Incorporating the fight against global poverty in fighting climate change is a matter of climate justice. As Shue (2014) argues, fighting climate change has two complementary solutions: the first is to keep the economic development of the developing countries as clean as possible and the second is to reduce the emissions of the developed countries by an amount greater than the amount of emissions produced by poor nations (Shue, 2014, p. 51). Fighting climate change and fighting poverty are two sides of the same medal. Successfully addressing both sides is certainly not only in the interest of the global developing nations, it is in the interest of the developed countries as well: The smaller the amount of additional emissions that are necessary for the poor countries to rise out of poverty, the smaller the necessary reduction in emissions for the industrialized countries (Shue, 2014, p. 51). An extensive analysis on the road to zero emissions will have to broaden the perspective to a developing-countries context.

The aim of this section was to illustrate that the direction for collective action with regard to zero-carbon emissions in general and with regard to decarbonising the electricity system in particular is far from clear. Current renewable technologies fail to be serious large-scale alternatives to fossil electricity generation for various reasons. Apart from uncertain cost competitiveness those reasons are bad reputation and high-level risk (nuclear energy), limited capacity (tidal, geothermal and hydro energy) and limited reliability (solar and wind power). Solar and wind power are certainly the most promising for large-scale application – if there was only a way to better deal with their high intermittencies. I introduced three instruments for grid stabilization, which, at last right now, fail to balance wind and solar intermittencies: There is insufficiently integrated interconnections (long and short distance), smart grids in the development and testing stage as well as power storage, which is either still lacking cost competitiveness (batteries) or requires sufficiently integrated long-distance electricity interconnections (hydropower). Broadening the view to power-to-X technologies in rather general terms, seems to be a promising pathway. Geoengineering might be used as a complementary measure. Fast and effective international research and development is required in all areas. The promotional task for scientists is certainly to concentrate existing know-how, financial means and efforts to push the technological progress in general.

Still, I do not intend to create the impression that this promotional task is only a task for natural science and engineering. I do not dare to speak for other disciplines, rather I wish to address those scientists concerned with climate ethics and maybe even applied ethics more broadly. Steigleder (2016c) claims that "[c]limate ethics must be to a great extent risk ethics and energy ethics" (Steigleder, 2016c, p. 121). In view of the massive transformation that is required to limit warming to below 2 °C, he argues that the transformation "not only involves many technological, scientific, economic and political, but also numerous moral questions" (Steigleder, 2016c, p. 122). Further, he emphasizes that "[a] significant number of these [moral] questions go along with the evaluation and weighing up of different kinds of risks which require considerations from risk ethics." (Steigleder, 2016c, p. 122). Yet, both energy and risk ethics are underdeveloped areas of normative ethics (Steigleder, 2016c, p. 122).[162] Discussing and clarifying these questions is a promotional duty of scientists concerned with climate ethics.

Though the decarbonization of the electricity system is centrally important to climate mitigation, electricity generation is not the only sector that requires transformation to zero-emissions. Global emissions from agriculture, forestry and land use as well as all sectors using energy as end-users including industrial processes, buildings and the transportation sector provide every reason for concern (Edenhofer et al., 2013, pp. 64–92). With regard to the Polluters, transportations will play a major role in the following chapter.

162 Starting from a very broad understanding of ethics, Frigo (2018) provides a literature overview on energy ethics covering social science, religious studies, energy justice and interdisciplinary approaches. Shrader-Frechette (2017) provides a rough overview of possible arguments related to energy ethics. Steigleder's (2016c) work on energy ethics with regard to CCS may be used as a template for further energy ethics research.

6.5 The Polluters: Mainstreaming a Sustainable Lifestyle

Given that the Polluters are weakly collectively responsible for climate change, the Polluters' task as a collectivity is to organise themselves to ensure mitigation. The Polluters are probably the most controversially discussed weak collectivity. With reference to per capita carbon budgets, chapter 5.2.4 claimed that individual emission levels are considerably higher than the level required to sustain the climate system. The Polluters are caught in a structure that heavily relies on the burning of fossil energy sources. In view of chapter 6.4, current technological solutions are rather insufficient to spare the Polluters fundamental changes in almost all areas of their lives. The huge challenge with regard to the Polluters' promotional duty is that members ought to promote measures that restrict their performing their lifestyle.

That given, the primary promotional task for members of the Polluters is promotion through the social channel in order to create a sense of social responsibility and to support political regulations. Yet, that is not all they can do. This chapter mainly elaborates the Polluters' promotional duties with regard to the political channel. I have claimed several times that structural changes are required. That is in order to achieve required emission thresholds at all and to decrease individual costs of changing the Polluters' lifestyle. With regard to promotion through the political channel, my claim is that members of the Polluters ought to call for a framework that allows for mainstreaming a green, sustainable lifestyle across their own collectivity. However, what are members supposed to be calling for exactly?

Focussing on the three main sources of individual carbon emissions, the first part of this chapter aims to draw an overview of possible options (chapter 6.5.1). Following Gewirth, I have defended a limited, secondary individual duty to mimic alongside promotion. The second part of this chapter takes up on this and examines what it means to live up to what one promotes as a member of the Polluters. There, I mainly aim to conceptualise ways of thinking that members of the Polluters may adopt when they try to mimic as regular individual agents (chapter 6.5.2).

6.5.1 *Promoting a Setting for Carbon-Free Consumption: Housing, Mobility and Food*

With regard to the Polluters' lifestyle there are two connected challenges: Right now, the Polluters' lifestyle is not at all compatible with projected climate targets. That is a serious problem. Yet, the way more serious problem is that the Polluters' lifestyle is projected to develop in adverse direction to what is actually required for effective mitigation. Instead of collectively performing less of those actions that contribute to climate change, polluting activity increases and will continue to increase in the next decades. While international climate policy aims to control the Polluters' emissions, the Polluters – though without intention – continue to increase their share of global carbon emissions.

But, one step at a time. The largest share of individual emissions is from three domains: (1) housing, (2) mobility, and (3) food. These three domains make up 70 to 80% of consumption-related life-cycle environmental impacts in industrialized countries (Mont, 2014; Tukker & Jansen, 2006). With regard to housing, the primary source is the heating system. Regarding mobility, the key sources are car use and air travel. And with respect to food, beverages, meat and dairy are the biggest contributors. The discussion below largely sticks to these main sources and examines (1) heating systems representative for housing, (2) tourist air travel as an example for mobility and (3) food consumption in rather general terms.

Connecting to top-down and bottom-up change with regard to collective climate action, consumer research differentiates three different approaches to change consumer patterns: (1) The voluntary approach that relies entirely on the consumers' interest, understanding and feeling of responsibility. (2) The soft bottom-up approach that aims to provide incentives to reconsider daily consumption behaviour for example by social marketing and nudging.[163] And (3) the hard top-down approach that enforces behavioural changes through state regulation (Higham, Cohen, Cavaliere, Reis, & Finkler, 2016, p. 337). With regard to different domains of private consumption, studies suggest that the voluntary approach, used on its own, is insufficient to effectively change individual behavioural patterns (Higham et al., 2016). Generally, sustainable choices are found to be easier in, what I refer to as, the domain of food consumption, than in the tourism and leisure context (Cohen, Higham, & Reis, 2013; Higham et al., 2016). Linked to that, the success of nudging as a soft bottom-up approach depends on the domain where nudging is used. Nudging may work well for food consumption but it remains largely unsuccessful in tourism and leisure (Barr, Gilg, & Shaw, 2011).

I assume the following: First, the Polluters promote against their own individual interests. In that regard, I consider it risky to rely on voluntary changes. Second, the more often individuals choose to perform certain actions, the greater the likelihood that changes may be initiated by the soft bottom-up

163 Examples for the soft bottom-up approach are social media campaigns on ocean plastic waste with reference to plastic particles in cosmetic and care products or offering options for private emission offsetting in order to visualise the true costs of polluting actions and to nudge people to fly less.

approach. Yet, and this is my third assumption, in view of rapid climate change, the soft-bottom-up approach is more of a substitute than a stand-alone solution. Based on that, it is possible to roughly specify the required ways that the Polluters ought to promote in order to conquer themselves and their lifestyle. I suggest that the focus with regard to tourist air travel is rather on top-down regulations, regarding food nudging, it may serve as a substitute. Since the choices with respect to housing are usually once in a lifetime choice with lasting impacts, appropriate means are rather top-down than soft-bottom up approaches. These assumptions do not argue against a primary promotional duty to target the social channel. As I emphasised elsewhere, any political measure will require social backing. Rather, the matter that I am trying to specify is what kind of political measures are appropriate. Correspondingly, my aim is to narrow down the broad range of options that promotional action may call for.

Housing and Heating Systems

For regular individual agents, the choice of a heating system is not a typical daily consumer decision. People either live in rented housings, which leaves the choice of the heating system to their owners or they own property themselves. And owning property is usually a once in a lifetime choice. Moreover, replacing heating systems in the context of energy conserving renovation of residential housing is mainly a question of financial ability. Many governments already provide private subsidized loans for sustainable housing. In that regard, the duty of the Polluters is possibly to call for politicians to extend amount and duration of state subsidies. Yet, beyond nudging, promotional action could also call for putting more ambitious standards for new buildings into place. There is a variety of alternative heating systems that do not rely on fossil fuels such as pellet heating, micro combined heat and power units (micro-CHP) or heat pump technologies. However, choosing between different forms of heating very much boils down to choosing between generating energy from fossil sources or green alternatives. Chapter 6.4 elaborated promotional duties regarding the road to zero emission electricity supply. I claimed that the clarification in this regard to be a promotional task for scientists as members of the Able. Heating is just another facet of global energy supply. Thus, the decision regarding the most effective and efficient form of heating is mainly in the hands of the Able.

Mobility and Air Travel

Chapter 4.4 elaborated whether the duty to do one's part of the collective scheme with regard to the limitation of tourist air travel is too demanding form a moral perspective. Readers may ask why I take up air travel one more time.

When I previously analysed air travel I aimed to provide guidance on the assessment of costs and, more specifically, I aimed to clarify whether the limitation of air travel is at all reasonable from a moral perspective. Now, I reconsider air travel and ask what members of the Polluters ought to promote for in the absence of morally required collective measures that were to limit air travel. This chapter also substantiates my claim that, in order to bring aviation in line with urgently needed, ambitious mitigation targets, the volume of air travel requires top-down limitation until large-scale technological innovations allow for environmentally friendly aviation. The following discusses air travel as a representative for mobility.

The number of worldwide tourist arrivals has more than doubled in the past twenty years. In 2017, there were 41.9 million scheduled commercial flights worldwide, which is more than 114,000 flights per day (ATAG, 2018, p. 5). With almost four billion people travelling around the globe in 2017, aviation is a serious issue when it comes to climate protection (World Bank, 2019). Even though the number of flights for tourist purposes is a smaller share, these numbers speak for themselves. So, the first finding is that current air travel is very high in volume.

Assuming optimistic per capita carbon budgets of around 2 t CO_2 per year, even one international flight per year per person exceeds the budget. According to an online carbon calculator provided by Atmosfair, a return flight from Frankfurt to New York City produces 3.6 t CO_2 per person, Frankfurt to Tokyo and return causes about 6 t CO_2 per person and Frankfurt to Perth causes more than 10 t CO_2 per person per return flight (status: March 2019).[164] The second finding is therefore that air travel is very high in emissions – in fact, too high.

Tourism in general (including transport, accommodation, attractions etc.) accounts for about 5% of global carbon emissions (P. Peeters & Dubois, 2010, p. 447; UNWTO et al., 2008, p. 13). Transport generates the largest share of these emissions: 75% with 40% coming from air travel alone (UNWTO et al., 2008, p. 33). To put these number into perspective, global carbon emission cuts will have to be between 60 and 80% by 2050, depending on scenario and proposed concentration of atmospheric carbon. Readers may object that 5% of a total of sixty or eighty does not count as a high potential for mitigation. Given that these 5% come from a single sector that is almost entirely connected to leisure and luxury, I consider that quite a high share. But perhaps, readers are more convinced when considering projections. Based on 3.5% compound annual growth rate for aviation industry, the International Air Transport Association (IATA) forecasts the number of passengers to reach 8.2 billion in

164 For carbon calculator see Atmosfair (2019).

2037, which is a doubling within twenty years (IATA, 2018). Forecasts predict that the share of global emissions generated by tourism in 2050 is about 40% (Grote, Williams, & Preston, 2014; Higham et al., 2016). Aviation's contribution to the share generated by tourism is very likely to increase disproportionally. That is for at least three reasons: First, the annual distance travelled is growing (P. Peeters & Landré, 2012). Second, stays seems to get shorter but more frequent (Alegre & Pou, 2006). Third, in contrast to air travel, the remaining sectors of tourism (e.g. accommodation) as well as remaining transport modes are very likely to pursue emissions reduction (Edwards, Dixon-Hardy, & Wadud, 2016; Filimonau, Dickinson, Robbins, & Huijbregts, 2011). Assuming that even critics agree that a doubling in the number of passengers within twenty years is a huge challenge to climate change, the third finding is that growth rates of air travel are largely inconsistent with mitigation targets. Adding another finding for tourism in particular, the fourth finding is that the biggest share of tourism is from aviation.

Critics might object that air travel is not the only problem with regard to mobility in general and international tourism in particular. Indeed, that is true. Land transportation and car use in particular are huge challenges to climate change. Also, the shipping industry has the potential to become at least as incompatible with climate targets as aviation (Capros et al., 2016, pp. 79–80).[165] With regard to land transportation and car use there are possible green or at least greener alternatives already available. Though they should be critically discussed with regard to electricity generation and extraction of rare resources (such as platinum used in fuel-cells), electric and hybrid cars are technological options. These options hardly exist for aviation and maritime shipping. Critics may further object that I am too quick, overlooking improvements in fuel-efficiency and innovations such as biokerosene. Though there are efforts, these technologies will only slow growth in emissions but will probably not be sufficient to reduce them down to zero. Regarding aviation and cruising, Robert Vergeer (2016) found that half of the fuel savings in land transport until 2030 will be undone by planes and ships. Because maritime transport will become a topic with regard to food, I have limited the analysis in this section to aviation.

With regard to these developments and predictions, the common response of governments and institutions in general has been to encourage voluntary behavioural changes (Barr et al., 2011). In view of current and forecasted numbers, this approach quite obviously fails. Air travel, more than ever, characterizes

165 For maritime driven emissions, the EU predicts an increase by 38% until 2050 mainly due to a growing demand in transportation activity. Growth rates are predicted to be positive but slightly slower for aviation (mainly to fuel efficiency) (Capros et al., 2016, pp. 79–80).

the contemporary lifestyle of the Polluters. In this context, some scientists even refer to "binge flying" as a kind of behavioural addiction to excessive flying that is shaping the lifestyle in industrialised countries (Cohen, Higham, & Cavaliere, 2011, p. 1070). Studies further suggest that consumers know about aviation's environmental impact but continue to travel by plane regardless of their concerns about climate change (Barr, Shaw, Coles, & Prillwitz, 2010; Becken, 2007; Cohen et al., 2013; Hares, Dickinson, & Wilkes, 2010; S. Shaw & Thomas, 2010).

One reason for the continuation of polluting behaviour could be the lack of alternatives. But that is true for overseas travel only. Taking Germany as an example, planes are the number one travel option even for medium-distance travel: According to the German Travel Association (German: Deutscher Reiseverband, DRV), 72% of the German travellers spend their holiday outside of their own country. Only 8.5% of which is long-distance travel. One third of the rest of these trips falls on car travel to neighbourhood countries and a staggering two third involves air travel to holiday destinations in the Mediterranean region (DRV, 2018, p. 8). The Germans seem to simply prefer taking a plane even though options for land transportation exist. In this context, another growing trend is important: Stays get shorter while individuals travel more often and further away. Weekend-trips to Rome or Madrid seem to have become the new usual. Just to be clear, my claim is not that individual agents ought not to travel (see chapter 4.4). However, in view of climate change, transportation that is compatible with climate change is a must. I suppose that with regard to green(er) transportation such as a twenty-hours train ride one way, a *weekend* in Rome or Madrid does not seem to be too reasonable. Yet, instead of adjusting the length of the stay, members of the Polluters seem to conclude that air travel is the only viable alternative to reach their desired destination. Thus, with regard to the majority of air travel, I assume, arguments concerning the lack of alternatives hardly succeed.

Another reason for the continuation of polluting behaviour could be adverse incentives. In this context, Björn Ahaus, Ludger Heidbrink and Imke Schmidt (2011) consider various examples for so-called rebound effects: purchasing small, fuel-efficient cars and then driving more often, replacing old energy systems and then use more electricity or compensating for carbon emissions from flying with the effect to step onto the plane with less concern the next time (Ahaus et al., 2011, pp. 7–8). Are nudging measures in risk of becoming incentives? Maybe they can contribute. However, according to Myclimate, a major European offsetting organisation, only 1% of all flights in 2015 have actually been offset (Laubert, 2015). Thus, adverse incentives do not count as a reason either.

The thing about air travel, I guess, is that it is fast, convenient and cheap. Too cheap. The role of low-cost carriers illustrates the point: Ryanair is one of the cheapest European low-cost carriers and offers flights for as little as 9.90 Euro. In 2016, Ryanair Chief Executive Officer (CEO) Michael O'Leary announced that Ryanair intends to offer flights for free in the future with revenues coming from travel-related areas such as baggage fees, on-board service, car rental and hotel bookings (Topham, 2016). If flying is that cheap or at least way cheaper than a train ticket, a bus and even a drive, it bodes badly for sustainable consumer choices. With aviation generally being one of the dirtiest transportation modes, Ryanair clearly reveals the more fundamental problem: aviation (but also consumer goods and services in general) needs to reflect its environmental externalities. With 9.90 Euro per flight, that is not yet the case at all.

Despite business or sales strategies, the more institutionally-induced problems with regard to airfare are more fundamental. First, international plane tickets are still free of value added tax (VAT) (Gössling & Peeters, 2007). That is in sharp contrast to basically any other consumer good. Lack of VAT on airline tickets is a considerable benefit to consumers and in particular to those who fly frequently and long-distant. Second, based on the Chicago Convention (the international treaty on civil aviation), jet fuel is largely excluded from taxes in order to avoid double taxation for international aircrafts (Keen, Parry, & Strand, 2013). Even though fuel tax was not originally designed to serve environmental purposes, various authors provide empiric proof that aviation fuel taxes can help reduce fuel consumption and thus carbon emissions from aviation (Fukui & Miyoshi, 2017; Hofer, Dresner, & Windle, 2010; Sterner, 2007). Critics may argue that introducing taxes will result in revenue shortfalls for airline companies, which will impede growth of an industry that drives economic prosperity. Even though that might be true, I rather regard the lack of taxation as a subsidy for a well-developed industry that is ever growing and contributing to climate change more than any other transportation option. With regard to policy on aviation and climate change, Alice Bows and Kevin L. Anderson diagnose a policy clash (Bows & Anderson, 2007, p. 103). While governments aim to achieve internationally agreed mitigation goals, they fail to address high level emissions of the aviation industry right now and, even worse so, they fail to bring the aviation industry in line with governments' future emission targets. While many other industries are almost naturally subject to policy measures, aviation has been an exception for too long. And that is even though there is no way of overlooking that growth forecasts are fundamentally incompatibility with climate protection. I have no intention at all to penalise or even harass the aviation industry as such. I have emphasised earlier that the aviation industry provides a large amount of jobs worldwide (chapter 4.4) The point is simple: in view of climate change and urgently required mitigation, climate policy should leave no potential unused.

However, taxation is not the only means to limit air travel and thereby to reduce emissions. In general, there are two possible types: technological means and economic means. I have mentioned biokerosene and energy efficiency already. With regards to efficiency, one could add more specific measures such as optimising weight of aircrafts, seat occupancy rate and optimising of routes. But, as I claimed above, technical means will hardly outweigh the increase in emissions from increased demand. With regard to economic means, taxes are certainly an option. However, considering the size of aviation industry, market-based solutions may be more effective.

The EU has been trying to reduce emissions from the aviation industry by integrating aviation into the European Union Emissions Trading Scheme (EU ETS) in 2012. The first response was international protest, which claimed the inclusion of carbon emissions coming from all flights from EU destinations (not just the components over EU airspace) to be an infringement on the sovereignty of non-EU states (Dickinson, Robbins, Filimonau, Hares, & Mika, 2013, p. 507). As a consequence, the EU narrowed the scope of the EU ETS to flights within the European Economic Area (EEA) until 2016 (Gössling, 2013, p. 433). However, the European air market only amounts for 8.5% of global emissions from aviation (Maertens, Scheelhaase, Grimme, & Jung, 2017, p. 592). The EU intended to allow the development of global measures by the International Civil Aviation Organization (ICAO), a UN Specialized Agency aiming to achieve a balance between growth and environmental impacts.[166] In October 2016, the ICAO agreed on a global market-based measure to reduce carbon emissions from 2021 onwards: The Carbon Offsetting and Reduction Scheme for International Aviation (CORSIA). CORISA seeks to guarantee a future zero-emission growth in aviation. Unlike the EU ETS as a cap-and-trade system, CORSIA is an offsetting system that builds on "Baseline&Credit" instead of a politically determined cap (Maertens et al., 2017, p. 589). In order to stabilize carbon emissions at 2020 levels (baseline), participants are required to monitor their carbon emissions. From 2021 onwards, the amount that deviates from the baseline-level in each year, is aviation's offsetting requirement for that year. Participants will have to purchase emission credits. Credits can for example be purchased by investments in climate protection projects or by participating in so-called Compliance Markets where credits can be traded among participants (Maertens et al., 2017, p. 589). Until 2026, trading is limited to international flights between countries voluntarily joining CORSIA: First

166 UN Specialized Agencies are autonomous but connected to the UN via agreements.

in a pilot phase (2021 to 2023) and then in a voluntary phase (2024 to 2026). As of 2027 and onwards, CORSIA applies to all countries whose operators represent more than 0.5% of global Revenue Ton Kilometres (RTK, measuring the share in international transport performances) or countries whose cumulative shares cover 90% of global RTK.[167] It still remains to be seen how to proceed with the EU ETS after CORSIA's entry into force (Scheelhaase, Maertens, Grimme, & Jung, 2018, pp. 59–60).[168]

Given the emission gap between countries' climate commitments on the one side, known as Nationally Determined Contributions (NDCs), and what is required to limit climate change to below 2 °C or even 1.5 °C on the other side, neither system seems to be sufficiently ambitious. The Emission Gap Report 2017 estimated a gap in 2030 of 16 to 19 Gt CO_2eq for 1.5 °C and 11 to 13.5 Gt CO_2eq for 2 °C (UNEP, 2017, p. 16). Using updated scenarios and data, the Emission Gap Report 2018 indicates that the gap in 2030 increased even further to 29 to 32 Gt CO_2eq for 1.5 °C and to 13 to 15 Gt CO_2eq for 2 °C (UNEP, 2018, p. 19).[169] In view of CORSIA, the Emission Gap Report 2017 indicated that overall global emissions from international aviation are projected to increase from 0.5 Gt CO_2eq in 2015 to around 1.1 Gt CO_2eq in 2030 despite emissions reductions from technical and operational improvement. Depending on the level of participation and the way rules are implied after 2020, CORSIA's contribution to saving emissions will be between 0 and 0.3 Gt CO_2eq per year on global overall emissions in 2030 (UNEP, 2017, p. 19). I leave the final analysis on CORSIA's efficiency and effectiveness to those with more expertise in climate economics. But what is safe to say is that a global, effective solution for aviation is urgently needed and CORISA can only be a beginning.

For the sake of completeness, I wish to mention that a carbon tax is a further market-based option to reduce emissions globally and it would also apply to aviation. In view of criticism regarding the EU ETS's ineffectiveness, several scholars have discussed the potential of a carbon tax as either a stand-alone solution or as a complementary option to the EU ETS (see e.g Helm, 2017,

167 Except for Least Developed Countries (LDC), Small Island Developing States (SIDS) and Landlocked Developing Countries (LLDC), whose participation is voluntary. For further analysis and comparison of EU ETS and CORSIA see e.g. Scheelhaase, Maertens, Grimme & Jung (2018).

168 Note that the EU ETS may be discussed critically.

169 The indicated range refers to the difference between conditional and unconditional NDCs. In contrast to NDCs without conditions attached (unconditional), the conditional NDCs are contingent commitments that depend on a number of factors such as "the ability of national legislatures to enact the necessary laws, ambitious action from other countries, realization of finance and technical support, or other factors" (UNEP, 2018, p. x).

pp. 57–58, 235–236; Hsu, 2011, pp. 15–17, 25–115; Patt, 2015, pp. 74–77, 77–79). Indeed, a carbon tax would probably be the most effective and most efficient means to increase the price of pollution and with it the price of pollution coming from aviation. Here, I limit myself to a very brief comment on a carbon tax in relation to the EU ETS. The EU's fiscal power is limited and fiscal policy is under competence of members states. The trouble comes with regard to the EU's secondary law in combination with the constitutional law of nation states. In order to avoid discrimination and to foster harmonisation between member states, the EU's secondary law sets stricter regulations for energy taxes than for consumer taxes. As a consequence, the UK, for example, has introduced a consumer tax – the so-called Climate Price Support. Using the UK's Climate Price Support as an example, Hartmut Kahl and Lea Simmel (2017) have, among other options for carbon pricing, assessed the scope for a carbon tax in German constitutional law and found that introducing a carbon tax as a consumer tax is not an option in Germany as it requires changing the constitution first. If a carbon tax imposes such challenges in the EU-framework, it probably bodes badly for a global carbon tax system. However, because my expertise is not in law, I leave the final assessment of possible tax options to those with expertise. Besides these legal challenges, strong industrial lobbying (e.g. from the aviation industry) probably does the rest to avoid the introduction of a carbon tax system either in Europe or anywhere in the world.

Just like figuring out a path to zero-emission electricity supply, finding the right choice of technical and economic means is a promotional duty of members of the Able but not a duty of members of the Polluters. It seems that figuring out the appropriate means is so difficult that promotional activity of the Polluters through the political channel is limited to rather general claims aiming to push forwards political action. In that case, the primary promotional duty with regard to housing and mobility, so it seems, is to target the social channel. But that is actually not bad in general. I have claimed above that members of the Polluters basically promote against their personal interests. In that respect, promotional action through the social channel, in order to create and secure social backing for any kind of coercive regulation, is a massive promotional task.

Food Market and the Lack of Information

Food is a paradox domain: The food market is one of the most globalized markets and is intensely shaped by international trade. Therefore, global food supply relies heavily on aviation and shipping. A first intuition could therefore be that changes in food must be subject to top-down measures only. Yet, with regard to social backing, the Polluters' willingness to switch towards more sustainable, greener choices seems to be considerably larger in the food domain than in housing or mobility (Cohen et al., 2013; Higham et al., 2016). That gives way to nudging as a measure to loosen up the coercive nature of top-down climate regulations. However, that does not mean leaving the final approval of sustainable choices to private consumers. The claim of this section is that individual consumers are currently at the very uninformed end of a long, complex and non-transparent supply chain. The main promotional task of the Polluters with regard to food is to decrease their own lack of information.

On the individual level, the Polluters' motivation to choose sustainably is probably diverse. Given that consumer choices in food are more frequent and considerably smaller in scope, personal impacts are not as restrictive as impacts with regard to mobility. My hypothesis is that 'doing one's bit' for the global fight against climate change is perceived to be easiest with regard to food. In addition, the fact that sustainable, green choices are considered to be generally healthier has a positive effect. Sustainable choices as healthy choices seem to be secured by prudential considerations. The problem is that the dominant moral narrative with regard to climate change is backed up by a narrative that consumption is a private matter when in fact, consumption is a public matter.

The first promotional task of the Polluters is to foster political consumerism through the social channel. Political consumerism is the idea of using one's money to "vote" for or against specific products and services (Brinkmann, 2004; Neilson, 2010; D. Shaw, Newholm, & Dickinson, 2006). Accordingly, political consumerism includes boycotting and buycotting. The former relates to punishing corporations for unfavourable or unethical behaviour. The latter refers to the opposite: supporting corporations for favourable and ethical practices. Even though buycotting is a measure in theory, boycotting has been the dominant tactic. In practice, examples for boycotting range from avoidance of use of plastic waist, production using palm oil, child labour, animal testing or genetic engineering and corporate tax avoidance to boycotting single products such as tinned tuna. Tinned tuna is in fact a good example for the impact that political consumerism may have. Tuna was first boycotted for dolphin bycatch in the late 1980s, which lead to a dolphin-safe label first in the US and following in the EU (Brown, 2005) and then a second time for coming from Bluefin Tuna as an endangered species. Yet, all campaigns that I have come across had been initiated by existing non-governmental organisations. However, that does not say anything against the Polluters' promotional duty. On the contrary, it actually underlines my claim that existing organisations and individual agents ought to use synergy effects in cooperation. With regard to climate change, the remaining question is what the Polluters ought to call for content-wise when calling for adherents to boycott or buycott certain products. What are the ultimate means with regard to consumer goods in food industry? Is it buying local, vegetarian, seasonal or rather non-plastic wrapping?

In order to decrease information-asymmetries, a number of food labels and certifications have been introduced. Eco- and Fair-trade labels or the Rainforest Alliance Certificate are only a few out of many different examples. Labels and certificates may be nudging tools too. Just as with any other nudging measure, the key assumption is that individuals reconsider their existing purchase patterns once they are empowered with enough information. Sweeping success with regard to nudging sustainable purchase patterns is missing (Grunert, Hieke, & Wills, 2014; van Amstel, Driessen, & Glasbergen, 2008). That is for at least four reasons: First, lack of clarity. The vast number of labels is often directly a source of confusion (Brécard, 2014). Or, if individuals make an effort to understand their meaning, the following paradox applies: The more information provided, the greater the knowledge and the greater the awareness on the consumer side. But the greater the awareness, the greater the ability to identify conflicting information, trade-offs and ambiguities. The feeling of not being able to truly choose sustainable options leads to insecurity and quite resignation. The second and third reason are lack of reliability and trustworthiness: Calculations are built upon standard methods without further explanation of what the standard methods actually incorporate. Only a reliable control procedure can reduce the risk of mislabelling. Various crises and scandals in the past have shown the risk of greenwashing and have done their bit to weaken consumers' confidence on labelling.[170] The fourth is lack of comprehensiveness in application. Labelling right now only allows for positive feedback. What is needed is a comprehensive application, which allows for positive and negative feedback. Right now, as the process of certification is rather resource intense, products without a label could either not have fulfilled the requirements, or products were simply not subject to the rather cost-intensive certification process. It seems that any tool of information becomes a matter of faith and conviction at some point.

With regard to climate change, one single label would be enough if it was introduced on a large scale across countries. In fact, there are a number of initiatives for carbon labels such as the Carbon Trust Label in the UK and, on a small scale, Thailand and Australia, the Climatop Approach in Switzerland

170 An example for greenwashing that has gained media attention in early 2018 is the scandal of the Marine Stewardship Council (MSC), an organization founded by Unilever and World Wide Fund for Nature (WWF) to certify fish that has been caught sustainably (see e.g. Beenen, 2018; Tomik, 2018).

or Carbon Neutral labels in the US and New Zealand (Schaefer & Blanke, 2014, pp. 76–78). Yet, almost all of them rely on positive feedback and none of them is used on large scale. The UK Carbon Trust Label for example is only on twenty single products sold at Tesco supermarkets. The Climatop label is used to indicate products that belong to the 20% range of least emission-intensive products at Migros supermarkets and Carbon Neutral labels indicate companies' willingness to conduct a comprehensive study with regard to particular products whose emissions are then offset in accordance (Schaefer & Blanke, 2014, pp. 76–78). In 2007, Tesco intended to introduce a carbon label for all their products. A few years later they announced the abandonment of the project because registration and certification procedures were extremely resource intensive and other players did not follow (Vaughan, 2012). The challenge to carbon labeling is, I think, that no company in isolation will be able to bear the costs of evaluation and certification without putting itself at a competitive disadvantage. That is just the same as with emission reductions of individual agents: taking costs on one's own is likely to result in overburdening. Thus, labeling requires a large-scale introduction not only to ensure that producers and companies are not disadvantaged for climate action, but also to ensure that consumers are provided with an opportunity for objective comparability. In analogy to food traffic lights that demonstrate aspects such as sugar and fat content, I think of a transnational product carbon footprint (e.g. introduced across the EU and North America) that provides insights on emissions from each part of the supply chain: production, processing, distribution, and sales. Here is the true problem: for parts of the supply chain – namely for the distribution part – reliable information does not exist.

The lack of information is best seen with regard to maritime transport. Consumer goods grown and produced across the globe have a high acceptance in parts of the world where they are not coming from due to fast, reliable and cheap merchandise shipping. Hence, it is not surprising that 80 to 90% of the world's trade includes shipping (ICS, 2018; IMO, 2018).[171] It stands to reason that life-cycle assessment of consumer goods includes the assessment of emissions from shipping. In fact, however, the scope of reliable and comprehensive assessment is limited to shipping activities in the northern parts of Europe (Brynolf, Baldi, & Johnson, 2016). For any other part of the world, emissions

171 This share demonstrates the exceptional role of maritime transport in global trade. The vast majority is carried by dry bulk carriers (e.g. for coal or iron) and tankers (e.g. for oil or chemicals). Assuming that consumer goods are mainly transported by container ships, its share is nevertheless significant (13.2% of global dead-weight tonnage) (UNCTAD, 2017, p. 24).

are top-down calculations from international sales statistics from bunker fuel, building upon so-called emission factors that represent average values for GHG from heavy fuel oil.[172] In contrast to top-down calculations, bottom-up calculations would start from activities of individual ships, which would be more reliable and comprehensive (Salo et al., 2016, p. 179). The problem is feasibility. The main reasons for that are found in the way emissions from shipping are monitored, verified and regulated. Alongside with aviation, shipping is the most unregulated, most polluting industry in the world.

The body in charge of regulating safety, security and pollution of maritime transport is the International Marine Organization (IMO).[173] As a UN specialized agency, the IMO refers to itself as the "global standard-setting authority" (IMO, 2018). In fact, it is rather a "forum in which States can discuss, negotiate and make common decisions regarding regulations of ships" than an authority (Linné & Svensson, 2016, p. 101). The IMO's members set so-called rules of reference, which establish the framework for actual rule-making and conventions. The actual setting of international rules is the UN Law of the Sea Convention (LOSC). The LOSC prescribes that states have the duty to protect marine environment by creating national regulations and enforcement (Linné & Svensson, 2016, p. 93).[174] The more ambitious the rules of reference set by the IMO are, the more ambitious and stricter the actual rules of the LOSC are. The crafting of agreements within the IMO can best be described as a contest between maritime and coastal interests: While coastal states have traditionally been the initiators of environmental protection conventions, maritime interests are primarily in the hands of so-called flag states. To understand the difference, a little bit more background information might be helpful: In order not to be a pirate ship, each ship has to have a flag under which it is flying. So-called open registries enable ship owners to fly under flags that are different from their

172 Bunker fuel is generally any type of fuel oil for ship use. The most common bunker fuel is heavy fuel oil, which is a product of refinery processes to crude oil. Lindgren et al. (2016, pp. 52–54) provide an overview on refining of crude oil.

173 More precisely, the forum in charge of all matters related to prevention and control of pollution from shipping is the Marine Environmental Protection Committee (MEPC), which is one of the five committees that conducts the IMO's work.

174 To be more precise, the LOSC does not actually create new rules for shipping, but it only refers to standards, which have been agreed upon within the IMO as a competence body. For ship-based pollution that is essentially an agreement referred to as MARPOL 73/78 (short for marine pollution; the agreement is from 1973 and was modified in 1978) (Linné & Svensson, 2016, p. 80). For the purpose of this paragraph the difference of the IMO as a body for setting rules of reference and the LOSC as the actual legal framework is sufficient. For a comprehensive overview on historic and present regulation of pollution from marine transport see e.g. Linné & Svensson (2016, pp. 78–116).

nationality. As a result, more than 70% of the global merchandise fleet was foreign-flagged in 2017 (UNCTAD, 2017, pp. 32–33). On the one hand the system of open registries provides opportunities for small island states and least developed countries. On the other hand, however, it has led to a great deal of problems connected to Flag of Convenience States. Some flag states compete for having the greatest advantages for ship owners mainly with regard to fewest regulations and weakest enforcement. Against this background it is no wonder that in 2017 the top three flag states were Panama, Liberia and the Marshall Islands. The 'top' refers to states with the greatest share of so-called dead-weight tonnage, a measure for ships' maximum loading capacity. Together with number four Hong Kong, five Singapore and six Malta, they represent more than 60% of the world's maritime trade (UNCTAD, 2017, p. 32).[175] Thus, the IMO's agreements are characterized by a contest between flag states with strong interest in few regulations and independent seaborn trade on the one hand and coastal states on the other hand, which have an interest in ambitious regulations and enforcement in order to protect their coastline ecosystem.

The problem with regulating pollution from shipping is at least threefold: First, coastal and therefore environmental interests are underrepresented in the IMO's Council that is responsible for convening the conference to adopt a new agreement in the first place (Linné & Svensson, 2016, p. 103).[176] Second, the IMO's budget is funded by its member states and calculated in accordance with their fleet (Linné & Svensson, 2016, p. 106). The majority of the IMO's budget therefore comes from flag states – a challenging precondition for concluding agreements against their interest. Third, as I mentioned above, the IMO is dependent on member states to define national regulations and ensure enforcement. The majority of maritime trade happens on the high seas. And on the high seas, flag states have the sole jurisdiction on the ships flying their flags. Thus, maritime transport on the high seas is largely unregulated and unmonitored and with it the pollution of shipping on the high seas.

Emission Controlled Areas (ECAs) defined by the IMO are limited to the North Sea and the Baltic Sea, a small part of the Californian coast, parts of the North American coastline (alongside of Canada and the US, including Hawaii) and parts of the Caribbean coast (IMO, 2019). Yet, emission control

175 Percentage expresses the share of world total dead-weight tonnage.

176 Note further that, just as in the case of a climate agreement, the negotiation process of new agreements on maritime transport adopted under the IMO is complex, slow and in order for the IMO to propose a new instrument, a "compelling need" for a new regulation is required (Linné & Svensson, 2016, p. 114). Linné & Svensson (2016) provide a detailed description on international agreements and challenges to IMO agreements in particular (Linné & Svensson, 2016, pp. 112–113, 114–117).

is limited to sulphur, nitrous oxide and particulate emissions (Salo et al., 2016, pp. 190–192). I do not intend to talk down the negative effects of these kinds of emissions at all. Nevertheless, in the face of climate change, ECAs with regard to CO_2 would be required too.

The need for strong regulation of the shipping industry is obvious: The primary target of regulations should not be market measures (introducing taxes, trade schemes, etc.) or operational measures (reducing emissions from fuel combustion, new criteria for marine fuel, development of new engines etc.) as for the aviation industry. These tasks are important too. Yet, the primary focus of regulation is way more fundamental: it is to ensure monitoring, verification and reporting of emissions and with it the accessibility of information. Only then, emissions reduction as well as market-based and operational measures might follow. There are two solutions for that: Either enforcing stricter obligations to regulate on flag states or strongly regulating open registries.[177]

In view of the lack of progress in the IMO over the years, the EU initiated its own scheme. Since January 2018, large ships using EU ports are required to monitor and report their verified carbon emissions and other information (e.g. on fuel combustion, amount of cargo and distance travelled) (EC, 2018). Nevertheless, just as in the case of aviation industry, the EU has declared itself in favour of a global agreement led by the IMO. In April 2018, the IMO finally agreed upon cutting emissions from shipping by at least 50% in 2050. The IMO's target is therefore not as ambitious as the EU's emission target of 70% but it is the right track (Saul & Chestney, 2018). In view of the fact that shipping (just as aviation) has not been covered by the Paris agreement in 2015, that is good news. Given that scientists have claimed the need for emission cuts of at least 85% in 2050, the IMO's target is probably not ambitious enough (Anderson & Bows, 2012).

However, with regard to what has been said above, the IMO misses the more fundamental point. The key challenge is that monitoring and reporting measures do not exist for big parts of the vessel on the high seas and that those actors who could ensure guidelines for monitoring, reporting and verifying for ship-based emissions have made it their business not to ensure them. That given, it bodes very poorly for providing accurate information on carbon emissions for consumer goods in the food domain.

177 The LOSC required a 'genuine link' (LOSC, Art. 91(1)) between the state and the ship in order to enforce strict obligations for flag states. Unfortunately, there is no conclusive and globally accepted definition on what that is (Zwinge, 2011). This way problems connected to open registries and Flag of Convenience States can occur at all.

So, what do members of the Polluters ought to do in this regard? In the beginning of this section, I claimed that individual consumers are currently at the very uninformed end of a very long, complex, and non-transparent supply chain. The case of maritime shipping underlines that, even if governments wanted to provide concrete information regarding the carbon footprint of consumer goods in the food domain (e.g. by a carbon label) in order to decrease information asymmetries, they would not be able to do so. Governance and structure of maritime shipping are very complex. For regular citizens as members of the Polluters, it is at least extremely difficult, if not impossible, to comprehend, review and call for the needed regulations in the shipping industry. That leaves the shipping industry in the hand of scientists and politics again. Yet, this section is not about promotional duties with regard to maritime transport. Even though the food market cannot be considered without the shipping industry, this section is about promotional duties with regard to the food market. It seems that willingness is greatest in the food domain and I suppose that many individual agents do not even want to purchase in the dark. With regard to collective mitigation, this has potential. Building on the Polluters' willingness, my hypothesis is that if individuals were provided with sufficient information, consumers would collectively have at least a limited capacity to choose consumer goods with smaller carbon footprints. In my view, a major promotional task of members of the Polluters in this regard is not to accept being at the very uninformed end of the consumer goods supply chain and to call for governments to tackle information asymmetries more effectively. Obviously, that requires politics to tackle the lack of information with regard to maritime transport too. Political consumerism is certainly one way to perform promotion. Another key task of the Polluters is to target the social channel in order to ensure support for any kind of political measure that the Able suggest as a means to reduce emissions of the Polluters down to zero.

In the introduction to this chapter, I claimed that members of the Polluters ought to call for a framework that allows for mainstreaming a green, sustainable lifestyle across their own collectivity. The motivation of this claim was the creation of a setting where members of the Polluters can do nothing but purchase sustainable, climate friendly consumer goods and services. With regard to housing, mobility and the food domain, it seems that major tasks are not in the hands of members of the Polluters. This chapter might leave readers with the impression that it has created more problems than it has solved. In fact, the challenge with mainstreaming a sustainable lifestyle of the Polluters' is extremely complex and this chapter has only revealed several, yet by far not all, factettes of the challenge. With regard to mobility, I said almost nothing about private car-use, which is still one of the major sources of individual carbon emissions and I hardly referred to the cruising industry, which is increasingly becoming a major challenge. With regard to the food market, I have only touched upon purchase in a very general way and left particular aspects such as meat consumption or dairy products untouched. Obviously, a lot more discussion is required.

Perhaps, my analysis simply underlines that the Polluters are caught in a structure that does not only rely on fossil fuels but in a structure that consists of practices whose success and efficiency inherently depends on the use of fossil energy. There is no easy way to switch the Polluters' lifestyle to zero-emissions. This chapter also underlines that the roots of the Polluters' dependency on fossil fuels will not be solved by mimicking even if the Polluters attempted to do so collectively. Nevertheless, transformation of almost all aspects that constitute the Polluters' lifestyle is mandatory. Clarifying the means for a sustainable lifestyle is the task of scientists. Fostering the creation of these means and implementing them is a task of political actors globally. Showing openness as well as demonstrating and ensuring support to any means that are to interfere with the Polluters' lifestyle is the major promotional task for members of the Polluters. And I believe, that is a huge task.

6.5.2 *Living Up to What One Promotes*

With regard to consistency, I have defended a limited, secondary individual duty for members of the Polluters to mimic alongside promotion. Even though I do not intend to make limited, secondary mimicking duties a primary matter in this book, I do not wish to leave it without a comment either. This chapter briefly examines what it means in practice to live up to what one promotes. The starting point is that members of the Polluters might find themselves in different roles in their daily lives. With regard to individual demandingness, these positions allow for various levels of ambitiousness. Or, put differently, the limits of demandingness, are narrower for some social roles than for others. In this respect, this chapter aims to conceptualise a way of thinking that members of the Polluters may adopt when they try to mimic as regular individual agents.

To see the point, consider an individual agent who aims to fulfil her promotional duty. For reasons of simplicity and illustration, suppose the agent is called Charlotte. Charlotte starts by promoting activity through the social channel, connects to others, raises awareness, tries to accelerate action in all areas and is very open to suggestions coming from science. Suppose scientists find that regarding air travel a global carbon tax system is an appropriate means to limit carbon emissions from aviation. So, Charlotte encourages others to promote a global carbon tax system with her through the political channel. Given that economic and technological means are probably insufficient to limit carbon emissions to a manageable amount, Charlotte aims to fulfil her secondary duty and is willing to adjust her own behaviour accordingly. She aims to mimic what is required under a collective scheme. Mimicking, though it is *not limited* to refraining from flying, *includes* refraining from flying as a limited, secondary duty parallel to her promotional activity (chapter 5.2.3).

Now suppose she, as an individual agent, fulfils different roles in life. Socially, Charlotte is a friend, a mother and a spouse. Professionally, she is a consultant for a company specialised in sustainability management. Organisationally, she is voluntarily engaging into a sports club as a chairwoman. And personally, Charlotte loves wildlife photography. Certainly, Charlotte's roles are connected, but that will not invalidate my point. Now, Charlotte is based in Germany. Suppose that her partner and children are dearly dreaming about a holiday in the tropics, the consultancy company assigns her to a project in New York City, the sports club is invited to a one-day charity event in London in order to raise money for an organisation supporting the poor and personally, she has always been dreaming about a wildlife safari in South Africa. Apart from the fact that Charlotte should probably strongly consider promoting the very idea of climate mitigation to her family, friends, colleagues and club members as her social setting (promotion through the social channel), I suppose that in order to fulfil her mimicking duty, Charlotte will be torn between considerations. Should she mimic by being a mother or spouse who is in the way of her family's dearest dreams, by being a consultant who refuses to travel to meet her client in person, by being a chairwoman who hinders the opportunity for a sporting competition that is at the same time helping the poorest or by being an individual that refuses to do what constitutes herself as a person? Most readers will probably agree that things are rather complicated. The conflict within Charlotte's family is foreseeable, the consulting company will probably not bear her refusing to do her job for forever and will threaten Charlotte with dismissal, club members will probably claim that she missed a great opportunity leading to heated discussions about her being the right cast for the chair and Charlotte herself will probably be torn between constituting her identity as a wildlife photographer or fulfilling her mimicking duty. My claim is that her mimicking duty is not equally weighty in all areas. But which conflict should Charlotte take? With regard to which role should she fulfil her secondary duty and refrain from flying? Assuming that all cases involve flying, in which setting may her individual costs exceed Charlotte's limits of demandingness?

Just to be clear, I am not discussing whether mimicking duties demand too much in the first place. The primary duty applies to the weak collectivity of the

Polluters and, in the absence of a collective scheme, members of the Polluters have a promotional duty in the first place. Based on Gewirth's rights-based theory, I have discussed demandingness and guidance on reasonable individual costs elsewhere (chapter 4.4). Mimicking alongside promotion may never become as demanding as promotion because, for reasons of effectiveness, efficiency, and fairness, mimicking is not the primary individual duty. Mimicking only supplements promotion as a matter of integrity and avoidance of hypocrisy. I rather aim to provide individual agents such as Charlotte with a tool that allows her to identify different potentials of mimicking and navigate between them in practice. Though exactly with regard to mimicking as a primary duty, Hohl (2017) suggests the use of "rules of thumb" in order to avoid demanding too much of individuals in practice (Hohl, 2017, p. 160, own translation). Even though that might be an option with regard to mimicking as a limited, secondary duty, I assume that 'rules of thumb' require some more specification. Otherwise, individuals may simply choose to refrain from those actions that suit their personal interests the most.

With regard to Charlotte's roles in life, I suggest a certain way of thinking that she might adopt. Though I do not intend to develop an elaborate theory of social roles, I attempt to provide some practical advice. Reflecting on Charlotte's roles, I suggest that roles vary with regard to three statements:

(1) The role inherently requires performing certain actions when living up to the role's purpose.
(2) The role is strongly connected to moral expectations on how role-specific actions ought to be performed.
(3) The role is performed as part of an authority-relationship between different roles.

My hypothesis is that the more the three statements apply to a certain role, the less potential there is to live up to what one promotes for.

Consider Charlotte as a mother and spouse: being a mother or wife does not inherently incorporate flying or enabling her family to spend a holiday in the tropics. Perhaps her family externally expects her to make a holiday in the sun possible, maybe she even, internally, expects herself as a loving mother to fulfil her children's dreams. Yet, these expectations have no moral ground at all. To see the point, consider a foreign minister in contrast. Certainly, a minister belongs to the Able and is not the typical regular individual agent. Yet, assuming that the role of the foreign minister is morally justifiable, the foreign minister may, at least from time to time, be expected to be in three different places in one day. Her role is connected to role-specific actions that ought to be performed in order to fulfil the role. However, that does not apply to Charlotte as a mother or wife. Also, in modern societies where the majority of the Polluters live, Charlotte's role as a mother or spouse is probably no subject to an authority-relationship. There are no consequences to be afraid of, consequences that might interfere with her rights. To see the importance of authority-relationships consider Charlotte's role in being a consultant in the following.

In many cases, being a consultant is inherently connected with travelling and air travel in particular. Maybe her colleagues and in particular her supervisor expect Charlotte to share their performance of the role by flying frequently. Yet, these expectations, it seems, do not have the same strong moral ground as in the case of the foreign minister. On the contrary, given that the consulting company focuses on sustainability, it would be surprising if members of the corporation would not align their own acting in accordance to the corporate focus. Yet, Charlotte performs her job as a consultant in the framework of an authority relationship. The relationship between her as an employee and the consultancy company as the employer is agreed by contract. To use Gewirth's applications once more, the employer-employee-relation is a voluntary association. Even though it is voluntary, Charlotte is required to fulfil the terms of the employment contract. And these terms will most likely include a paragraph on international travelling. If she does not fulfil the terms of the agreement, she will lose her job sooner or later, which results in losing income opportunities to support her family. The scope of action with regard to mimicking is rather limited if Charlotte does not want to risk losing her job.

In this regard, I wish to add a very brief comment on Charlottes promotional duty. Charlotte's promotional potential on the contrary is rather extensive. She could promote more sustainable business policies such as fewer flights and longer stays. She could also promote the idea of corporate offsetting. Great potential might also unfold by improving teleconferencing, which would avoid short-haul routes of aviation. And generally, she could promote the very idea of corporate environmental responsibility. She might be considered to be an inconvenient colleague, but I hope that inconvenience is no reason for termination.

Now consider Charlotte's role as a chairwoman. Being a chairwoman of a sporting club is not inherently connected to flying. Even if club members expect Charlotte to enable them with an opportunity and even if Charlotte would be eager to take the opportunity too, Charlotte's being a chairwoman does not require her to fulfil these expectations from a moral perspective. Charlotte may be in an authority relationship, but first of all, she is the primary authority figure and second, being the chairwomen of a sporting club on voluntary basis is not connected to anything morally relevant such as generating an income.

Charlotte could either promote the idea of travelling to the venue either by bus or train. Otherwise, she could initiate a similar charity event close to her place of residency.

Now, consider Charlotte as a wildlife photographer. In general, photography is not necessarily connected to flying. Maybe she has internal expectations about herself as a person. Certainly, constituting oneself as a person is morally required. But self-constitution is not inherently connected to flying. She could do wildlife photography right at her doorstep or travel somewhere close and thereby constitute herself as a person. Nevertheless, lifelong dreams probably have more moral weight than other actions of self-constitution.

Defenders of mimicking as a primary duty have a hard time defending the general statement that any kind of air travel ought to be prohibited. And that is for a reason. Some role, such as the minister or the consultant seem to require air travel in some situations or to some extent. In addition, I have stressed elsewhere that prominent scientists are members of the Able and members of the Polluters alike (chapter 5.2.3). For climate scientists polluting activities such as flying might – in strict limits – be legitimate in order to enable large-scale promotional activities. Even though regular individual agents are not in the position of a minister or a climate scientist, there might be some roles (e.g. in the professional context) where the limits of individual demandingness are narrower than for other roles.

Right now, with the dominant moral narrative in place, every individual agent subjectively chooses to do her bit wherever it suits her most. Many times, choosing is probably based on the criterion 'least interference with personal interests'. Some may feel that it is easy to have a vegetarian diet, some may feel that buying at the farmers market is easy, some others may feel that it is easy to take a bike, and so on. Some others may agree and again some others may not. Often, the consensus among individuals seems to be that one can have different opinions with regard to participating in global climate action. The approach with regard to social roles briefly introduced above attempts to provide a framework for individual evaluation in practice. It aims to transform the rather subjective discussion on agreeing or disagreeing on rather ineffective, direct mimicking actions to a more objective discussion regarding individual mimicking potential. My conviction is that mimicking, even as a secondary duty, is in great risk of demanding too much of individual agents. That is one reason why promotion is the primary duty. Yet, that does not mean that individual agents cannot be demanded anything from the moral perspective. As I stressed elsewhere, it is a thin line between demanding too much and demanding too little.

Perhaps, the point about mimicking is that demandingness strongly depends on the individual circumstances. That is also the case for mimicking as a limited, secondary duty next to promotion. The potential for mimicking, so it seems, always ends in some kind of subjective evaluation. I have attempted to provide a framework that individuals may consider when trying to mimic alongside promotion. Nevertheless, the aim of mimicking is to promote collective action. Recalling chapter 5.2.3, Cripps has stressed this point referring to the difference between private and public mimicking (Cripps, 2013, pp. 152, 153). She correctly concludes that refraining from polluting actions in silence is unlikely to initiate others to rethink their own lifestyle (Cripps, 2013, p. 153). Maybe subjective evaluation of individual circumstances is in the end not generally a bad thing as long as individuals attempt to evaluate their personal potential for (limited, secondary) mimicking as ambitiously as possible and as long as individuals promote the collective solution in the first place.

CHAPTER 7

Summary and Concluding Remarks

Climate change is the single biggest threat to human well-being and probably the biggest challenge of the 21st century. That given, this book asked: What ought regular individual agents do with regard to climate change? In this concluding chapter, I will first outline how I proceeded in this book by summarising the central findings of each chapter. Based on that, I will briefly explicate my response to the research question of this book. Next, I broaden the perspective to global climate policy by roughly asking what policy institutions can do to foster promotional activity. This chapter closes with two final remarks regarding political will and effectiveness.

Summary

Chapter 1 started by providing a rough overview of basic scientific facts about climate change by highlighting three aspects: the urgency to act, the impacts and risks, in particular with regard to human well-being, and the fact that climate-induced impacts will increasingly be felt in all regions of the planet. Next, I followed the dominant moral narrative and focused on the question whether individual agents have a duty to refrain from polluting actions directly. By providing a rough overview of the status quo of the philosophical discussion, I suggested that it is time to move away from accounts that try to prove causality of individual actions. Yet, I stressed that this does not mean either to forgo causality altogether or to leave individuals without moral duties at all.

Taking up on that, chapter 2 accelerated the challenge of identifying individual duties for cases of collective harm. Focusing on the avoidance of further climate-induced suffering, I emphasised that no individual agent will be able to prevent harm on her own. However, taking Gilbert's (2006, 2009) theory as an example, I showed that the central condition for assigning collective responsibility in the strong, conventional sense, namely intention in a rather broad understanding, is not fulfilled either. The challenge is that even though neither concepts of individual responsibility, nor standard concepts of collective responsibility are sufficient to justify either individual or collective duties; climate change requires action – collective action. Cripps (2013) has developed a theory that aims to solve this challenge.

Chapter 3 introduced Cripps' theory and I claimed that it is strongest for rethinking individuals as collectivities. That way, Cripps moves away from the intentionalist concept of collectivities. Building on an interest-based

© MENTIS VERLAG, 2020 | DOI:10.30965/9783957437426_008

normative foundation, Cripps defines mutual dependence as a criterion for non-intentional, weak collectivities. Accordingly, weak collectivities are assigned weakly collective duties. With regard to climate change mitigation, Cripps argues, the Young and the Able have a weakly collective duty to cooperate based on their collective self-interest (Young) and their ability to act (Able) and the Polluters have a weakly collective duty to organize themselves as necessary based on their weakly collective responsibility to prevent serious harm. Given that none of the three collectivities is yet able to act collectively, Cripps defends the primacy of individual promotional duties as derivatives of the weakly collective duties. Promotional duties are assigned to members of the weak collectivities and aim to initiate required collective action. In distinction to mimicking and direct duties, the key reasons for the primacy of promotion are connected to effectiveness, fairness, and efficiency. Discussing three objections, I argued that the demandingness-objection with regard to individual costs is the most serious one – even with regard to promotion. Following, I argued that Cripps is not fully able to cope with the strength of the demandingness objection, which is for two aspects: the lack of transparency with regard to the justification and the lack of a hierarchy of entitlements. In connection to that, I suggested the modification of the normative foundation (chapter 4). Further, I claimed that weak collectivities (chapter 5) as well as promotional duties (chapter 6) require discussion, advancement, and, if necessary, modification and defence.

Chapter 4 introduced Gewirth's (1978) rights-based moral theory as an alternative normative foundation. Based on rational considerations, Gewirth provides a substantial justification of a supreme moral principle, the PGC, claiming that all agents have equal moral rights to freedom and well-being. In application, the PGC provides an adequate framework to explicate and evaluate morally relevant actions and institutions in a consistent and conflict-solving way. Though Gewirth's theory varies from Cripps' normative foundation with regard to several aspects, I demonstrated that Cripps' idea of weak collectivities is compatible with a rights-based foundation. Drawing on Gewirth, I suggested the modification of the cost-expressions of the three guiding moral principles for the Young, the Able and the Polluters and emphasised that promotional action ought to initiate top-down *and* bottom-up change. Accordingly, climate action of individual agents is about pushing the effects of collaboration between individuals and institutions. Focusing on the Polluters, I further examined whether duties impose too much of an individual cost. In order to provide guidance on reasonable costs, I defined several fields of tension that shape the discussion. Following, I argued that demandingness is rather a challenge with regard to doing one's part in the presence of a collective scheme than a matter of promotional action as long as costs of promotion are distributed across the Polluters. Using air travel for tourist purposes as an example, I demonstrated the complexity of the discussion with regard to the limits of individual demandingness but stressed that the limitation of air travel must not be confused with a general prohibition of travel or mobility. Further, I argued that the limitation of air travel is rather unlikely to cause unreasonable costs for members of the Polluters as long as economic costs of collective mitigation and with it the establishment of relevant institutions that protect members' and non-members' rights is subject to adjustment across the collectivity.

Chapter 5 aimed to advance the nature of weakly collective duties. I started by defending weakly collective duties against several criticisms. With regard to the weakly collective duty of the Polluters, I defended predictability and avoidability as the two central requirements of weakly collective responsibility and stressed that raising social awareness is required to ensure that both conditions are fulfilled in the long-run. By discussing Hohl's (2017) approach that aims to justify individual duties to refrain from contribution, I defended Cripps' theory against a strong case for mimicking duties. Beside the fact that Hohl misses the true concept of weak collectivities right at the beginning of her theory, I defended the primacy of promotional duties over mimicking by considering Cripps' reasoning with regard to effectiveness from chapter 3. As long as there is no effective collective scheme in the first place, individual mimicking will not be able to effectively reduce climate-induced harm. Yet, taking accounts of mimicking seriously, I examined whether promotion incorporates mimicking at least to some extent. With reference to integrity and avoidance of hypocrisy, I have indeed made a case for a limited, secondary duty for mimicking alongside promotional duties. Lastly, I defended and roughly defined a threshold-level for individual emissions to decide who contributes to climate change in a morally relevant way as a member of the Polluters. The analysis showed that, on average, all but the world's poorest belong to the Polluters. Using average data, I was not able to exclude the unlikely case of the 'green German' as an individual who belongs to the Polluters on average but who personally succeeds to emit below the critical threshold. Expanding the case to a person who does not belong to any of the three weak collectivities but still has the option to promote, I suggested opening up the Able to all those who have the ability to do something against climate change in the broadest sense. With regard to the weakly collective duty of the Young, I focused on young parents as a subset. Based on the morally relevant relation between parents and their children and common membership in the Young, I claimed that young parents have a duty to mitigate in the interest of their children. That is in sharp contrast to one of

Cripps' (2017a) recent contributions. With reference to Shue (2010) and Lynas (2008), I argued that Cripps greatly underestimates the risk of climate-induced impacts by the middle as well as the end of the 21st century. Given that failure to act now will leave future children with no opportunity for effective climate action and assuming that the indirect relation between young parents today and future children matters, I argued that the duty of young parents extends to future descendants – at least to those born roughly by the end of the century. Further, I examined whether young parents have a duty to ensure mitigation in the long-run by raising their children to be climate-aware and responsible. Assuming that institutional education can hardly make up for failure to provide parental climate education, I claimed that parents owe their children this kind of education in order to empower them to fulfil their duties as adults and to ensure mitigation in their children's interest in the long-run.

Chapter 6 aimed to advance Cripps' idea of promotional duties. Conceptually, I defined two channels of promotional actions: the social channel and the political channel. Using Rainie and Wellmann's (2012) idea of networked individualism from social science, I conceptualised the transformation from individual agents finding themselves in the position to have a promotional duty to networked individuals to collective action. Considering Pulse of Europe and Pantsuit Nation as two practical examples for social change, I stressed that individuals are empowered to take action in their digital community. However, I also stressed that connected individuals require some kind of structure in order to develop possible solutions and to become a meaningful part of global climate action. Regarding the content of promotional action, I limited myself to presenting different facettes and areas of promotional action. Considering promotional action of the Young, I claimed that their key promotional tasks are to target the social channel and to promote large-scale climate education going beyond the kind of parental education defended in chapter 5. Considering the Able, I focused on scientists globally as a subset. My claim was that the path to zero-emissions is far from clear and that their main promotional duty is to figure out a way to get there. Taking electricity generation as an example, I demonstrated that alternatives are hardly able to compete with fossil sources for various reasons. Further, I discussed options such as smart grids, power storage as well as long-distant grids and briefly touched upon geoengineering. Even though I suggested that key questions regarding the electricity transition are subject to promotional actions of natural science and engineering, I stressed, with reference to Steigleder (2016c), that climate ethicists ought to accompany the process by engaging into energy ethics as well as risk ethics. Considering the Polluters, I acknowledged that members of the Polluters basically promote against their own lifestyle. For that reason, I suggested that the key promotional duty of the Polluters is to call for a framework that allows the mainstreaming of a sustainable, climate friendly lifestyle across the Polluters. Focussing on the three major sources of consumption-related emissions (housing, mobility, and food), I argued that major tasks with regard to housing and mobility are again in the hands of scientists. Promotional action of the Polluters is probably limited to the social channel, which is nevertheless a huge task. Only with regard to the food market, may nudging become an option and may be complemented by political consumerism. In the face of information asymmetries, I discussed the lack of available and reliable data from the shipping industry. Given that I defended a limited, secondary duty to mimic alongside with promotion, I lastly developed a tool that conceptualises a way of thinking that members of the Polluters may adopt. Building on a model of social roles, I stressed that individuals should assess their potentials for mimicking as ambitious as possible but that potentials remain a matter of subjective evaluation.

Responding to the Research Question

In response to what regular individual agents ought to do, the short answer is the following: individuals ought to promote a collective solution to climate change. The slightly more detailed response is this: Given that mimicking is not enough to reduce climate-induced harm, individuals ought to step out of their comfort zone and transform individual climate action into a public matter. A key component of promotional action is the promotion of the new moral narrative. Depending on the kind of membership, promotional action through the social channel ranges from connecting to others online and raising awareness in various ways to providing climate education. Promotional action through the social channel aims to ensure the required social backing for political, economic and technical means building the path to decarbonisation. Promotional action through the political channel aims to initiate the required regulations to limit emissions as soon, as comprehensively and as quickly as possible. Possible actions range from voting and writing e-mails to ambitious political candidates, political consumerism to focusing available resources on research and development. Next to promotional actions, individuals ought to evaluate their personal potential for mimicking alongside promotion as ambitiously as possible. Yet, even mimicking alongside promotion ought to be public and aim for promoting a collective solution. Generally, climate action of individual agents is about pushing the effects of collaboration between individuals and institutions. That given, I believe that individual agents are required to rethink their own relation to existing institutions. Climate change is about us as human beings – when did we start to believe that we must wait for existing institutions to make a change for the better?

Concluding Remarks

Focusing on the individual level, this book left the tasks of non-state actors, sub-state actors, but also state actors out of sight. In this concluding chapter, I wish to stress that pushing the collaboration between individuals and existing institutions is not a task that is entirely on the shoulders of regular individual agents. Turning to the opposing level, the following examines possible action that the UNFCCC together with its members states might take to foster promotional action of individual agents.

I wish to start with a few remarks regarding the UNFCCC's self-understanding. Until the late 2000s, the UNFCCC rather strictly followed a top-down approach of imposing mitigation targets on its members as defined in the Kyoto Protocol from 1997. The monocentric governance structure left little space for non-state actors to either put pressure on governments or influence the negotiations' outcome (van Asselt & Zelli, 2014). As a consequence, member states did not follow up on defined emission reduction goals, leaving a "normative and institutional vacuum on energy" (Karlsson-Vinkhuyzen, 2010, p. 175). After a disastrous Climate Conference in Copenhagen in 2009, the main task of the conferences in the following two years in Cancun and Durban was basically to keep up the process of climate negotiations at all. Agreements during these years lacked ambition, specificity and commitment. In the follow-up of Durban and in preparation for the Paris Agreement, governments were asked to determine their own contributions by including action of non-state and sub-state actors to increase ambitions (van Asselt & Zelli, 2014). The result was the formulation of the Intended Nationally Determined Contributions (INDCs), which were renamed as NDCs after the Paris Agreement.[178] The idea that climate policy is exclusively a matter of the UNFCCC is quite resilient (Cole, 2015). I would not go as far as to say that the UN's self-understanding has effectively shifted from monocentric to polycentric. However, the fact that the Paris Agreement actively calls for bottom-up change is a good sign even though, as Sander Chan, Clara Brandi and Steffen Bauer (2016) point out, the agreement does not specify the role of non-state actors other than emphasising their importance for global climate action and asking member states to foster their action (Chan et al., 2016, p. 242).

---

178 Mbeva and Pauw (2016) provide a detailed analysis of the role of the INDCs.

Evidence for the changing of the UNFCCC's self-understanding might come from two examples: First, in the course of the Climate Conference in Lima in 2014, the UNFCCC launched a networking platform, the so-called Non-State Actor Zone for Climate Action (NAZCA). To date more than 12,000 different initiatives have been registered so far (status: March 2019), ranging from companies and cities to non-market organisations and NGOs. However, none of them are an initiative formed by regular individual agents. Second, the UNFCCC now acknowledges nine so-called constituencies, which basically group NGOs according to their purpose or interest. Among them are NGOs with a focus on environmental protection, research, farming and agriculture but also Youth NGOs. According to the UNFCCC, they represent civil society (UNFCCC, 2019). A selected number of them are therefore granted observer status at conferences. Yet, it is one thing to have observer status and it is quite another to be able to actively shape the outcome of the negotiation process as representatives of society. Elaborating the role of constituencies with regard to the UNFCCC greatly exceeds the scope of this chapter. My point is simply that the UNFCCC's self-understanding is changing, and international climate policy is trying to figure a way to represent civil society.

Nevertheless, encompassing the action of already organised collectivities as representing civil society is different from encompassing climate action of regular individuals. I claimed that in order to become an effective part of global climate action, groups of individual agents as networked individuals require some kind of structure at some point (chapter 6.2). As groups of agents with some kind of structure, their action might be embedded in global climate action. Yet, my claim is that in order to foster social backing for required national reduction targets that are secured under the UNFCCC, the UNFCCC and its member states would be well advised to find a way to integrate regular individuals in some way. My argument is neither that regular individuals should be able to actively shape the outcome of climate negotiations nor that anyone should be able to attend official conferences. However, I wonder how individual agents are supposed to participate in global mitigation if they have no access at all to the arena that sets global mitigation pathways. In order to encourage individual agents to make a meaningful difference, climate policy has to become accessible for regular individuals. That is true for any level of climate policy.

Jumping from regular individual agents to the UNFCCC and back to regular individuals leaves quite a challenging gap. Representatives of climate policy may ask how they are supposed to enable access to climate policy to 7.7 billion people worldwide. My very brief answer is that I agree with Harro van Asselt and Friborz Zelli (2014) who stress that the new role of the UNFCCC ought to focus on providing information, network opportunities and funding as well as on supporting technological progress, innovation and capacity building (van Asselt & Zelli, 2014, pp. 138–139). With regard to promotional duties of the Young, I have briefly discussed the role of the UNESCO with regard to climate education. The UNESCO attempts to foster national climate education for example by providing training materials or by supporting initiatives across the planet in order to reach citizens who may be in the position to accelerate the scope and pace of climate education. Analogous to this, my idea is that the UNFCCC could provide information, networking opportunities, and so on for groups of individual agents who are thrilled to take climate action. Even though social media provides great networking opportunities, networking must not be limited to these platforms. In spring 2019, following the example of students and pupils marching under the heading Fridays for Future and Youth for Future, different sets of individual agents formed groups such as Scientists for Future and Teachers for Future. Why not use this momentum to establish a platform for initiatives to connect with each other globally?

Readers might further ask why I focused on the level of international climate policy but not on the national, regional or local level of political action to encourage promotional activities of individual agents. Given that the key arena of climate policy still is the UNFCCC, I aimed to show that there is some potential and scope to push mutual collaboration even on this global level. National, regional and local governments as well as the media will have to play a major role in transferring information from the lowest level of action (individuals) to the highest level of action (UN) and vice versa. The urgency of climate action requires challenging our common understanding of governance. Perhaps, connecting climate action across levels will turn out not to be as difficult as it may seem at first glance.

In addition, I wish to make two final comments: First, integrating individual agents in climate policy generally requires willingness on the political side. I have claimed several times throughout this book that, in my view, climate action is rather a matter of means and acceptance than of political will. Reflecting on the lack of ambitiousness with regard to individual politicians and even with regard to some national governments, I have to admit that it is from time to time tempting to diagnose a general lack of political will. However, I doubt that the matter is that simple. Perhaps, the trouble is that a number of stakeholders continue to pretend that there is an alternative to either interim mitigation targets or zero emissions by 2050. In fact, the relevant question is neither if measures are taken nor how ambitiously they are taken. Mitigation targets are set and mandatory. The only relevant question is how to ensure that they are reached and sustainably secured. It would indeed be extremely helpful if political actors internalized this question and focused political discussions on formulating an answer. Yet, even though the message might not have come across to all politicians individually, I would not diagnose a lack of political will to a whole party or even a whole government. Likewise, I would not judge either a community of states or the UNFCCC to lack political will because single members lack political ambitiousness.

The second comment is rather a note to fellow philosophers. The argumentative strength of this book stems – though not exclusively – from focussing on effectiveness. Perhaps, some scholars will claim that effectiveness is not a matter of philosophy. After all, philosophy is a theoretical discipline of humanities. I have a deep conviction that philosophy is able to formulate urgently required answers to questions that many other disciplines do not even know that philosophy is capable of. With climate change, I have dedicated my book to a challenge that is urgent, devasting and huge. In the face of this, I consider every year that scientists (not only philosophers) work on this topic as a privilege. This privilege comes with the requirement to propose effective solutions.

The solution to climate change is about putting together a very ambitious, fair, effective and efficient package of political, technological, economic, and social means. Perhaps, that is not only the biggest challenge of the 21st century but one of the biggest challenges in the history of mankind. As such, climate action is at constant risk of leaving agents with feelings of fear, panic, helplessness, and resignation. However, if we – collectively – succeed in putting the four components together, I am convinced that change is possible. I have formulated my book as a plea for the promotion of a collective solution to climate change and more generally, as a plea for changing the dominant moral narrative. That was for a purpose. From a moral perspective, it is the only way to capture individual moral duties for climate change as a collective problem. But more importantly, my hope is that readers will turn this last page with a feeling of confidence and empowerment to finally be able to make a meaningful difference to the global fight against climate change.

References

A

Adler, M. D. (2007). Corrective justice and liability for global warming. *University of Pennsylvania Law Review*, (115), 1859–1867.

Aengenheyster, M., Feng, Q. Y., van der Ploeg, F., & Dijkstra, H. A. (2018). The point of no return for climate action: Effects of climate uncertainty and risk tolerance. *Earth System Dynamics*, *9*(3), 1085–1095.

Ahaus, B., Heidbrink, L., & Schmidt, I. (2011). Der verantwortliche Konsument: Wie Verbraucher mehr Verantwortung für ihren Alltagskonsum übernehmen können [The responsible consumer: How consumers may take more responsibility in their daily life]. *Working Papers, Center for Responsibility Research, 1*(10). Retrieved 14 March 2019 from https://duepublico.uni-duisburg-essen.de/servlets/DerivateServlet/Derivate-25913/WP_10_Der_verantwortliche_Konsument.doc.pdf.pdf

Alegre, J., & Pou, L. (2006). The length of stay in the demand for tourism. *Tourism Management, 27*(6), 1343–1355.

Allen, M. R., Babiker, M., Chen, Y., de Coninck, H., Connors, S., van Diemen, R., ... Zickfeld, K. (2018). *Global warming of 1.5 °C: An IPCC Special Report on the impacts of global warming of 1.5 °C above pre-industrial levels and related global greenhouse gas emission pathways, in the context of strengthening the global response to the threat of climate change, sustainable development, and efforts to eradicate poverty. Summary for policymakers*. Incheon, Korea: Intergovernmental Panel on Climate Change (IPCC).

Allen, M. R., Frame, D. J., Huntingford, C., Jones, C., D., Lowe, J. A., Meinshausen, M., & Meinshausen, N. (2009). Warming caused by cumulative carbon emissions towards the trillionth tonne. *Nature, 458*, 1163–1166.

Allison, I., Carrasco, J., Kaser, G., Kwok, R., Mote, P., Murray, T., ... Zhang, T. (2013). Oberservations: Cryosphere. In T. F. Stocker, G.-K. Plattner, M. Tignor, S. K. Allen, J. Boschung, A. Nauels, ... P. M. Midgley (Eds.), *Climate change 2013: The physical science basis. Contribution of Working Group I to the Fifth Assessment Report of the Intergovernmental Panel on Climate Change* (pp. 317–382). Cambridge (UK) and New York: Cambridge University Press.

Allison, W. (2015). *Nuclear is for life: A cultural revolution*. Wade Allison Publishing.

Almassi, B. (2012). Climate change and the ethics of individual emissions: A response to Sinnott-Armstrong. *Perspectives: International Postgraduate Journal of Philosophy, 4*, 4–21.

Anderson, K. L., & Bows, A. (2012). Executing a Scharnow turn: Reconciling shipping emissions with international commitments on climate change. *Carbon Management, 3*(6), 615–628.

Andreae, P. (2017). Pulse of Europe: 'Sie haben das Potential, Wahlen zu beeinflussen' [Pulse of Europe: They have the potential to influence elections]. *Frankfurter Allgemeine Zeitung*. Retrieved 14 March 2019 from http://www.faz.net/aktuell/rhein-main/frankfurt/pulse-of-europe-sie-haben-das-potential-wahlen-zu-beeinflussen-14919190.html

Archard, D. (2010). The obligations and responsibilities of parenthood. In D. Archard & D. Benatar (Eds.), *Procreation and parenthood: The ethics of bearing and rearing children*. Oxford: Oxford University Press.

Arneson, R. (2004). Moral limits on the demands of beneficence. In D. K. Chatterjee (Ed.), *The ethics of assistance: Morality and the distant needy* (pp. 33–58). Cambridge (UK): Cambridge University Press.

Arpino, B., Gumà, J., & Albert, J. (2017). The demography of grandparenthood: The role of family histories. RECSM Working Paper Number. (50), 1–31.

Ashford, E. (2006). The inadequacy of our traditional conception of the duties imposed by human rights. *Canadian Journal of Law and Jurisprudence, 19*(2).

ATAG. (2018). *Aviation: benefits beyond borders. Powering global economic growth, employment, trade links, tourism and support for sustainable development through air transport, October 2018*. Retrieved 14 March 2019 from Air Transport Action Group (ATAG) website: https://aviationbenefits.org/media/166344/abbb18_full-report_web.pdf.

Atmosfair. (2019). CO_2-Fußabdruck meines Flugs berechnen [Calculating my flight's carbon dioxide footprint]. Retrieved 14 March 2019, from nachdenken – klimabewusst reisen – atmosfair website: https://www.atmosfair.de/de/kompensieren/flug/

Attfield, R. (2009). Mediated responsibility, global warming, and the scope of ethics. *Journal of Social Philosophy, 40*(2), 225–236.

Audunsson, S. (2017). Micro-Influencer sorgen für mehr Engagement [Micro influencers ensure more involvement]. Retrieved 8 October 2018, from https://www.adzine.de/2017/02/micro-influencer-sorgen-fuer-mehr-engagement/.

Aufrecht, M. (2011). *Climate change and structural emissions: Moral obligations at the individual level. 25*(2), 201–213.

B

Baatz, C. (2014). Climate change and individual responsibility to reduce GHG emissions. *Ethics, Policy and Environment, 17*(1), 1–19.

Baatz, C. (2016). Reply to my critics: Justifying the fair share argument. *Ethics, Policy & Environment, 19*(2), 160–169.

Baltzer, U. (2002). Joint action of large groups. In G. Meggle (Ed.), *Social facts and collective intentionality* (pp. 1–18). Frankfurt: Dr. Hänsel-Hohenhausen AG.

Barr, S., Gilg, A., & Shaw, G. (2011). 'Helping people make better choices': Exploring the behaviour change agenda for environmental sustainability. *Applied Geography, 31*(2), 712–720.

Barr, S., Shaw, G., Coles, T., & Prillwitz, J. (2010). 'A holiday is a holiday': Practicing sustainability, home and away. *Journal of Transport Geography, 18*(3), 474–481.

Barry, C., & Øverland, G. (2015). Individual responsibility for carbon emissions: Is there anything wrong with overdetermining harm? In J. Moss (Ed.), *Climate Change and Justice* (pp. 165–183). Cambridge (UK): Cambridge University Press.

Bayne, T., & Kolers, A. (2003). Toward a pluralistic account of parenthood. *Bioethics, 17*(3), 221–242.

Becken, S. (2007). Tourists' perception of international air travel's impact on the global climate and potential climate change policies. *Journal of Sustainable Tourism, 15*(4), 351–368.

Beenen, J. (2018). Streit um die Glaubwürdigkeit des MSC-Siegels für Fisch [Dispute over the trustworthiness of MSC labels for fish]. *Süddeutsche*. Retrieved 14 March 2019 from https://www.sueddeutsche.de/wirtschaft/fisch-im-supermarkt-streit-um-die-glaubwuerdigkeit-des-msc-siegels-fuer-fisch-1.3905424

Bertsch, V., Hall, M., Weinhardt, C., & Fichtner, W. (2016). Public acceptance and preferences related to renewable energy and grid expansion policy: Empirical insights for Germany. *Energy, 114*, 465–477.

Beyleveld, D. (1991). *The dialectical necessity of morality: An analysis and defense of Alan Gewirth's argument to the Principle of Generic Consistency*. Chicago: University of Chicago Press.

Bittner, J., Brost, M., Krupa, M., Ladurner, U., Pinzler, P., Raether, E., ... Schirmer, S. (2017). Europe wakes up. *Die Zeit*. Retrieved 14 March 2019 from http://www.zeit.de/politik/2017-03/european-union-politics-crisis-populism-future-economics/komplettansicht.

Blackburn, S. (2008). *Oxford Dictionary of Philosophy* (2nd ed. rev.). New York: Oxford University Press.

Blustein, J. (1997). Procreation and parental responsibility. *Journal of Social Philosophy, 28*(2), 79–86.

Booth, C. (2012). Bystanding and climate change. *Environmental Values, 21*(4), 397–416.

Bowman, P. (2014). Fair shares and decent lives. *Ethics, Policy & Environment, 17*(1), 24–26.

Bows, A., & Anderson, K. L. (2007). Policy clash: Can projected aviation growth be reconciled with the UK Government's 60% carbon-reduction target? *Transport Policy, 14*(2), 103–110.

Braham, M., & Holler, M. J. (2009). Distributing causal responsibility in collectivities. In R. Gekker & T. Boylan, *Economics, rational choice and normative philosophy* (pp. 145–163). London, New York: Routledge.

Braham, M., & van Hees, M. (2009). Degrees of causation. *Erkenntnis, 71*(3), 323–344.

Braham, M., & van Hees, M. (2012). An anatomy of moral responsibility. *Mind, 121*(483), 601–634.

Brake, E. (2005). Fatherhood and child support: Do men have a right to choose? *Journal of Applied Philosophy, 22*(1), 55–73.

Brake, E. (2010). Willing parents: A voluntarist account of parental role obligation. In D. Archard & D. Benatar, *Procreation and parenthood* (pp. 151–177). Oxford: Oxford University Press.

Brake, E., & Millum, J. (2018). Parenthood and procreation. In *Stanford Encyclopedia of Philosophy Archive: Spring 2018 Edition*. Retrieved 14 March 2019 from https://plato.stanford.edu/archives/spr2018/entries/parenthood/.

Bratman, M. (1999). *Faces of intention: Selected essays on intention and agency.* Cambridge (UK): Cambridge University Press.

Brécard, D. (2014). Consumer confusion over the profusion of eco-labels: Lessons from a double differentiation model. *Resource and Energy Economics, 37*, 64–84.

Brenan, M., & Saad, L. (2018). *Global warming concern steady despite some partisan shifts*. Retrieved 14 March 2019 from Gallup Inc. website: https://news.gallup.com/poll/231530/global-warming-concern-steady-despite-partisan-shifts.aspx.

Brighouse, H. (2002). *School choice and social justice*. Oxford: Oxford University Press.

Brighouse, H., & Swift, A. (2014). *Family values: The ethics of parent-child relationships.* New Jersey: Princeton University Press.

Brinkmann, J. (2004). Looking at consumer behavior in a moral perspective. *Journal of Business Ethics, 51*(2), 129–141.

Broome, J. (2012). *Climate matters: Ethics in a warming world.* New York: W. W. Norton & Company.

Brown, J. (2005). An account of the dolphin-safe tuna issue in the UK. *Marine Policy, 29*(1), 39–46.

Brynolf, S., Baldi, F., & Johnson, H. (2016). Energy efficiency and fuel changes to reduce environmental impact. In K. Andersson, J. F. Lindgren, S. Brynolf, & M. Wilewska-Bien (Eds.), *Shipping and the environment: Improving environmental performance in marine transport* (pp. 295–339). Berlin, Heidelberg: Springer.

Budolfson, M. B. (2009). *Collective action, climate change, and the ethical significance of futility*. Retrieved 14 March 2019 from https://b1bc346f-a-62cb3a1a-s-sites.googlegroups.com/site/budolfson/papers/BudolfsonFutility.pdf.

Buffoli, M., Odone, A., & Signorelli, C. (2016). Not in my back yard (NIMBY), an endemic syndrome influencing environmental policies. *Epidemiology Biostatics and Public Health, 13*(4).

Bundesnetzagentur. (2017). *Quartalsbericht zu Netz- und Systemsicherheitsmaßnahmen: Erstes Quartal 2017* [*Quarterly report regarding grid- and system safety measures: first quarter 2017*]. Retrieved 26 January 2018 from Bundesnetzagentur für Elektrizität, Gas, Telekommunikation, Post und Eisenbahnen (Federal Network Agency) website: https://www.bundesnetzagentur.de/SharedDocs/Downloads/DE/Allgemeines/Bundesnetzagentur/Publikationen/Berichte/2017/Quartalsbericht_Q1_2017.pdf.

Burtt, S. (2003). The scope of parental authority: Why we don't owe children an 'open future'. In S. Macedo & I. M. Young (Eds.), *Child, family and state* (pp. 243–270). New York: New York University Press.

C

Campbell, I., Kalanki, A., & Sachar, S. (2018). *How to counter the climate threat from room air conditioners*. Retrieved 14 March 2019 from Rocky Mountain Institute website: https://www.rmi.org/wp-content/uploads/2018/11/Global_Cooling_Challenge_Report_2018.pdf.

Caney, S. (2005). Cosmopolitan justice, responsibility, and global climate change. *Leiden Journal of International Law, 18*(4), 747–775.

Caney, S. (2006). Cosmopolitan justice, rights and global climate change. *Canadian Journal of Law & Jurisprudence, 19*(2), 255–278.

Caney, S. (2010). Climate change, human rights and the moral thresholds. In S. Gardiner, S. Caney, D. Jamieson, & H. Shue (Eds.), *Climate ethics: Essential readings* (pp. 163–177). Oxford: Oxford University Press.

Caney, S. (2012). Just emissions. *Philosophy & Public Affairs, 40*(4), 255–300.

Caney, S. (2014). Two kinds of climate justice: Avoiding harm and sharing burdens. *Journal of Political Philosophy, 22*(2), 125–149.

Capros, P., Höglund-Isaksson, L., Winiwarter, W., Purohit, P., Gomez-Sanabria, A., Frank, S., ... Kesting, M. (2016). *EU Reference Scenario 2016: Energy, transport and GHG emissions. Trends to 2050*. Retrieved 14 March 2019 from European Commission (EC), Directorate-General for Energy, Directorate-General for Climate Action and Directorate-General for Mobility and Transport website: https://ec.europa.eu/energy/sites/ener/files/documents/20160713%20draft_publication_REF2016_v13.pdf.

Caws, P. (2005). The distributive structure of social groups. *Journal of Social Philosophy, 36*(2), 218–232.

Chan, S., Brandi, C., & Bauer, S. (2016). Aligning transnational climate action with international climate governance: The road from Paris. *Review of European Comparative & International Environmental Law, 25*(2), 238–247.

CIEL. (2011). *Climate change & human rights: A primer*. Retrieved 14 March 2019 from Center for International Environmental Law (CIEL) website: http://www.ciel.org/Publications/CC_HRE_23May11.pdf.

Claassen, R., & Düwell, M. (2013). The foundation of capability theory: comparing Nussbaum and Gewirth. *Ethical Theory and Moral Practice, 16*, 493–510.

Clifton, J. (2016). Top 5 Influencer Marketing trends and implications. In *Edelmann Digital: 2017 Digital Trends*. Retrieved 14 March 2019 from http://edelmandigital.com/wp-content/uploads/2016/12/2017-Edelman-Digital-Trends-Report.pdf.

Cohen, S. A., Higham, J. E. S., & Cavaliere, C. T. (2011). Binge flying: Behavioural addiction and climate change. *Annals of Tourism Research, 38*(3), 1070–1089.

Cohen, S. A., Higham, J. E. S., & Reis, A. (2013). Sociological barriers to developing sustainable discretionary air travel behaviour. *Journal of Sustainable Tourism, 21*(7), 982–998.

Cole, D. C. (2015). Advantages of a polycentric approach to climate change policy. *Nature Climate Change, 5*(2), 114–118.

Collins, M., Knutti, R., Arblaster, J., Dufresne, J.-L., Fichefet, T., Friedlingstein, P., … Wehner, M. (2013). Long-term climate change: Projections, commitments and irreversibility. In T. F. Stocker, D. Qin, G.-K. Plattner, M. M. B. Tignor, S. K. Allen, J. Boschung, … P. M. Midgley (Eds.), *Climate change 2013: The physical science basis. Contribution of Working Group I to the Fifth Assessment Report of the Intergovernmental Panel on Climate Change* (pp. 1029–1136). Cambridge (UK) and New York: Cambridge University Press.

Connolly, K., & Oltermann, P. (2017, March 12). Sunday of protest in 40 cities to show growth of pro-EU movement. *The Guardian.* Retrieved 14 March 2019 from https://www.theguardian.com/world/2017/mar/12/sunday-of-protest-in-40-cities-to-show-growth-of-pro-eu-movement

Cook, J., Oreskes, N., Doran, P. T., Anderegg, W. R. L., Verheggen, B., Maibach, E. W., … Rice, K. (2016). Consensus on consensus: A synthesis of consensus estimates on human-caused global warming. *Environmental Research Letters, 11*(4).

Cooper, M. (2014). Small modular reactors and the future of nuclear power in the United States. *Energy Research and Social Science, 3*, 161–177.

Cripps, E. (2011a). Climate change, collective harm and legitimate coercion. *Critical Review of International Social and Political Philosophy, 14*(2), 171–193.

Cripps, E. (2011b). Collectivities without intention. *Journal of Social Philosophy, 42*(1), 1–20.

Cripps, E. (2013). *Climate change and the moral agent: Individual duties in an interdependent world.* Oxford: Oxford University Press.

Cripps, E. (2017a). Do parents have a special duty to mitigate climate change? *Politics, Philosophy & Economics, 16*(3), 308–325.

Cripps, E. (2017b). Justice, integrity and moral community: Do parents owe it to their children to bring them up as good global climate citizens? *Proceedings of the Aristotelian Society, 117*(1), 41–59.

Cruz, J., & Atcheson, M. (2016). *Floating offshore wind energy: The next generation of wind energy.* Springer International Publishing.

Cullity, G. (2015). Acts, omissions, emissions. In J. Moss (Ed.), *Climate change and justice* (pp. 148–64). Cambridge (UK): Cambridge University Press.

Cuomo, C. J. (2011). Climate change, vulnerability, and responsibility. *Hypatia, 26*(4), 690–714.

Czisch, G. (2005). *Szenarien zur zukünftigen Stromversorgung: Kostenoptimierte Variationen zur Versorgung Europas und seiner Nachbarn mit Strom aus erneuerbaren*

Energien [Scenarios for future electricity supply: optimal cost variations towards supplying Europe and its neighbours with renewable electricity] (Universität Kassel). Retrieved 9 January 2018 from https://kobra.bibliothek.uni-kassel.de/bitstream/urn:nbn:de:hebis:34-2006041195961/1/DissVersion0502.pdf.

D

Dalton, G. (2018). *Selling the science of climate change.* Retrieved 14 March 2019 from https://www.climateone.org/audio/selling-science-climate-change.

Davison, L., Littleford, C., & Ryley, T. (2014). Air travel attitudes and behaviours: The development of environment-based segments. *Journal of Air Transport Management, 36,* 13–22.

Dessouky, M. O. (2013). The environmental impact of large scale solar energy projects on the MENA deserts: Best practices for the DESERTEC initiative. *Eurocon 2013,* 784–788. Zagreb: IEEE Xplore.

Dickinson, J. E., Robbins, D., Filimonau, V., Hares, A., & Mika, M. (2013). Awareness of tourism impacts on climate change and the implications for travel practice: A Polish perspective. *Journal of Travel Research, 52*(4), 506–519.

DRV. (2018). *Der Deutsche Reisemarkt: Zahlen und Fakten 2017* [*The German travel sector: Numbers and facts 2017*]. Retrieved 4 April 2018 from Deutscher Reiseverband (DRV) website: https://www.drv.de/securedl/106/0/0/1522935434/966f8cbd41b0b65f29854576 04cac11516928aae/fileadmin/user_upload/Fachbereiche/Statistik_und_Marktforschung/Fakten_und_Zahlen/18-03-01_DRV_ZahlenFakten-2018_ITB_final.pdf.

Du, Y., & Parsons, J. E. (2009). *Update on the cost of nuclear power.* Cambridge (USA): MIT Press.

Dunbar, R. I. M. (2016). Do online social media cut through the constraints that limit the size of offline social networks? *Open Science, 3*(1).

Duus-Otterström, G. (2014). Individual climate obligations and non-subsistence emissions. *Ethics, Policy & Environment, 17*(1), 27–30.

E

EC. (2016). Energy to smart grid. Retrieved 14 March 2019, from European Commission – CORDIS – Bringing you the results of EU research and innovation framework programmes since 1990: Horizon 2020, FP7 and earlier website: https://cordis.europa.eu/project/rcn/201959/factsheet/en

EC. (2018). *Reducing emissions from the shipping sector.* Retrieved 14 March 2019 from European Commission (EC) website: https://ec.europa.eu/clima/policies/transport/shipping_en.

Edenhofer, O., Pichs-Madruga, R., Sokona, Y., Kadner, S., Minx, J. C., Brunner, S., … Cruz-Núnez, X. (2013). Technical summary. In T. F. Stocker, D. Qin, G.-K. Plattner,

M. M. B. Tignor, S. K. Allen, J. Boschung, ... P. M. Midgley (Eds.), *Climate change 2014: Mitigation of climate change. Contribution of Working Group III to the Fifth Assessment Report of the Intergovernmental Panel on Climate Change* (pp. 35–94). Cambridge (UK) and New York: Cambridge University Press.

Edwards, H. A., Dixon-Hardy, D., & Wadud, Z. (2016). Aircraft cost index and the future of carbon emissions from air travel. *Applied Energy, 164*, 553–562.

Erikson, E. H., & Erikson, J. M. (1998). *The life cycle completed.* W. W. Norton.

Ernst & Young. (2010). *Cost of and financial support for wave, tidal stream and tidal range generation in the UK: A report for the Department of Energy and Climate Change and the Scottish Government.* Ernst & Young LLP.

F

Feinberg, J. (1970). *Doing and deserving: Essays in the theory of responsibility.* Princeton: Princeton University Press.

Feinberg, J. (1980). *Rights, justice and the bounds of liberty.* Princeton: Princeton University Press.

Feinberg, J. (1984). *Harm to others.* New York: Oxford University Press.

Ferreira, H. L., Garde, R., Fulli, G., Kling, W., & Lopes, J. P. (2013). Characterisation of electrical energy storage technologies. *Energy, 53*, 288–298.

Field, C., Barros, V. R., Mach, K. J., Mastrandrea, M. D., van Aalst, M. K., Adger, W. N., ... Yohe, G. W. (2013). Technical summary. In C. Field, V. R. Barros, D. Dokken, K. J. Mach, M. D. Mastrandrea, T. E. Bilir, ... L. White (Eds.), *Climate change 2014: Impacts, adaptation, and vulnerability. Part A: Global and sectoral aspects. Contribution of Working Group II to the Fifth Assessment Report of the Intergovernmental Panel on Climate Change* (pp. 35–94). Cambridge (UK) and New York: Cambridge University Press.

Field, C., Barros, V. R., Mastrandrea, M. D., Mach, K. J., Abdrabo, M. A. K., Adger, W. N., ... Yohe, G. W. (2014). Summary for policymakers. In C. Field, V. R. Barros, D. Dokken, K. J. Mach, M. D. Mastrandrea, T. E. Bilir, ... L. White (Eds.), *Climate change 2014: Impacts, adaptation, and vulnerability. Part A: Global and sectoral aspects. Contribution of Working Group II to the Fifth Assessment Report of the Intergovernmental Panel on Climate Change* (pp. 1–32). Cambridge (UK) and New York: Cambridge University Press.

Filimonau, V., Dickinson, J., Robbins, D., & Huijbregts, M. a. J. (2011). Reviewing the carbon footprint analysis of hotels: Life Cycle Energy Analysis (LCEA) as a holistic method for carbon impact appraisal of tourist accommodation. *Journal of Cleaner Production, 19*(17/18), 1917–1930.

Foit, S. R., Vinke, I. C., de Haart, L. G. J., & Eichel, R.-A. (2017). Power-to-syngas: An enabling technology for the transition of the energy system? *Angewandte Chemie International Edition, 56*(20), 5402–5411.

Fox, M. H. (2014). *Why we need power: The environmental case*. New York: Oxford University Press.

Fragnière, A. (2014). Climate change, neutrality and the Harm Principle. *Ethical Perspectives, 21*(1), 73–99.

Fragnière, A. (2015). Review on: Elizabeth Cripps' 'Climate change and the moral agent: Individual duties in an interdependent world'. *Environmental Values, 24*(5), 698–700.

Fragnière, A. (2016). Climate change and individual duties. *Wiley Interdisciplinary Reviews: Climate Change, 7*, 798–814.

Freberg, K., Graham, K., McGaughey, K., & Freberg, L. A. (2011). Who are the social media influencers? A study of public perceptions of personality. *Public Relations Review, 37*(1), 90–92.

Frid, C., Andonegi, E., Depestele, J., Judd, A., Rihan, D., Rogers, S. I., & Kenchington, E. (2012). The environmental interactions of tidal and wave energy generation devices. *Environmental Impact Assessment Review, 32*(1), 133–139.

Frigo, G. (2018). Energy ethics: A literature review. *Relations: Beyond anthropocentrism, 6*(2), 177–214.

Fukui, H., & Miyoshi, C. (2017). The impact of aviation fuel tax on fuel consumption and carbon emissions: The case of the US airline industry. *Transportation Research Part D: Transport and Environment, 50*, 234–253.

Fuscaldo, G. (2006). Genetic Ties: Are they morally binding? *Bioethics, 20*(2), 64–76.

G

Gardiner, S. M. (2010). Is 'arming the future' with geoengineering really the lesser evil? In S. M. Gardiner, S. Caney, D. Jamieson, & H. Shue (Eds.), *Climate ethics: Essential readings* (pp. 284–312). Oxford: Oxford University Press.

Gardiner, S. M. (2011). *A perfect moral storm: The ethical tragedy of climate change*. Oxford, New York: Oxford University Press.

Gardiner, S. M. (2017). Geoengineering: Ethical questions for deliberate climate manipulation. In S. M. Gardiner & A. Thompson (Eds.), *The Oxford Handbook of Environmental Ethics* (pp. 501–514). New York: Oxford University Press.

Garvey, J. (2011). Climate change and causal inefficacy: Why go green when it makes no difference? *Royal Institute of Philosophy Supplement, 69*, 157–174.

German Parliament. (2017). *Minutes of plenary proceedings 221*. Retrieved 14 March 2019 from German Parliament website: https://www.bundestag.de/blob/496808/7e4a9dbc60b4a2184a1c78b63965a946/18221-data.txt.

Gewirth, A. (1978). *Reason and morality*. Chicago, London: The University of Chicago Press.

Gewirth, A. (1982). *Human rights: Essays on justification and application*. Chicago, London: The University of Chicago Press.

Gewirth, A. (1990). Two types of Cost-Benefit Analysis. In D. Scherer (Ed.), *Upstream/ downstream: Issues in environmental ethics* (pp. 205–232). Philadelphia: Temple University Press.

Gewirth, A. (1996). *The community of rights*. Chicago: University of Chicago Press.

Gheaus, A. (2018). Biological parenthood: Gestational, not genetic. *Australasian Journal of Philosophy*, *96*(2), 225–240.

Gilbert, M. (1989). *On social facts*. London: Routledge.

Gilbert, M. (1997). What is it for us to intend? In G. Holmstrom-Hintikka & R. Tuomela, *Contemporary action theory, vol. 2: The philosophy and logic of social action*. Dordrecht: Kluwer Academic Publishers.

Gilbert, M. (2006). *A theory of political obligation: Membership, commitment, and the bonds of society*. Oxford, New York: Oxford University Press.

Gilbert, M. (2009). Shared intention and personal intentions. *Philosophical Studies*, *144*(1), 167–187.

Global Footprint Network. (2018). National Footprint Accounts 2018 edition. Retrieved 17 October 2018 from http://data.footprintnetwork.org/#/.

Goldberg, S. M., & Rosner, R. (2011). *Nuclear reactors: Generation to generation*. Cambridge (USA): American Academy of Arts and Sciences.

Gössling, S. (2013). National emissions from tourism: An overlooked policy challenge? *Energy Policy*, *59*, 433–442.

Gössling, S., & Peeters, P. (2007). 'It does not harm the environment!': An analysis of industry discourses on tourism, air travel and the environment. *Journal of Sustainable Tourism*, *15*(4), 402–417.

Götz, M., Lefebvre, J., Mörs, F., McDaniel Koch, A., Graf, F., Bajohr, S., … Kolb, T. (2016). Renewable power-to-gas: A technological and economic review. *Renewable Energy*, *85*, 1371–1390.

Gould, C. C. (1988). *Rethinking democracy: Freedom and social cooperation in politics, economy and society*. Cambridge: Cambridge University Press.

Greenwood, S., Perrin, A., & Duggan, M. (2016). Demographics of social media users in 2016. Retrieved 6 October 2018, from Pew Reserach Center Internet & Technology website: http://www.pewinternet.org/2016/11/11/social-media-update-2016/.

Grote, M., Williams, I., & Preston, J. (2014). Direct carbon dioxide emissions from civil aircraft. *Atmospheric Environment*, *95*, 214–224.

Grubler, A. (2010). The costs of the French nuclear scale-up: A case of negative learning by doing. *Energy Policy*, *38*(9), 5174–5188.

Grunert, K. G., Hieke, S., & Wills, J. (2014). Sustainability labels on food products: Consumer motivation, understanding and use. *Food Policy*, *44*, 177–189.

Gubernator, S. (2017). Reicht es heute, einfach nur für Europa zu sein? [Is it enough to only support the European idea?] *WeltN24*. Retrieved 14 March 2019 from https://

www.welt.de/politik/deutschland/article163279367/Reicht-es-heute-einfach-nur-fuer-Europa-zu-sein.html.

Gullberg, A. T., Ohlhorst, D., & Schreurs, M. (2014). Towards a low carbon energy future: Renewable energy cooperation between Germany and Norway. *Renewable Energy, 68*, 216–222.

H

Haberl, H., Sprinz, D., Bonazountas, M., Cocco, P., Desaubies, Y., Henze, M., … Searchinger, T. (2012). Correcting a fundamental error in greenhouse gas accounting related to bioenergy. *Energy Policy, 45*, 18–23.

Hagenbring, A., Brodowski, A. K., & Winter, P. (2018). *Lufthansa Group: Annual report 2017*. Retrieved 14 March 2019 from Deutsche Lufthansa AG website: https://investor-relations.lufthansagroup.com/fileadmin/downloads/en/financial-reports/annual-reports/LH-AR-2017-e.pdf.

Hale, B. (2011). Nonrenewable resources and the inevitability of outcomes. *The Monist, 94*(3), 369–390.

Hall, B. (1999). The origin of parental rights. *Public Affairs Quarterly, 13*(1), 73–82.

Hansen, J., Kharecha, P., Sato, M., Masson-Delmotte, V., Ackerman, F., Beerling, D. J., … Zachos, J., C. (2013). Assessing 'dangerous climate change': Required reduction of carbon emissions to protect young people, future generations and nature. *Public Library of Science (PLoS) ONE, 8*(12), 1–26.

Hansen, J., Sato, M., Russell, G., & Kharecha, P. (2013). Climate sensitivity, sea level and atmospheric carbon dioxide. *Philosophical Transaction of the Royal Society A, 371*.

Hardin, R. (1988). *Morality within the limits of reason*. Chicago: University of Chicago Press.

Hares, A., Dickinson, J., & Wilkes, K. (2010). Climate change and the air travel decisions of UK tourists. *Tourism and Climate Change, 18*(3), 466–473.

Hawkins, E., Ortega, P., Suckling, E., Schurer, A., Hegerl, G., Jones, P., … van Oldenborgh, G. J. (2017). Estimating changes in global temperature since the preindustrial period. *Bulletin of the American Meteorological Society, 98*(9), 1841–1856.

Hellenkemper, M. (2018). The perfect fit: Wie jedes Unternehmen passgenaue und hochwertige Influencer identifiziert [The perfect fit: How each business identifies custom-fit and high-quality influencers]. In A. Schach & T. Lommatzsch (Eds.), *Influencer Relations: Marketing und PR mit digitalen Meinungsführern* (pp. 201–224). Wiesbaden: Springer Fachmedien Wiesbaden.

Helm, D. (2015). *The carbon crunch: How we're getting climate change wrong – and how to Fix It*. New Haven: Yale University Press.

Helm, D. (2017). *Burn out: The endgame for fossil fuels*. New Haven and London: Yale University Press.

Hennessy, K., Fitzharris, B., Bates, B. C., Harvey, N., Howden, S. M., Hughes, L., ... Warrick, R. (2007). Australia and New Zealand. In M. L. Parry, O. F. Canziani, J. P. Palutikof, P. J. van der Linden, & C. E. Hanson (Eds.), *Climate change 2007: Impacts, adaptation and vulnerability. Contribution of Working Group II to the Fourth Assessment Report of the Intergovernmental Panel on Climate Change* (pp. 507–540). Cambridge (UK): Cambridge University Press.

Higham, J., Cohen, S. A., Cavaliere, C. T., Reis, A., & Finkler, W. (2016). Climate change, tourist air travel and radical emissions reduction. *Special Volume: Sustainable Tourism: Progress, Challenges and Opportunities, 111*, 336–347.

Hill, J. L. (1991). What does it mean to be a 'parent'? The claims of biology as a basis for parental rights. *New York University Law Review, 66*, 353–420.

Hiller, A. (2011). Climate change and individual responsibility. *The Monist, 94*(3), 349–368.

Hofer, C., Dresner, M. E., & Windle, R. J. (2010). The environmental effects of airline carbon emissions taxation in the US. *Transportation Research Part D, 15*, 37–45.

Hohfeld, W. N. (1919). *Fundamental legal conceptions*. New Haven: Yale University Press.

Hohl, S. (2017). *Individuelle Verantwortung für kollektiv verursachte Übel [Individual responsibility for collectively caused harms]*. Münster: mentis Verlag.

Horton, K. (2004). International aid: The fair shares factor. *Social Theory and Practice, 30*(2), 161–174.

Hourdequin, M. (2010). Climate, collective action and individual ethical obligations. *Environmental Values, 19*(4), 443–464.

Hsu, S.-L. (2011). *The case for a carbon tax: Getting past our hang-ups to effective climate policy*. Washington, D.C.: Island Press.

Hultman, N. E., Koomey, J., & Kammen, D. M. (2007). What history can teach us about the future costs of U.S. nuclear power. *Environmental Science Technology, 41*(7), 2088–2093.

I

IAEA. (2018a). *Deployment indicators for small modular reactors: Methodology, analysis of key factors and case studies*. Vienna: International Atomic Energy Agency (IAEA).

IAEA. (2018b). *Fast reactors and related fuel cycles: Next generation nuclear systems for sustainable development (FR17)*. Retrieved 14 March 2019 from https://www-pub.iaea.org/MTCD/Publications/PDF/STIPUB1836web.pdf.

IAEA. (2019). Nuclear share of electricity generation in 2017. Retrieved 14 March 2019, from International Atomic Energy Agency (IAEA) PRIS – Power Reactor Information System website: https://www.iaea.org/PRIS/WorldStatistics/NuclearShareofElectricityGeneration.aspx.

IATA. (2018). IATA forecast predicts 8.2 billion air travelers in 2037. Retrieved 14 March 2019, from International Air Transport Association (IATA) website https://www.iata.org/pressroom/pr/Pages/2018-10-24-02.aspx.

ICS. (2018). *Shipping facts: Shipping and world trade*. Retrieved 14 March 2019 from International Chamber of Shipping (ICS) website: http://www.ics-shipping.org/shipping-facts/shipping-and-world-trade.

IEA. (2017a). *Energy access outlook 2017: From poverty to prosperity*. Retrieved 14 March 2019 from International Energy Agency (IEA) website: https://www.iea.org/publications/freepublications/publication/WEO2017SpecialReport_EnergyAccessOutlook.pdf.

IEA. (2017b). *Key world energy statistics 2017*. Retrieved 14 March 2019 from International Energy Agency (IEA) website: https://www.iea.org/publications/freepublications/publication/KeyWorld2017.pdf.

IEA. (2018a). *Key world energy statistics 2018*. Retrieved 14 March 2019 from International Energy Agency (IEA) website: https://webstore.iea.org/download/direct/2291?fileName=Key_World_2018.pdf.

IEA. (2018b). Share of electricity generation by fuel – European Union-28. Retrieved 14 March 2019, from International Energy Agency (IEA) website: https://www.iea.org/statistics/?country=EU28&year=2016&category=Electricity&indicator=undefined&mode=chart&dataTable=ELECTRICITYANDHEAT.

Ignaciuk, A., Vöhringer, F., Ruijs, A., & van Ierland, E. C. (2006). Competition between biomass and food production in the presence of energy policies: a partial equilibrium analysis. *Energy Policy*, *34*(10), 1127–1138.

ILO, & WHO. (2017). *International Chemical Safety Cards (ICSCs), ICSC 0291, Methane*. Retrieved 14 March 2019 from International Labour Organization (ILO) website: https://www.ilo.org/dyn/icsc/showcard.display?p_lang=en&p_card_id=0291&p_version=2.

IMO. (2018). *Introduction to IMO*. Retrieved 14 March 2019 from International Maritime Organisation (IMO) website: http://www.imo.org/en/About/Pages/Default.aspx.

IMO. (2019). Emission Control Areas (ECAs) designated under regulation 13 of MARPOL Annex VI. Retrieved 14 March 2019, from International Maritime Organisation (IMO) website http://www.imo.org/en/OurWork/Environment/PollutionPrevention/AirPollution/Pages/Emission-Control-Areas-(ECAs)-designated-under-regulation-13-of-MARPOL-Annex-VI-(NOx-emission-control).aspx.

Ioelovich, M. (2015). Recent findings and the energetic potential of plant biomass as a renewable source of biofuels: A Review. *BioResources*, *10*(1), 1879–1914.

J

Jackson, F. (1987). Group morality. In J. J. C. Smart, P. Pettit, R. Sylvan, & J. Norman (Eds.), *Metaphysics and morality: Essays in honour of J.J.C. Smart*. Blackwell.

Jackson, R. B., Vengosh, A., Carey, J. W., Davies, R. J., Darrah, T. H., O'Sullivan, F., & Pétron, G. (2014). The environmental costs and benefits of fracking. *Annual Review of Environment and Resources, 39*(1), 327–362.

Jamieson, D. (2007a). The moral and political challenges of climate change. In L. Dilling & S. C. Moser (Eds.), *Creating a climate for change: Communicating climate change and facilitating social change* (pp. 475–482). Cambridge: Cambridge University Press.

Jamieson, D. (2007b). When utilitarians should be virtue theorists. *Utilitas, 19*(2), 160–183.

Jamieson, D. (2010). Climate change, responsibility, and justice. *Science & Engineering Ethics, 16*(3), 431–445.

Jamieson, D. (2011). Energy, ethics, and the transformation of nature. In D. G. Arnold (Ed.), *The Ethics of Global Climate Change* (pp. 16–37).

Jamieson, D. (2014). *Reason in a dark time: Why the struggle against climate change failed – and what it means for our future*. Oxford: Oxford University Press.

Johnson, B. (2003). Ethical obligations in a tragedy of the commons. *Environmental Values, 12*(3), 271–287.

Johnson, B. (2011). The possibility of a joint communique: My response to Hourdequin. *Environmental Values, 20*(2), 147–156.

Johnson, E. (2009). Goodbye to carbon neutral: Getting biomass footprints right. *Environmental Impact Assessment Review, 29*(3), 165–168.

K

Kagan, S. (2011). Do I make a difference? *Philosophy & Public Affairs, 39*(2), 105–141.

Kahl, H., & Simmel, L. (2017). Europa- und verfassungsrechtliche Spielräume einer CO2-Bepreisung in Deutschland [Legal scopes for carbon pricing in Germany considering European and constitutional law]. *Würzburger Studien zum Umweltenergierecht, 6*. Retrieved 14 March 2019 from https://stiftung-umweltenergierecht.de/wp-content/uploads/2017/10/stiftung_umweltenergierecht_wuestudien_06_co2_bepreisung.pdf.

Kamm, F. M. (2004). The new problem of distance in morality. In D. K. Chatterjee (Ed.), *The ethics of assistance: Morality and the distant needy* (pp. 59–74). Cambridge (UK): Cambridge University Press.

Kampton, W., & Tomic, J. (2005). Vehicle-to-grid power fundamentals: Calculating capacity and net revenue. *Journal of Power Sources, 144*, 268–279.

Karl, T. R., Melillo, J. M., & Paterson, T. C. (2009). *Global climate change impacts in the United States: A state of knowledge report* (U.S. Global Change Research Program, Ed.). Cambridge (UK) and New York: Cambridge University Press.

Karlsson-Vinkhuyzen, S. I. (2010). The United Nations and global energy governance: Past challenges, future choices. *Global Change, Peace & Security, 22*(2), 175–195.

Katsman, C., Sterl, A., Beersma, H. W., van den Brink, H., Church, J. A., Hazeleger, W., … Weise, R. (2011). Exploring high-end scenarios for local sea level rise to develop flood protection strategies for a low-lying delta – the Netherlands as an example. *Climatic Change, 109*(3–4), 617–645.

Kawall, J. (2011). Future harms and current offspring. *Ethics, Policy and Environment, 14*(1), 23–26.

Keen, M., Parry, I., & Strand, J. (2013). Planes, ships and taxes: charging for international aviation and maritime emissions. *Economic Policy, 28*(76), 701–749.

Kernohan, A. (2000). Individual acts and accumulative consequences. *Philosophical Studies, 97*(3), 343–366.

Kietzmann, J. H., Hermkens, K., McCarthy, I. P., & Silvestre, B. S. (2011). Social media? Get serious! Understanding the functional building blocks of social media. *Business Horizons, 54*(3), 241–251.

Kim, S. Y., & Wolinsky-Nahmias, Y. (2014). Cross-national public opinion on climate change: The effects of affluence and vulnerability. *Global Environmental Politics, 14*(1), 79–106.

Kingston, E., & Sinnott-Armstrong, W. (2018). What's wrong with joyguzzling? *Ethical Theory and Moral Practice, 21*(1), 169–186.

Kirchmeier, R. (2018). Bloggen und Kooperationen: Aus der Perspektive von Mikro-Influencern [Blogging and cooperation: The perspective of micro-influencers]. In A. Schach & T. Lommatzsch (Eds.), *Influencer Relations: Marketing und PR mit digitalen Meinungsführern* (pp. 303–314). Wiesbaden: Springer Fachmedien Wiesbaden.

Klein, R. J. T., Midgley, G. F., Preston, B. L., Alam, M., Berkhout, F. G. H., Dow, K., … Thomas, A. (2014). Adaptation opportunities, constraints, and limits. In C. Field, V. R. Barros, D. Dokken, K. J. Mach, M. D. Mastrandrea, T. E. Bilir, … L. White (Eds.), *Climate change 2014: Impacts, adaptation, and vulnerability. Part A: Global and sectoral aspects. Contribution of Working Group II to the Fifth Assessment Report of the Intergovernmental Panel on Climate Change* (pp. 899–943). Cambridge (UK) and New York: Cambridge University Press.

Knight, K. W. (2016). Public awareness and perception of climate change: A quantitative cross-national study. *Environmental Sociology, 2*(1), 101–113.

Kolers, A., & Bayne, T. (2001). "Are you my mommy?" On the genetic basis of parenthood. *Journal of Applied Philosophy, 18*(3), 273–285.

Kortschak, D. (2015). Tschechien bremst deutschen Wind-Strom [Czech Republic thwarts German wind electricity]. *Frankfurter Rundschau.* Retrieved 14 March 2019 from http://www.fr.de/wirtschaft/energie/oekostrom-aus-deutschland-tschechien-bremst-deutschen-wind-strom-a-455579.

Kovats, S., Valentini, R., Bouwer, L. M., Georgopoulou, E., Jacob, D., Martin, E., ... Soussana, J.-F. (2015). Europe. In V. R. Barros, C. Field, D. Dokken, M. Mastrandrea, K. J. Mach, T. E. Bilir, ... L. White (Eds.), *Climate change 2014: Impacts, adaptation, and vulnerability. Part B: Regional aspects. Contribution of Working Group II to the Fifth Assessment Report of the Intergovernmental Panel on Climate Change* (pp. 1267–1326). Cambridge (UK) and New York: Cambridge University Press.

Kumar, V., & Mirchandani, R. (2013). Increasing the ROI of social media marketing. *IEEE Engineering Management Review, 41*(3), 17–23.

Kuper, S. (2018). How Pantsuit Nation founder Libby Chamberlain took on Trump. *Financial Times*. Retrieved 9 January 2019 from https://www.ft.com/content/51fcd900-cc1c-11e8-b276-b9069bde0956.

Kutz, C. (2000). *Complicity: Ethics and law for a collective age*. Cambridge: Cambridge University Press.

L

Lammers, M. (2018). Wie Unternehmen aus Mico-Influencern Co-Marketer machen [How businesses transform micro-influencers into co-marketers]. In M. Jahnke, R. Brix, A. Bruce, & T. Fuchs (Eds.), *Influencer Marketing: für Unternehmen und Influencer: Strategien, Plattformen, Instrumente, rechtlicher Rahmen: mit vielen Beispielen*. Wiesbaden: Springer Gabler.

Laubert, B. (2015). Kompensation für Klimaschäden: In Luft aufgelöst [Compensation for climate change: Vanishing into thin air]. *Spiegel Online*. Retrieved 14 March 2019 from http://www.spiegel.de/wirtschaft/service/atmosfair-co-co2-emissionen-kompensieren-was-bringt-das-a-1044894.html.

Lawford-Smith, H. (2016). Difference-making and individuals' climate-related obligations. In C. Hayward & D. Roser (Eds.), *Climate justice in a non-ideal world* (pp. 64–82). Oxford: Oxford University Press.

Lawrence, M. G., Schäfer, S., Muri, H., Scott, V., Oschlies, A., Vaughan, N. E., ... Scheffran, J. (2018). Evaluating climate geoengineering proposals in the context of the Paris Agreement temperature goals. *Nature Communications, 9*(1), 3734.

Le Quéré, C., Andrew, R. M., Friedlingstein, P., Sitch, S., Pongratz, J., Manning, A. C., ... Zhu, D. (2018). Global carbon budget 2017. *Earth System Science Data Discussions, 10*, 405–448.

Lee Phillips, M. (2011). The mind at midlife. *Monitor on Psychology, 42*(4).

Leiserowitz, A., Maibach, E., Roser-Renouf, C., Rosenthal, S., Cutler, M., & Kotcher, J. (2018). *Climate change in the American mind: March 2018*. New Haven: Yale University and George Mason University.

Leiserowitz, A., Maibach, E., Roser-Renouf, C., & Smith, N. (2011). *Climate change in the American mind: Americans' global warming beliefs and attitudes in May 2011*. New Haven: Yale University and George Mason University.

Lenton, T. M., Held, H., Kriegler, E., Hall, J. W., Lucht, W., Rahmstorf, S., & Schellnhuber, H. J. (2008). Tipping elements in the Earth's climate system. *Proceedings of the National Academy of Sciences, 105*(6), 1786–1793.

Leopold, T., & Skopek, J. (2015). The demography of grandparenthood: An international profile. *Social Forces, 94*(2), 801–832.

Lewis, A., Estefen, S., Huckerby, J., Musial, W., Pontes, T., & Torres-Martinez, J. (2011). Ocean energy. In O. Edenhofer, R. Pichs-Madruga, Y. Sokona, K. Seyboth, P. Matschoss, & S. Kadner (Series Ed.), *Renewable energy sources and climate change mitigation* (pp. 497–534). Cambridge: Cambridge University Press.

Lichtenberg, J. (2010). Negative duties, positive duties, and the "new harms". *Ethics, 120*(3), 557–578.

Lin, H.-C., Bruning, P. F., & Swarna, H. (2018). Using online opinion leaders to promote the hedonic and utilitarian value of products and services. *Business Horizons, 61*(3), 431–442.

Lindberg, M. (2012). *Possibilities for electricity exchange between Norway and Germany.* Retrieved 14 March 2019 from Zero Emission Resource Organization (ZERO) website: https://www.zero.no/wp-content/uploads/2016/05/Possibilities-for-electricity-exchange-between-Norway-and-Germany.pdf.

Lindgren, J. F., Salo, K., Brynolf, S., Andersson, K., Svensson, E., Zetterdahl, M., ... Magnusson, M. (2016). The natural environment and human impacts. In K. Andersson, S. Brynolf, J. F. Lindgren, & M. Wilewska-Bien (Eds.), *Shipping and the environment: Improving environmental performance in marine transport* (pp. 29–74). Berlin, Heidelberg: Springer.

Linné, P., & Svensson, E. (2016). Regulating pollution from ships. In K. Andersson, S. Brynolf, J. F. Lindgren, & M. Wilewska-Bien (Eds.), *Shipping and the environment: Improving environmental performance in marine transport* (pp. 75–121). Berlin, Heidelberg: Springer.

Lovering, J. R., Yip, A., & Nordhaus, T. (2016). Historical construction costs of global nuclear power reactors. *Energy Policy, 91*, 371–382.

Lowe, J. A., Howard, T., A, P., Tinker, J., Holt, J., Wakelin, S., ... Bradley, S. (2009). *UK Climate projections science report: Marine and coastal projections*. Met Office Hadley Centre.

Lynas, M. (2008). *Six degrees: Our future on a hotter planet.* London: Harper Perennial.

Lynas, M. (2013). *Nuclear 2.0: Why a green future needs nuclear power.* Cambridge (UK): UIT Cambridge.

M

Mackie, J. L. (1965). Causes and conditions. *American Philosophical Quarterly, 2*(4), 245–264.

Maertens, S., Scheelhaase, J., Grimme, W., & Jung, M. (2017). Klimaschutz im Luftverkehr: vom EU-Emissionshandel zu CORSIA [Climate protection in aviation: from EU emissions trading to CORISA]. *Wirtschaftsdienst, 97*(8), 588–595.

Maltais, A. (2013). Radically non-ideal climate politics and the obligation to at least vote green. *Environmental Values, 22*(5), 589–608.

Mander, J. (2016). *GWI social summary: GlobalWebIndex's quarterly report on the latest trends in social networking.* Retrieved 14 March 2019 from GlobalWebIndex (GWI) website: https://www.amic.media/media/files/file_352_955.pdf.

Margolis, R. (2016). The changing demography of grandparenthood. *Journal of Marriage and Family, 78*(3), 610–622.

Mason, P. M., Glover, K., Smith, J. A. C., Willis, K. J., Woods, J., & Thompson, I. P. (2015). The potential of CAM crops as a globally significant bioenergy resource: moving from 'fuel or food' to 'fuel and more food'. *Energy & Environmental Science, 8*(8), 2320–2329.

Matthews, H. D., Zickfeld, K., Knutti, R., & Allen, M. R. (2018). Focus on cumulative emissions, global carbon budgets and the implications for climate mitigation targets. *Environmental Research Letters, 13*(1).

May, L. (1992). *Sharing responsibility.* University of Chicago Press.

May, L., & Hoffman, S. (1991). *Collective responsibility: Five decades of debate in theoretical and applied ethics.* Rowman & Littlefield Publishers.

May, L., & Tuomela, R. (Eds.). (2007). Introduction (to special issue on collective responsibility). *Journal of Social Philosophy, 38*(3), 365–368.

Mbeva, K. L., & Pauw, P. (2016). Self-differentiation of countries' responsibilities: addressing climate change through intended nationally determined contributions. Discussion paper, (4). Bonn: German Development Institute / Deutsches Institut Für Entwicklungspolitik (DIE).

McCright, A. M., & Dunlap, R. E. (2011). The politicization of climate change and polarization in the American public's views of global warming, 2001–2010. *The Sociological Quarterly, 52*(2), 155–194.

McGarne, S. (2017, April 19). Unite, unite Europe! A protest in favor of the European Union. *The New Yorker.* Retrieved 14 March 2019 from https://www.newyorker.com/culture/culture-desk/unite-unite-europe-a-protest-in-favor-of-the-european-union

McKinnon, C. (2014). Climate change: Against despair. *Ethics and the Environment, 19*(1), 31–48.

Meinshausen, M., Meinshausen, N., Hare, W., Raper, S. C., Frieler, K., Knutti, R., ... Allen, M. R. (2009). Greenhouse-gas emission targets for limiting global warming to 2 °C. *Nature, 458*, 1158–1162.

Mellema, G. (1997). *Collective responsibility.* Amsterdam, Atlanta: Rodopi.

Miller, D. (2011). Taking up the Slack? Responsibility and justice in situations of partial compliance. In C. Knight & Z. Stemplowska (Eds.), *Responsibility and distributive justice* (pp. 230–245). Oxford: Oxford University Press.

Miller, R. B. (2004). Moral closeness and world community. In D. K. Chatterjee (Ed.), *The Ethics of assistance: Morality and the distant needy* (pp. 101–122). Cambridge (UK): Cambridge University Press.

Millum, J. (2010). How do we acquire parental rights? *Social Theory and Practice, 36*(1), 112–132.

Mishan, E. J. (1976). *Cost-Benefit Analysis*. New York: Praeger.

Mont, O. (2014). Consumption and climate change. *Klimat | Fokus,* (11).

Morgan-Knapp, C., & Goodman, C. (2015). Consequentialism, climate harm and Individual obligations. *Ethical Theory and Moral Practice, 18*(1), 177–190.

Moser, S. C. (2016). Reflections on climate change communication research and practice in the second decade of the 21st century: what more is there to say? *Wiley Interdisciplinary Reviews: Climate Change, 7*(3), 345–369.

Murphy, L. B. (1993). The demands of beneficence. *Philosophy and Public Affairs, 22*(4), 267–292.

Murphy, L. B. (2000). *Moral demands in non-ideal theory*. New York: Oxford University Press.

N

Nagel, T. (1991). *Equality and partiality* (Vol. 43). New York and Oxford: Oxford University Press.

Nasruddin, Alhamid, M. I., Daud, Y., Surachman, A., Sugiyono, A., Aditya, H. B., & Mahlia, T. M. I. (2016). Potential of geothermal energy for electricity generation in Indonesia: A review. *Renewable and Sustainable Energy Reviews, 53*, 733–740.

Nefsky, J. (2012). *The morality of collective harm*. Berkeley: UC Berkeley.

Nefsky, J. (2017). How you can help, without making a difference. *Philosophical Studies, 174*, 2743–2767.

Neilson, L. A. (2010). Boycott or buycott? Understanding political consumerism. *Journal of Consumer Behaviour, 9*(3), 214–227.

Neuteleers, S. (2014). Individual environmental duties: questions from an institution-oriented perspective. *Ethics, Policy & Environment, 17*, 20–23.

Nolt, J. (2011). How harmful are the average American's greenhouse gas emissions. *Ethics, Policy and Environment, 14*(1), 3–10.

Nolt, J. (2017). Future generations in environmental ethics. In S. M. Gardiner & A. Thompson (Eds.), *The Oxford Handbook of Environmental Ethics* (pp. 344–354). New York: Oxford University Press.

Nussbaum, M. C. (1992). Human functioning and social justice: In defense of Aristotelian essentialism. *Political Theory, 20*(2), 202–246.

Nussbaum, M. C. (2000). *Women and human development*. Cambridge (UK): Cambridge University Press.

Nussbaum, M. C. (2006). *Frontiers of justice*. Harvard University Press.

O

Oberdiek, J. (2008). What's wrong with infringements (insofar as infringements are not wrong)?: A Reply. *Law and Philosophy*, *27*, 293–307.

O'Connor, C. (2017). Forbes top influencers: Meet the 30 social media stars of fashion, parenting and pets (Yes, Pets). Retrieved 8 October 2018, from https://www.forbes.com/sites/clareoconnor/2017/09/26/forbes-top-influencers-fashion-pets-parenting/#7b4f7e997683.

O'Neill, O. (1979). Begetting, bearing, and rearing. In O. O'Neill & W. Ruddick, *Having children: philosophical and legal reflections on parenthood* (pp. 25–38). New York: Oxford University Press.

Otsuka, M. (1991). The paradox of group beneficence. *Philosophy & Public Affairs*, *20*(2), 132–149.

Oxford Dictionaries. (2019). 'middle age', Definition of middle age in English by Oxford Dictionaries Online. Retrieved 14 March 2019, from Oxford Living Dictionaries, English website: https://en.oxforddictionaries.com/definition/middle_age.

P

Paas, L., & Goodman, D. (2016). *Action for climate empowerment: Guidelines for accelerating solutions through education, training and public awareness; 2016*. Retrieved 14 March 2019 from UNESCO and UNFCCC website: http://unesdoc.unesco.org/images/0024/002464/246435e.pdf.

Palizban, O., & Kauhaniemi, K. (2016). Energy storage systems in modern grids: Matrix of technologies and applications. *Journal of Energy Storage*, *6*, 248–259.

Pantsuit Nation. (2019). About Pantsuit Nation: Mission, Vision, & Team. Retrieved 14 March 2019, from Pantsuit Nation – Your voice matters website: https://www.pantsuitnation.org/mission.html.

Parfit, D. (1984). *Reasons and persons*. Oxford: Oxford University Press.

Patt, A. (2015). *Transforming energy*. New York: Cambridge University Press.

Peeters, P., & Dubois, G. (2010). Tourism travel under climate change mitigation constraints. *Tourism and Climate Change*, *18*(3), 447–457.

Peeters, P., & Landré, M. (2012). The emerging global tourism geography: An environmental sustainability perspective. *Sustainability*, *4*(1), 42–71.

Peeters, W., De Smert, A., Diependaele, L., & Sterckx, S. (2015). *Climate change and individual responsibility: Agency, moral disengagement and the motivational gap*. Palgrave Macmillan.

Pellegrino, G. (2018). Robust individual responsibility for climate harms. *Ethical Theory and Moral Practice*, *21*(4), 811–823.

Pérez-Diaz, J., Chazarra, M., Garcia-González, J., Cavazzini, G., & Stoppato, A. (2015). Trends and challenges in the operation of pumped-storage hydropower plants. *Renewable and Sustainable Energy Reviews*, *44*, 767–784.

Pidgeon, N. F. (2012). Public understanding of, and attitudes to, climate change: UK and international perspectives and policy. *Climate Policy, 12*, 85–106.

Planton, S. (2013). Annex III: Glossary. In T. F. Stocker, D. Qin, G.-K. Plattner, M. Tignor, S. K. Allen, J. Boschung, ... P. M. Midgley (Eds.), *Climate change 2013: The physical science basis. Contribution of Working Group I to the Fifth Assessment Report of the Intergovernmental Panel on Climate Change* (pp. 1447–1465). Cambridge (UK) and New York: Cambridge University Press.

Posner, D., & Weisbach, E. A. (2010). *Climate change justice.* Princeton: Princeton University Press.

Poullikkas, A. (2013). A comparative overview of large-scale battery systems for electricity storage. *Renewable and Sustainable Energy Reviews, 27*, 778–788.

Prusak, B. G. (2011). The costs of procreation. *Journal of Social Philosophy, 42*(1), 61–75.

Pugliese, A., & Ray, J. (2009). A heated debate: Global attitudes toward climate change. *Harvard International Review, 31*(3), 64–68.

Pulse of Europe. (2017). Pulse of Europe: Über uns [Pulse of Europe: About us]. Retrieved 22 December 2017, from https://pulseofeurope.eu/ueber-uns/.

R

Rainie, L., & Wellmann, B. (2012). *Networked: The new social operating system.* Cambridge (USA): MIT Press.

Ratter, B. M. W., Philipp, K. H. I., & von Storch, H. (2012). Between hype and decline: recent trends in public perception of climate change. *Environmental Science & Policy, 18*, 3–8.

Rayner, T., & Minns, A. (2015). The challenge of communicating unwelcome climate messages. *Tyndall Working Paper, 162*.

Raz, J. (1986). *The morality of freedom*. Oxford: Clarendon Press.

Rehman, S., Al-Hadhrami, L. M., & Alam, M. M. (2015). Pumped hydro energy storage system: A technological review. *Renewable and Sustainable Energy Reviews, 44*, 586–598.

Rentmeester, C. (2010). A Kantian look at climate change. *Essays in Philosophy, 11*(1), 76–86.

Reser, J. P., Bradley, G. L., Glendon, A. I., Ellul, M. C., & Callaghan, R. (2011). *Public risk perceptions, understandings and responses to climate change and natural disasters in Australia and Great Britain.* Retrieved 14 March 2019, from http://hdl.handle.net/10462/comp/7116.

Revkin, A. C. (2010). China sustains blunt 'you first' message on CO_2. *New York Times Blog*. Retrieved 14 March 2019 from https://dotearth.blogs.nytimes.com/2010/09/02/china-sustains-blunt-you-first-message-on-co2/.

Ribeiro, F., Ferreira, P., Araújo, M., & Braga, A. C. (2014). Public opinion on renewable energy technologies in Portugal. *Energy, 69*, 39–50.

Rignot, E., Mouginot, J., Scheuchl, B., van den Broeke, M., van Wessem, M. J., & Morlighem, M. (2019). Four decades of Antarctic ice sheet mass balance from 1979–2017. *Proceedings of the National Academy of Sciences.*

Röder, M., Whittaker, C., & Thornley, P. (2015). How certain are greenhouse gas reductions from bioenergy? Life cycle assessment and uncertainty analysis of wood pellet-to-electricity supply chains from forest residues. *Biomass and Bioenergy*, *79*, 50–63.

Röös, B. (2017). Die Forderungen sind zu abstrakt [Claims are too abstract]. *General Anzeiger Bonn*. Retrieved 14 March 2019 from https://www.general-anzeiger-bonn.de/news/politik/%E2%80%9EDie-Forderungen-sind-zu-abstrakt%E2%80%9C-article3511202.html.

Rosillo-Calle, F. (2016). A review of biomass energy: Shortcomings and concerns. *Journal of Chemical Technology & Biotechnology*, *91*(7), 1933–1945.

S

Salama, S., & Aboukoura, K. (2018). Role of emotions in climate change communication. In W. Leal Filho, E. Manolas, A. M. Azul, U. M. Azeiteiro, & H. McGhie (Eds.), *Handbook of climate change communication: Vol. 1: Theory of climate change communication* (pp. 137–150). Cham: Springer International Publishing.

Salo, K., Zetterdahl, M., Johnson, H., Svensson, E., Magnusson, M., Gabrielii, C., & Brynolf, S. (2016). Emissions to the air. In K. Andersson, S. Brynolf, J. F. Lindgren, & M. Wilewska-Bien (Eds.), *Shipping and the environment: Improving environmental performance in marine transport* (pp. 169–227). Berlin, Heidelberg: Springer.

Sandberg, J. (2011). My emissions make no difference. *Environmental Ethics*, *33*(3), 229–248.

Sandler, R. (2010). Ethical theory and the problem of inconsequentialism: Why environmental ethicists should be virtue-oriented ethicists. *Journal of Agricultural and Environmental Ethics*, *23*, 167–183.

Sannino, G., & Cavicchioli, C. (2013). *Overcoming research challenges for ocean renewable energy*. Luxembourg: European Union.

Saul, J., & Chestney, N. (2018). U.N. shipping agency reaches deal to cut CO2 emissions. *Reuters*. Retrieved 14 March 2019 from https://uk.reuters.com/article/us-imo-emissions/u-n-shipping-agency-reaches-deal-to-cut-co2-emissions-idUKKBN1HK20F.

Schaefer, F., & Blanke, M. (2014). Opportunities and challenges of carbon footprint: Climate or CO_2 labelling for horticultural products. *Erwerbs-Obstbau*, *56*(2), 73–80.

Scheelhaase, J., Maertens, S., Grimme, W., & Jung, M. (2018). EU ETS versus CORSIA – A critical assessment of two approaches to limit air transport's CO_2 emissions by market-based measures. *Journal of Air Transport Management*, *67*, 55–62.

Scheffler, S. (1995). Individual responsibility in a global age. *Social Philosophy and Policy*, *12*(1), 219–236.

Schellnhuber, H. J. (2009). Tipping elements in the Earth system. *Proceedings of the National Academy of Sciences of the United States of America*, *106*(49), 20561–20563.

Schlothfeldt, S. (2009). *Individuelle oder gemeinsame Verpflichtung? Das Problem der Zuständigkeit bei der Behebung gravierender Übel* [*Individual or shared responsibility? The problem of accountability for removing severe harms*]. Paderborn: mentis Verlag.

Schurer, A. P., Mann, M. E., Hawkins, E., Tett, S. F. B., & Hegerl, G. C. (2017). Importance of the pre-industrial baseline in determining the likelihood of exceeding the Paris Limits. *Nature Climate Change*, *7*(8), 563–567.

Schwenkenbecher, A. (2014). Is there an obligation to reduce one's individual carbon footprint? *Critical Review of International Social and Political Philosophy*, *17*(2), 168–188.

Scruggs, L., & Benegal, S. (2012). Declining public concern about climate change: Can we blame the great recession? *Global Environmental Change*, *22*(2), 505–515.

Seager, T. P., Selinger, E., & Spierre, S. (2011). Determining moral responsibility for CO_2 emissions: a reply to Nolt. *Ethics, Policy and Environment*, *14*(1), 39–42.

Segura, E., Morales, R., Somolinos, J. A., & López, A. (2017). Techno-economic challenges of tidal energy conversion systems: Current status and trends. *Renewable and Sustainable Energy Reviews*, *77*, 536–550.

Sewill, B. (2009). *Airport jobs: false hopes, cruel hoax*. Retrieved 14 March 2019 from Aviation Environment Federation website: http://www.aef.org.uk/downloads/Airport_jobs_false_hopes_cruel_hoax_March2009_AEF.pdf.

Shaw, D., Newholm, T., & Dickinson, R. (2006). Consumption as voting: an exploration of consumer empowerment. *European Journal of Marketing*, *40*(9/10), 1049–1067.

Shaw, S., & Thomas, C. (2010). Discussion note: Social and cultural dimensions of air travel demand: Hyper-mobility in the UK? *Journal of Sustainable Tourism*, *14*(2).

Shockley, K. (2013). Collective responsibility. In H. LaFollette (Ed.), *International Encyclopedia of Ethics*. Oxford: Whiley-Blackwell.

Shockley, K. (2017). Individual and contributory responsibility for environmental harm. In *The Oxford Handbook of Environmental Ethics* (pp. 265–275). New York: Oxford University Press.

Shrader-Frechette, K. (2017). Ethical energy choices. In *The Oxford Handbook of Environmental Ethics* (pp. 391–403). New York: Oxford University Press.

Shue, H. (1993). Subsistence emissions and luxury emissions. *Law and Policy*, *15*(1), 39–59.

Shue, H. (2010). Deadly delays, saving opportunities. Creating a more dangerous world? In S. M. Gardiner, S. Caney, D. Jamieson, & H. Shue (Eds.), *Climate ethics: Essential readings* (pp. 146–162). New York: Oxford University Press.

Shue, H. (2014). *Climate justice: Vulnerability and protection.* Oxford: Oxford University Press.

Shue, H. (2015a). Last opportunities: Future human rights generate urgent present duties. In M. Di Paola & D. Kamal, *Climate change and human rights: The 2015 Paris Conference and the tasks of protecting people on a warming planet.* Retrieved 14 March 2019 from https://www.globalpolicyjournal.com/blog/26/11/2015/last-opportunities-future-human-rights-generate-urgent-present-duties.

Shue, H. (2015b). Transboundary damage in climate change: Criteria for allocating responsibility. In A. Nollkaemper & D. Jacobs (Eds.), *Distribution of responsibility in International Law* (pp. 321–340). Cambridge (UK): Cambridge University Press.

Singer, P. (1972). Famine, affluence, and morality. *Philosophy & Public Affairs, 1*(3), 229–243.

Singer, P. (2002). *One world: The ethics of globalization.* Yale University Press.

Sinnott-Armstrong, W. (2005). It's not my fault: Global warming and individual moral obligations. *Perspectives on Climate Change: Science, Economics, Politics, Ethics, 5,* 293–315.

Small, M. J., Stern, P. C., Bomberg, E., Christopherson, S. M., Goldstein, B. D., Israel, A. L., … Zielinska, B. (2014). Risks and risk governance in unconventional shale gas development. *Environmental Science & Technology, 48*(15), 8289–8297.

Smartgrids Salzburg. (2019). Projekte [Projects]. Retrieved 14 March 2019, from SMARTGRIDS Modellregion Salzburg website: http://www.smartgridssalzburg.at/content/website_smartgrids/de_at/LaufendeProjekte/leafs.html.

Smith, A. (2014). What people like and dislike about Facebook. Retrieved 6 October 2018, from Pew Research Center website: http://www.pewresearch.org/fact-tank/2014/02/03/what-people-like-dislike-about-facebook/

Sommerfelt, N., & Zschäpitz, H. (2017). Strompreis-Kollaps durch 'Herwart' offenbart Wahnsinn der Energiewende [Collapsing electricity prices caused by ‚Herwart' revealed madness of the energy transition]. *WeltN24.* Retrieved 14 March 2019 from https://www.welt.de/wirtschaft/energie/article170189224/Strompreis-Kollaps-durch-Herwart-offenbart-Wahnsinn-der-Energiewende.html

Sovacool, B. K. (2014). Cornucopia or curse? Reviewing the costs and benefits of shale gas hydraulic fracturing (fracking). *Renewable and Sustainable Energy Reviews, 37,* 249–264.

Steigleder, K. (1992). *Die Begründung des moralischen Sollens: Studien zur Möglichkeit einer normativen Ethik [The justification of moral obligation: Studies on the possibility of normative ethics].* Tübingen: Attempto Verlag Tübingen GmbH.

Steigleder, K. (1997). Gewirth und die Begründung der normativen Ethik [Gewirth and the foundation of normative ethics]. *Zeitschrift Für Philosophische Forschung, 51,* 250–266.

Steigleder, K. (1999). *Grundlegung der normativen Ethik: Der Ansatz von Alan Gewirth* [*Groundwork of normative ethics: The theory of Alan Gewirth*]. Verlag Karl Alber.

Steigleder, K. (2016a). Climate risks, climate economics, and the foundation of rights-based risk ethics. *Journal of Human Rights, 15*(2), 251–271.

Steigleder, K. (2016b). Deontologische Theorien der Verantwortung [Deontologic theories of responsibility]. In L. Heidbrink, C. Langbehn, & J. Sombetzki (Eds.), *Handbuch Verantwortung* (pp. 171–188). Wiesbaden: Springer.

Steigleder, K. (2016c). The tasks of climate related energy ethics: The example of Carbon Capture and Storage. *Jahrbuch Für Wissenschaft Und Ethik, 21*, 121–145.

Steinmeier, F.-W. (2017). *Bundespräsident Frank-Walter Steinmeier nach seiner Vereidigung zum Bundespräsidenten im Deutschen Bundestag am 22.03.2017* [*Federal president Frank-Walter Steinmeier after his swearing in to the German parliament on 22 March 2017*]. Presented at the Berlin. Retrieved 14 March 2019 from https://www.bundespraesident.de/SharedDocs/Reden/DE/Frank-Walter-Steinmeier/Reden/2017/03/170322-Vereidigung.html.

Stern, N. (2006). *Stern Review: The economics of climate change. Executive summary*. Retrieved 21 February 2017 from http://www.wwf.se/source.php/1169157/Stern%20Report_Exec%20Summary.pdf.

Stern, N. (2007). *The economics of climate change: The Stern Review*. Cambridge: Cambridge University Press.

Sternberg, A., & Bardow, A. (2015). Power-to-what? Environmental assessment of energy storage systems. *Energy & Environmental Science, 8*(2), 389–400.

Sterner, T. (2007). Fuel taxes: An important instrument for climate policy. *Energy Policy, 35*, 3194–3202.

Steward, F. (1996). Basic needs, capabilities, and human development. In Avner Offer (Ed.), *In Pursuit of the Quality of Life* (pp. 46–65). New York: Oxford University Press.

Stocker, T. F., Plattner, G.-K., Alexander, L. V., Allen, S. K., Bindoff, N. L., Bréon, F.-M., … Xie, S.-P. (2013). Technical summary. In D. Qin, T. F. Stocker, G.-K. Plattner, S. K. Allen, J. Boschung, A. Nauels, … P. M. Midgley (Eds.), *Climate change 2013: The physical science basis. Contribution of Working Group I to the Fifth Assessment Report of the Intergovernmental Panel on Climate Change* (pp. 33–115). Cambridge (UK) and New York: Cambridge University Press.

Sverdlik, S. (1987). Collective responsibility. *Philosophical Studies: An International Journal for Philosophy in the Analytic Tradition, 51*(1), 61–76.

T

Tan, K.-C. (2015). Individual duties of climate justice under non-ideal conditions. In J. Moss (Ed.), *Climate Change and Justice* (pp. 129–147). Cambridge: Cambridge University Press.

Thomson, J. J. (1986). *Rights, restitution, and risk: essays in moral theory* (W. Parent, Ed.). Cambridge (USA) and London (UK): Harvard University Press.

Tollefsen, D. P. (2002). Collective intentionality and the social sciences. *Philosophy of the Social Sciences*, *32*(1), 25–50.

Tollefson, J. (2015). Global-warming limit of 2 °C hangs in the balance. *Nature*, *52*, 14–15.

Tomik, S. (2018). Tod trotz Ökosiegel: Warum Delfine beim Fischfang sterben müssen [Death despite eco-label: Why dolphins must die during fishing]. *Frankfurter Allgemeine Zeitung*. Retrieved 14 March 2019 from https://www.faz.net/1.5556048.

Topham, G. (2016). Many Ryanair flights could be free in a decade, says its chief. *The Guardian*. Retrieved 14 March 2019 from https://www.theguardian.com/business/2016/nov/22/ryanair-flights-free-michael-oleary-airports.

Tranter, B., & Booth, K. (2015). Scepticism in a changing climate: A cross-national study. *Global Environmental Change*, *33*, 154–164.

Trolliet, M., Barbier, T., & Jacquet, J. (2019). From awareness to action: Taking into consideration the role of emotions and cognition for a stage toward a better communication of climate change. In W. Leal Filho, B. Lackner, & H. McGhie (Eds.), *Addressing the challenges in communicating climate change across various audience* (pp. 47–64). Cham: Springer International Publishing.

Tukker, A., Bulavskaya, T., Giljum, S., de Koning, A., Lutter, S., Simas, M., ... Wood, R. (2016). Environmental and resource footprints in a global context: Europe's structural deficit in resource endowments. *Global Environmental Change*, *40*, 171–181.

Tukker, A., & Jansen, B. (2006). Environmental impacts of products: A detailed review of studies. *Journal of Industrial Ecology*, *10*(3), 159–182.

Tuomela, R. (1992). Group beliefs. *Synthese*, *91*, 285–318.

Tuomela, R. (1995). *The importance of us: A Philosophical study of basic social notions*. Stanford University Press.

Tuomela, R. (2002). *The philosophy of social practices: A collective acceptance view*. Cambridge: Cambridge University Press.

U

UC San Diego. (2018). A daily record of atmospheric carbon dioxide from Scripps Institution of Oceanography at UC San Diego. Retrieved 14 March 2019, from The Keeling Curve website: https://scripps.ucsd.edu/programs/keelingcurve/.

Uihlein, A., & Magagna, D. (2016). Wave and tidal current energy: A review of the current state of research beyond technology. *Renewable and Sustainable Energy Reviews*, *58*, 1070–1081.

UN DESA. (2018). *World urbanization prospects: The 2018 revision*. Retrieved 14 March 2019 from United Nations, Department of Economics and Social Affairs (DESA), Population Division website: https://population.un.org/wup/Download/

UNCTAD. (2017). *Review of maritime transport.* Retrieved 14 March 2019 from United Nations Conference on Trade and Development (UNCTAD) website: http://unctad.org/en/PublicationsLibrary/rmt2017_en.pdf.

UNEP. (2017). *The Emissions Gap Report 2017: A UN Environment synthesis report.* Retrieved 14 March 2019 from United Nations Environment Programme (UNEP) website: https://wedocs.unep.org/bitstream/handle/20.500.11822/22070/EGR_2017.pdf.

UNEP. (2018). *The Emissions Gap Report 2018.* Retrieved 14 March 2019 from United Nations Environment Programme (UNEP) website: http://wedocs.unep.org/bitstream/handle/20.500.11822/26895/EGR2018_FullReport_EN.pdf?sequence=1&isAllowed=y.

UNFCCC. (2015a). *Adoption of the Paris Agreement: Draft decision -/CP.21.* Retrieved 14 March 2019 from United Nations Framework Convention on Climate Change (UNFCCC) website: https://unfccc.int/resource/docs/2015/cop21/eng/l09.pdf.

UNFCCC. (2015b). *Paris Agreement.* Retrieved 14 March 2019 from United Nations Framework Convention on Climate Change (UNFCCC) website: https://unfccc.int/sites/default/files/english_paris_agreement.pdf

UNFCCC. (2019). Admitted NGOs. Retrieved 14 March 2019, from United Nations Framework Convention on Climate Change (UNFCCC) – Parties & Non-Party Stakeholders website: https://unfccc.int/process-and-meetings/parties-non-party-stakeholders/non-party-stakeholders/admitted-ngos#constituencies

University of Oxford. (2019). TrillionthTonne.org: Explaining the need to limit cumulative emission of carbon dioxide. Retrieved 14 March 2019, from http://www.trillionthtonne.org

UNWTO. (2017). *2016 Annual Report.* Retrieved 14 March 2019 from United Nations World Tourism Organization (UNWTO) website: http://cf.cdn.unwto.org/sites/all/files/pdf/annual_report_2016_web_0.pdf

UNWTO, UNEP, & WMO. (2008). *Climate change and tourism: Responding to global challenges.* Retrieved 14 March 2019 from http://sdt.unwto.org/sites/all/files/docpdf/climate2008.pdf

V

van Amstel, M., Driessen, P., & Glasbergen, P. (2008). Eco-labeling and information asymmetry: a comparison of five eco-labels in the Netherlands. *Journal of Cleaner Production, 16*(3), 263–276.

van Asselt, H., & Zelli, F. (2014). Connect the dots: managing the fragmentation of global climate governance. *Environmental Economics and Policy Studies, 16*(2), 137–155.

van de Poel, I., Nihlén Fahlquist, J., Doorn, N., Zwart, S., & Royakkers, L. (2012). The problem of many hands: Climate change as an example. *Science and Engineering Ethics, 18*(1), 49–67.

van der Linden, S., Maibach, E., & Leiserowitz, A. (2015). Improving public engagement with climate change: Five "best practice" insights from psychological science. *Perspectives on Psychological Science, 10*(6), 758–763.

van Zyl, L. (2002). Intentional parenthood and the nuclear family. *Journal of Medical Humanities, 23*(2), 107–118.

Vanderheiden, S. (2008). *Atmospheric justice: A political theory of climate change.* New York: Oxford University Press.

Vassilev, S. V., Vassileva, C. G., & Vassilev, V. S. (2015). Advantages and disadvantages of composition and properties of biomass in comparison with coal: An overview. *Fuel, 158*, 330–350.

Vaughan, A. (2012). Tesco drops carbon-label pledge. *The Guardian.* Retrieved 14 March 2019 from https://www.theguardian.com/environment/2012/jan/30/tesco-drops-carbon-labelling.

Vergeer, R. (2016). *The share of aviation and maritime transport in the EU's transport related fossil fuel demand.* Retrieved 14 March 2019 from CE Delft website: https://www.transportenvironment.org/sites/te/files/publications/2016_11_CE_Delft_7G93_Share_of_aviation_and_maritime_transport_DEF.pdf.

W

Wallace, K. A. (forthcoming). A Kantian perspective on individual responsibility for sustainability. In K. Shockley & A. Light, *Ethics in the Anthropocene.* Cambridge (USA): MIT Press.

Wetzel, D. (2012). Polen macht die Grenze für deutschen Strom dicht [Poland closes its borders to German electricity]. *WeltN24.* Retrieved 14 March 2019 from https://www.welt.de/wirtschaft/article112279952/Polen-macht-die-Grenze-fuer-deutschen-Strom-dicht.html.

Whitmarsh, L. (2011). Scepticism and uncertainty about climate change: Dimensions, determinants and change over time. *Global Environmental Change, 21*, 690–700.

Wolfson, C. (2017). Macro vs. micro influencers. Retrieved 8 October 2018, from https://www.revolutiondigital.com/article/macro-vs-micro-influencers/.

World Bank. (2018a). World Bank open data on CO_2 emissions (metric tons per capita). Retrieved 18 October 2018, from https://data.worldbank.org/indicator/EN.ATM.CO2E.PC?name_desc=false.

World Bank. (2018b). World Bank open data on life expectancy at birth, total (years). Retrieved 27 December 2018, from https://data.worldbank.org/indicator/SP.DYN.LE00.IN?locations=RU.

World Bank. (2019). World Bank open data on air transport, passengers carried. Retrieved 10 March 2019, from https://data.worldbank.org/indicator/IS.AIR.PSGR.

World Energy Council. (2016). *World Energy Resources, Geothermal 2016.* Retrieved 8 January 2018 from World Energy Council website: https://www.worldenergy.org/wp-content/uploads/2017/03/WEResources_Geothermal_2016.pdf.

Wright, R. W. (1985). Causation in tort law. *California Law Review, 73*(6), 1735–1828.

Wright, R. W. (1988). Causation, responsibility, risk, probability, naked statistics, and proof: Pruning the bramble bush by clarifying the concepts. *Iowa Law Review, 73,* 1001–1070.

Z

Zhang, C., Wei, Y.-L., Cao, P.-F., & Lin, M.-C. (2018). Energy storage system: Current studies on batteries and power condition system. *Renewable and Sustainable Energy Reviews, 82,* 3091–3106.

Zoller, D. (2015). Moral responsibility for distant collective harms. *Ethical Theory and Moral Practice, 18,* 995–1010.

Zwinge, T. (2011). Duties of flag states to implement and enforce international standards and regulations – and measures to counter their failure to do so. *Journal of International Business and Law, 10*(2), 297–323.

Index